QC753.2 .B85 1986
Burke, Harry E.
3 4369 00006692

W9-BLA-774

# HANDBOOK
# of
# MAGNETIC PHENOMENA

# HANDBOOK of MAGNETIC PHENOMENA

Harry E. Burke

VNR VAN NOSTRAND REINHOLD COMPANY
*New York*

Library of Congress Catalog Card Number: 85-2409
ISBN: 0-442-21184-8



Manufactured in the United States of America

Published by Van Nostrand Reinhold Company Inc.
115 Fifth Avenue
New York, New York 10003

Van Nostrand Reinhold Company Limited
Molly Millars Lane
Wokingham, Berkshire RG11 2PY, England

Van Nostrand Reinhold
480 La Trobe Street
Melbourne, Victoria 3000, Australia

Macmillan of Canada
Division of Canada Publishing Corporation
164 Commander Boulevard
Agincourt, Ontario M1S 3C7, Canada

15 14 13 12 11 10 9 8 7 6 5 4 3 2

Library of Congress Cataloging in Publication Data

Burke, Harry E.
Handbook of magnetic phenomena.

Includes index.
1. Magnetism. 2. Electronics. I. Title.
QC7532.B85 1985 538 85-2409
ISBN 0-442-21184-8

*Dedicated to*

*Jeffrey Michael Burke*

*and to his great great grandfather,*

*Herman Daniel Burke (1842–1888)*

*who was wounded both at Gettysburg and in the Wilderness and later by Indians on his way to California*

*and to his great great great grandfather,*

*William Cleminson Armstrong (1815–1874)*

*who, as a civilian, was dragged from a sick bed to spend the Civil War in a Union prison.*

# PREFACE

The general theory of magnetism and the vast range of individual phenomena it embraces have already been examined in many volumes. Specialists hardly need help in charting their way through the maze of published information. At the same time, a nonspecialist might easily be discouraged by this abundance. Most texts are restricted in their coverage, and their concepts may well appear to be disorganized when the uninitiated attempt to consider them in their totality. Since the subject is already thoroughly researched with very little new information added year by year, this is hardly a satisfactory state of affairs. By now, it should be possible for anyone with even a minimum of technical competence to feel completely at home with all of the basic magnetic principles.

The present volume addresses this issue by stressing simplicity—simplicity of order and simplicity of range as well as simplicity of detail. It proposes a pattern of logical classification based on the electronic consequences that result whenever any form of matter interacts with any kind of energy. An attempt has been made to present each phenomenon of interest in its most visually graphic form while reducing the verbal description to the minimum needed to back up the illustrations. This might be called a *Life* magazine type of approach, in which each point is principally supported by a picture. The illustrations make use of two (perhaps unique) conventions. First, small dots are used to indicate slow-moving (cold) electronic carriers, whereas large dots stand for fast-moving (hot) ones. This difference in size helps dramatize the relationship between thermal and electronic phenomena. Second, the small spinning charged particles that represent magnetized bodies—microminiature magnets—are referred to as "magnetons." (All submolecular particles spin, whether charged or uncharged.) The various ways magnetons relate to each other in terms of both their electric charge and their magnetic polarization constitute the bases for many magnetic phenomena.

Having recognized first that all magnetic phenomena are ultimately based on Ampere's Law, which relates a magnetic field to a moving electric charge, the discussion proceeds to the Biot-Savart fields that surround the linear paths of all electric currents and then to the toroidal fields that surround both current-carrying coils and spinning electric charges. Current-carrying coils are used to construct electromagnets, whereas spinning charged bodies are responsible for permanent magnetic phenomena.

All told, there are more than four times as many different magnetic phenomena detailed here than can be found in any other publication known to the author.

This book is divided into five parts. Part I explores the relationships between moving electrically charged particles and the various energy components of the environments through which they move. Part II investigates the effects of magnetic field changes on these particles. The next three parts discuss what happens magnetically to moving magnetons. Part III discusses the consequences that may be expected when magnetons move under tight constraints, as in a solid or liquid. Part IV discusses the consequences when they are able to move without much interference from, or interaction with, other particles in the same environment, as in a vacuum or gas. In conclusion, Part V considers the quite different consequences when magnetons are able to move through crystalline structures without significant energy interchange.

HARRY E. BURKE

# CONTENTS

# HANDBOOK
# of
# MAGNETIC PHENOMENA

# 1
# INTRODUCTION

Curiously enough, from an entry-level engineering viewpoint at least, today's literature covering magnetism and its various manifestations is frequently incomplete, often inaccurate, and certainly lacking in structural discipline. This discussion sets out to correct that situation primarily for the benefit of electronic engineers.

Here answers are provided to the basic question: "What are the electronic consequences when material and energy interact in the presence of magnetic fields?" The term *MAGNETOTRANSDUCTION* has been coined to encompass the possibilities. Descriptions are formulated using the particle model to represent material structures. The particle model is used because it best communicates the nature of the largest number of phenomena to the largest audience. Although the particle model performs yeoman service in this matter, it will not stand up in all circumstances. Particularly, it serves no useful purpose in relation to phenomena that occur near absolute zero. Such a deficiency is only to be expected as no model ever represents exactly the system for which it was developed.

Some readers might prefer an energy model, for this is the model used by most universities now teaching material science. Its use would restrict the audience, however, to those few who have already given the subject a great deal of attention and who would learn little more from these pages. Then, too, the particle and energy models complement each other, and those seeking comprehension in greater depth would do well to compare these explanations with those provided elsewhere.

In short, this document is prepared so that the average technically oriented individual, when contemplating a particular device, can know what to expect from its use—both its performance and its limitations.

## 1.1 MAGNETOTRANSDUCTION

The basic concept of electrotransduction—the useful electric consequences of material/energy interactions—serves as a convenient focus for phenomenological consideration of those phenomena of interest to technicians of an electronic bent.

The definition

> an *ELECTROTRANSDUCER* is a device capable of producing a voltage analogous to a stimulating phenomenon, where this analog may be expressed in terms of frequency, amplitude, phase, or time-duration

directs the discussion toward those phenomena that at least have the potential of activating useful devices where the "stimulating phenomena" are expressed as some form of energy. Although all forms of energy have a single common denominator, this discussion will consider the following five subforms: chemical, thermal, mechanical, electrical, and radiant (electromagnetic). Each of these energy subforms is converted into an electric equivalent by an electrotransductive process.

In addition, because of the relation between electric and magnetic fields, electrotransductive responses are influenced by magnetic fields. Thus we have the definition,

> a *MAGNETOTRANSDUCER* is a specialized form of electrotransducer that performs its function in the presence of a magnetic field.

This book will cover those magnetic phenomena that illustrate electronic consequences and are therefore of interest to electronic engineers.

The implications developed here, however, extend well beyond a mere listing of phenomena. For, although electrotransduction itself is limited to the process of converting a phenomenon of interest into an electric analog, today's analog-to-digital conversion techniques easily provide quantified information once a transducer generates its appropriate analog. Each principal area of magnetotransductive interest has been emphasized, therefore, by underscoring both the number and title of its section head (see, for example, Sec. 3.27).

Here, then, is outlined a series of fundamental concepts that make it possible to use basic phenomena to generate information with precision and dispatch.

## 1.2 DISCUSSION BOUNDARIES

This is a *PHENOMENOLOGICAL* DISCUSSION. To the degree that it is practical, principles of transduction are presented as separate simplified considerations in a pattern of logical classification. The mechanical devices that are often a part of a transducer assembly are described only as an aid to understanding the underlying principles, and rigorous explorations of theory are left to the references.

The definition bounding the scope of this presentation implies that transduction involves one or more of the following: a voltage is generated; current flow is varied by changes in circuit parameters; the coupling between two circuits is altered; or a time delay is introduced in some fashion.

A study of transductive principles includes a study of various "effects." These effects are often ascribed to the individuals who discovered them during the course of history as scientific progress expanded the level of man's comprehension of the world about him. The successful design of a useful device is achieved by a relative isolation of one of these effects from all of the others, while the limitations to transductive accuracy are second-order effects, which are reducible only to some acceptable minimum.

This discussion is intended as a simple exposition of pertinent phenomena. As such, it is not subject to obsolesence as are theoretical dissertations, lists of existing transducers, or current recomendations on how to make specific measurements. Either the interpretation, or utilization, of a phenomenon might change with time, but the fact of its existence is immutable. Although new phenomena doubtlessly remain to be discovered, those listed here should remain constant to the degree the author's objectives have been achieved. A definite attempt has been made to avoid editorial comments by eliminating words such as "good," "useful," "large," "effective," and the like, or other expressions of value judgment whose meanings are subject to change with time and whose use is generally restrictive to comprehensive processes.

The text does contain a limitation that is a direct result of editorial decision. Here the choice has been made, from a very large number of observable phenomena, to delineate only those with transductive possibilities. Admittedly this is inconsistent with the stated objectives. Although every kink in a transfer function might have transductive possibilities under some circumstances, only those that have titillated the author's interest have been included in the interest of ultimately completing this document for publication.

## 1.3 MODEL

The increasing complexity of technological specialties tends to make them incomprehensible to the nonspecialist. Unfortunately, many specialists encourage this situation by their cultivation of in-vocabularies and their tendency to publish only for the benefit of their peers. In particular, page-after-page of complex formulae, utilizing incomprehensible symbology, is enough to discourage all but the most erudite.

Admittedly, it does take a mathematical approach to describe the universe as it is now conceived in all of its detail. Such an approach is not required, however, for either the level of understanding or the audience intended here. Nevertheless, if this discussion is to extend beyond a mere listing of phenomena, some reasonable concept of the constitution of matter must be offered. To be effective, this concept should conform to accepted theory as far as it goes, and it should provide a step from which those who have more interest can progress.

Such eminent scientists as Doctors Einstein and Infeld concur in the reasonableness of such an approach, as evidenced by the following quotation:

> The purpose of any physical theory is to explain as wide a range of phenomena as possible. It is justified insofar as it does make events understandable.

Surely, these gentlemen would not object to the addition, ". . . to a specified audience." The problem, then, lies in the choice of a comprehensive model suitable for a chosen audience.

Actually, there is very little latitude in this matter. The vast majority of mankind treats objects as objects and is most comfortable with models constructed from objects. In this audience, energy levels, quantized waves, and the like are admitted to exist but they relate to objects only as secondary considerations and they do not constitute entities in themselves.

This discussion is structured around the interaction of matter and energy—what kind of energy, what kind of matter, and what kind of interaction. The concern is primarily with electric carriers—how they are created, the forms of energy that cause them to move after they are created, and the forces that restrict their motion.

As a consequence, a convenient model has been constructed from electrons and nuclei of various kinds. In this model, electrons spin on their axes, orbit nuclei, pass through crystalline labyrinths whose walls are constructed from wildly swinging nuclei, and so forth.

Furthermore, both electrons and nuclei function as constituents of various material structures so that the directional characteristics of both magnetic and electric dipoles influence the positional relationships of nuclei in gases as well as in crystals.

The logic used in the development of any model has four phases: postulating, testing, correcting, and, finally, recognizing deficiencies.

It would be nice, but not absolutely necessary, for a model to be able to stand rigorous test. Certainly, rigor is a cornerstone in the foundation of all science, and there is no intent here to ignore that fact. Nevertheless, a simple understanding of a subject can be easily lost in a maze of nonessential detail and qualifying explanations. At the very least, all of the factors that have led to the existence of a specific phenomenon do not necessarily contribute to a useful understanding of the way in which it performs.

One of the goals here is simplification. At the same time, if a subject is oversimplified, its real meaning becomes confused and may even be lost. This state of affairs presents rather neat matters for editorial judgment. An attempt has been made to find an illuminating path between what can be two absurd extremes. The reader may judge for himself the degree to which success has been achieved.

## 1.4 AUDIENCE

This text is intended as a reference for electronic engineers, but it should be of use to a much broader audience.

Students approaching magnetic phenomena for the first time will find here a simple, comprehensive, graphic presentation that complements more theoretical tomes. The provision of an illustration for each phenomenon of interest is designed to accelerate the learning process by offering a perspective of the subject that can be achieved with minimal effort. If the illustrations manage to communicate some sense of the beautiful simplicity involved in the construction of matter (rather than its complexity), their purpose has been achieved.

Technicians concerned with the installation of specific transducers and who thus need to understand their functions in order to avoid their limitations; engineers from other disciplines who must specify measurements by using catalog devices; and members of management—all should find this information useful.

In fact, all interested parties can find a series of answers to the basic question: "How does it work?" Furthermore, since the text lists a large number of phenomena not currently found elsewhere in one reference, designers and other established specialists can use this document as a source that will allow their expanding curiosity to push on to the more detailed and theoretical information their needs require.

## 1.5 THEME

Charles Litton once said:

> If you wish to understand how an electronic device works, make yourself very small, enter the device and look around. If you are observant, you can see what makes it function.

This presentation is aimed at furthering the development of the "intelligent intuition" implied by Mr. Litton's comment.

# PART I
# ENVIRONMENTS EXPERIENCED BY MOVING ELECTRIC CHARGES

Any very small particle carrying an electrostatic charge has electrostatic relationships with any other electrostatically charged particle to which it may be exposed in any given environment. Like charges repel, unlike attract, particles with plus and minus charges tend to pair (and in the process cancel each other's macro effects), and accumulations of particles with like charges create electrostatic gradients.

In this context, the presence of many electrostatically charged particles in various configurations creates the various environments in which all of the particles move. For, in fact, all atomic-scaled particles are always in motion. The paths of motion may be linear, curvilinear, oscillatory, or even random, but in all circumstances all such particles spin. Because of this spin, all particles also have direction, the particular direction being determined by the orientation of their spin axis.

In addition, moving electric charges are accompanied by magnetic fields whose direction is determined by the direction of motion. Both spin fields and path fields are associated with each particle, and the fields created by one particle interact with the fields created by every other particle. Fields oppositely directed cause a force of attraction between the particles creating these fields and in the process reduce each other's field strength, whereas similarly directed fields cause particles to repel each other and in the process strengthen each other's effect.

One curious result occurs in the circumstance of two spinning electrons whose axes are oppositely directed. The associated, oppositely directed magnetic fields cause a force of attraction between the two electrons, while their like electrostatic charges create a force of repulsion. The result is a tendency for electrons to pair at a distance where the attractive and the repulsive forces are in balance. Each such electrostatic pair, while maintaining its own relationships, then tends to be attracted to a particle with a plus charge as exemplified by a positive ion or a positive atomic nucleus.

In this relationship, the negative particles are forced to circle the positive particle in a path whose velocity increases as the diameter of traverse decreases. In other words, as the force of electrostatic attraction brings the particles closer together, the mechanical force of centrifugal action tends to drive them apart. The result is a tendency for electrons to circle a positive ion in a path where the electrostatic forces are in balance with the mechanical forces.

What is more, all of these particles are moving at different, constantly changing speeds and in different, constantly changing directions. These speed changes tend to create, or absorb, electromagnetic energy, whereas the mean effect of all motion is a manifestation of thermal energy.

The total energy represented by any given enviroment, therefore, is made up of the sum of its electrostatic, mechanical, thermal, electromagnetic, and chemical (positional) components. Although the total energy content may be constant, the constituents that make up this total may be in a constant state of interchange. Part I of this book explores some of the relationships that are established between moving electrically charged particles (moving magnetons) and the various energy components of the environments through which they move.

# 2
# BASIC LAWS AND DEFINITIONS

The foundation of the subject of "magnetics," as is true of most subjects, rests on a very few, fairly simple concepts. At least these concepts are simple in terms of "what" they are if not "why" they are. Once accepted as given, they can be used as the basic building blocks from which the entire discipline can be developed. They also serve as benchmarks when trying to interpret an observed phenomenon.

First, there is such a thing as an "electron." Not only do electrons exist, there are countless numbers of them everywhere you look. An electron is an object with a miniscule mass that carries one unit of negative charge and spins on its axis at some constant rate. Electric currents are one manifestation of moving electrons; that is, electric currents are "carried" by electrons.

Second, there is such a thing as a "field" that can be used to transmit a force across what would otherwise be empty space. In this context there are three basic types of fields: gravitational, electric, and magnetic.

Third, according to Ampere, a magnetic field surrounds each moving electron. Since spinning electrons are moving electrons, a magnetic field surrounds each electron as a result of its spin. As a consequence, each electron acts as a microminiature permanent magnet.

Fourth, according to Lorentz, an electric charge experiences a force when it moves through a magnetic field. This force is a result of the interaction between an ambient field and an Ampere field.

Finally, all matter is held together by the forces of attraction that exist between particles that have electric fields because of their electric charge and magnetic fields because of their spin. Chapter 2 expands these concepts in more detail.

## 2.1 THE THREE TYPES OF FIELDS

Two objects separated by empty space can exert a force upon each other. For instance, an unsupported body falls toward the earth; electrons repel each other when in close proximity; bar magnets exchange either repulsive or attractive forces depending on their relative orientation.

A *FIELD* is the mechanism by which force is transmitted across space. Any object that can exert a force upon another object at a distance is surrounded by a field of some kind. A field extends throughout space, although, for all practical purposes, its effects may not extend very far.

A field can only be identified as a relationship between two objects. The two objects cannot exert a force upon each other unless each is surrounded by its own field. Consequently, a force interchange over a distance of open space is based on an interaction of two fields.

Three different kinds of fields exist, as follows: electric, magnetic, and gravitational. Only fields of the same kind can interact to transmit force.

A field is described in terms of the strength and direction of the force it can provide at every point in space. Vector arrows that indicate both magnitude and direction provide one means of field representation. A number of such arrows can be sprinkled liberally throughout a region in which a field is evident to indicate its shape.

Although all objects are surrounded by their own characteristic fields, these fields do not have constant, unchanging patterns since the interaction of two of them distorts the shape of both fields. However, the field normally surrounding a single object can be evaluated in terms of the force it would exert on a hypothetical second object if that second object were present. This second object is imagined to be a point source acompanied by a field of unit magnitude. In this case, the field of the second object is assumed not to distort the field of the first.

## 2.2 DEFINITION OF A MAGNETIC FIELD

A magnetic field can be defined in either of two ways, as follows:

> A *MAGNETIC FIELD* is that region in the neighborhood of moving electric charges in which magnetic forces are observable.

or,

> Any region in which an electrically charged body experiences a force by virtue of its motion is considered to contain a *MAGNETIC FIELD*.

## 2.3 ELECTRICALLY CHARGED PARTICLES

An *ELECTRON* is a very small, electrically charged particle. Each electron is identical in every way to every other electron.

A *PROTON,* which is much larger than an electron, carries an electric charge of exactly the same magnitude as that of an electron, but of opposite polarity. The concept of "opposite polarity" is derived from the contrary phenomena in which an electron and a proton experience a force of attraction toward each other, whereas either two electrons or two protons repel each other.

A convention derived from the experiments of Benjamin Franklin labels the electron's charge as "negative" and the proton's charge as "positive." Since all other electrically charged bodies carry electric charges, either positive or negative, that are exact, integral multiples of the electron's charge, the electron's charge is used to define the "unit magnititude" for this phenomenon.

## 2.4 MOVING ELECTRICALLY CHARGED PARTICLES

An electrically charged particle is surrounded by an *ELECTRIC FIELD.* A *moving* electrically charged particle is also accompanied by a *magnetic* field. *AMPERE'S LAW* establishes the relationship between moving charges and magnetic fields, as follows:

AMPERE'S LAW $$dH = kI(dl/r^2)\sin\Theta \qquad (2.1)$$

where $dH$ represents the strength of the magnetic field that results from charge movements over the short distance $dl$ when the moving charge is described in terms of a current $I$ ; $r$ represents the distance from path $dl$ to the point where the field strength is of interest; $\Theta$ represents the angle between the path of movement and a line extending from the center of $dl$ to the point of interest; and $k$ represents the constant of proportionality.

If many small electrically charged particles continually traverse a common path at a constant speed, the net effect of all of their individually moving magnetic fields is a stationary magnetic field, known as *BIOT-SAVART'S FIELD.* As shown by Fig. 2.1, this stationary field assumes a cylindrical configuration about a linear path.

Biot-Savart's specialized version of Ampere's Law establishes the magnetic field strength at some distance from an infinitely long, straight conductor carrying an electric current as follows:

BIOT-SAVART'S LAW $$H = kI/r \qquad (2.2)$$

where the magnetic field strength $H$ at a particular point in space surrounding a conductor is proportional to the current $I$ flowing through that

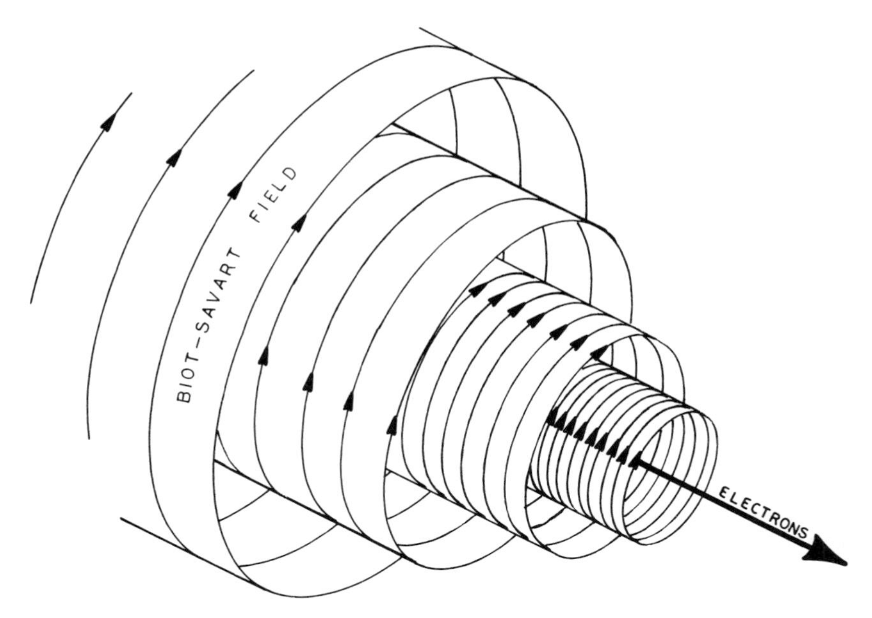


Fig. 2.1. Biot-Savart's Field: The shape of the magnetic field surrounding an electric current flowing uniformly along a linear path.

conductor and inversely proportional to the distance $r$ from that conductor and where $k$ is the constant of proportionality.

A magnetic field has strength. The stronger the moving electric charge, the stronger the resulting magnetic field. The faster the electric charge moves, moreover, the stronger the magnetic field. A magnetic field also has direction. The direction depends on the direction of the movement of the electric charge and on the polarity of that charge.

No magnetic field is associated with a stationary electric charge. In fact, a magnetic field cannot exist apart from the movements of an electric charge.

## 2.5 FLUX LINES

An electron is an object with a "unit" negative charge. A proton provides the equivalent unit positive charge. If one of these particles is placed in an electric field, it experiences an accelerating force that attempts to move it through space.

The *UNIT POLE* is a useful magnetic concept analogous to the electric charges mentioned above. In this case, the terms "north" and "south" carry the positive and negative connotations of the analogy. Just as there is a force of repulsion between two electrons and a force of attraction between an electron and a proton, so is there a force of repulsion between two "north" poles and a force of attraction between a "north" pole and a "south" pole.

*COULOMB'S LAW* establishes standards for the strength of this force as follows:

COULOMB'S LAW $$F = m_1 m_2 / k r^2 \quad (2.3)$$

where "F" is the force of attraction or repulsion between two magnetic poles of strengths $m_1$ and $m_2$, $r$ the distance between these poles, and $k$ is a constant.

Magnetic fields can be described in terms of "lines of magnetic flux." This concept is derived from the hypothetical action of a "unit north pole" moving in an ambient magnetic field. If there were such a pole in such a field, it would tend to move in the direction of the field at every point in space. In this motion, it would trace out the paths called "lines of force," or simply, *FLUX LINES*. A "unit south pole" moves along flux lines in a direction opposite to that of a "unit north pole."

The movement of a unit pole along flux lines is a response to a Coulomb force where the effect of one of the two unit poles is replaced by an equivalent magnetic field. The force on the one pole is a result of its own localized field's interaction with the broader ambient field. Although the ambient field strength is felt by the pole, the location of the source of the ambient field need not be known as far as the force on the pole is concerned. The ambient field simply has an effect on the pole at the point in space occupied by the pole. A measure of the reaction of a unit pole to an ambient field provides the standard of strength against which that ambient field is judged.

Inasmuch as magnetic fields are not necessarily uniform throughout the volume they occupy, this concept is pictorially useful in describing a field's curvature and intensity as well as its direction. In this convention, one flux line indicates both curvature and direction, whereas a number of parallel flux lines per unit area indicates intensity.

## 2.6 MAGNETIC POLES

Both electric and magnetic fields can be pictorialized in general by utilizing a flux-line concept. Electric unit charges tend to move along electric flux lines, whereas magnetic unit poles tend to move along magnetic flux lines. There are fundamental differences, however, between these two concepts.

For instance, there are two types of electrically charged particles: positive and negative. Each type acts as a source of electric flux. If both types are present, electric flux lines originate with one and terminate on the other. Under these circumstances, all electric flux lines have beginnings and endings as well as directions. If only one type of electrically charged particle is present, however, electric flux lines extend between that particle and infinity. In this case, an electric flux line has a beginning and a direction but no ending.

On the other hand, although a magnetic flux line has a direction, it has neither a beginning nor an ending. Magnetic flux lines are always continuous. As a result, there is no such particle as a magnetic "unit pole" analogous to the "unit charge" concept of either an electron or a proton. Although the concept of "north" and "south" magnetic "unit poles" is useful in evaluating magnetic fields, such particles cannot exist in nature. As shown by the sketches of Figs. 2.5, 3.1, and others, however, it is possible for magnetic flux lines to emanate from one end of a body and enter the other end. Under these circumstances, that body is said to be *MAGNETICALLY POLARIZED*. Similarly, a body is said to be *ELECTRICALLY POLARIZED* when electric flux lines exit one end and enter the other.

In electric polarization, an electric flux line originates at some point within a polarized body. The end of a flux line is identified with a specific electron or a specific proton. With magnetic polarization, on the other hand, a magnetic flux line simply passes through a body, and there is no point within that body where it either begins or ends.

As an example, consider the external field surrounding a bar magnet. This magnetic field appears strongest at the bar's two ends. Superficially, this would seem to indicate a source of magnetism inside of each end. "North" magnetism seems to emanate from one end, and "south" magnetism emerges from the other end. This is apparent only from an external point of view, however, for the field strength is actually greatest near the center of the metal bar rather than at either end. The magnetic poles thus represent no more than points of entry or exit, not points of origin or termination.

The "north-south" connotation was established through historical association. The magnetic field of the earth is oriented in a manner that places the "magnetic poles" in close physical juxtaposition with the geographic poles. In fact, a compass needle actually points toward the north geographic pole from many places on earth. These two entirely different concepts merge into one in some people's minds.

Even when using the north–south convention, there is a certain amount of confusion recognizing the difference between a north-seeking pole, which is a "true" north pole, and a "south" magnetic pole, which is the type that would be associated with the geographic north pole if there was such a thing as a magnetic "unit pole." In short, although a body can be polarized so that magnetic flux lines emanate from one end and enter the other, there is no such thing as a magnetic "unit polar" body.

## 2.7 FORCES ON MOVING ELECTRICALLY CHARGED PARTICLES

If a charged particle moves in an ambient magnetic field, the magnetic field accompanying that moving particle and the ambient field interact to exert a force on that particle. This force seeks to change the direction of particle movement.

A single moving particle carrying an electric charge is accompanied by a Biot-Savart magnetic field. Although the Biot-Savart field, strictly speaking, results only from an infinitely long conductor carrying many charged particles, a cross section of the magnetic field around the path of one particle, taken through that particle, has the same general, circular shape. A Biot-Savart field is stationary in relation to both space and time, however, whereas the field of a single particle changes in space with the passage of that particle.

*LORENTZ'S LAW* describes the force on moving electrically charged particles as follows:

LORENTZ'S LAW $$F = kQB\, dx/dt \tag{2.4}$$

where $Q$ is the magnitude of a particle's electric charge; $B$, the strength of the ambient field through which that particle moves; $dx/dt$, the velocity of particle movement; $F$, the resulting force on the particle; and $k$, the constant of proportionality.

A Lorentz force acts at right angles both to the direction of the ambient field and the direction of movement. It is a "sidewise" thrust exerted on charged particles as they move at right angles to flux lines.

A magnetically charged body in an ambient magnetic field experiences a force that attempts to move it from a position where its presence strengthens the ambient field toward a position where its presence weakens the ambient field. This is a manifestation of the principle that all systems seek minimum energy conditions.

The magnetic field surrounding an electron's path is clockwise, facing the direction in which the electron approaches the viewer. Under the circumstances illustrated by Fig. 2.2, the magnetic field surrounding a moving electron opposes the ambient field toward the bottom and strengthens the ambient field toward the top. Both of these circumstances cause a down-

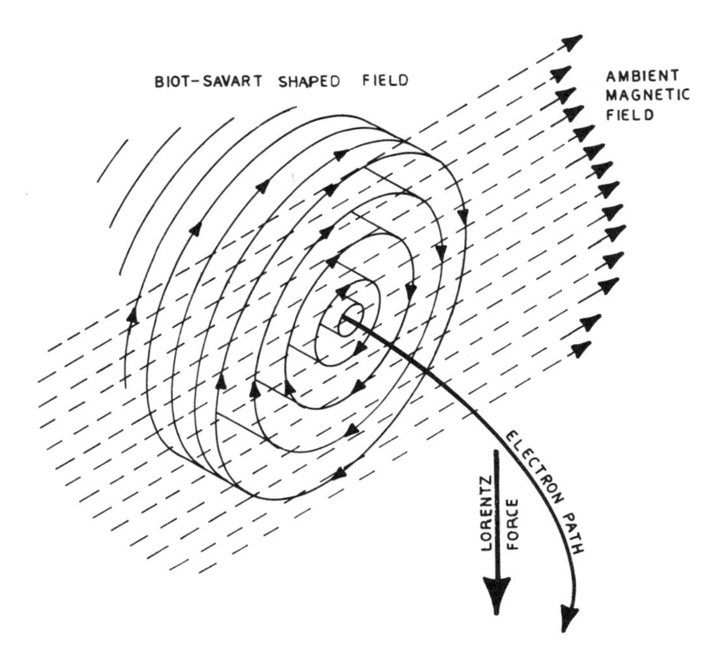


Fig. 2.2. Forces on the path of a negatively charged particle moving at right angles to the direction of a magnetic field.

ward force on the electron. Along the axis of the ambient field, the electron's field is at right angles to the ambient field. There is no force of any kind created by field interactions in a perpendicular configuration. In short, if a negatively charged particle moves from left to right across Fig. 2.2 and an ambient field is directed into it, the Lorentz force on that particle extends from the top to the bottom.

## 2.8 PATHS FOLLOWED BY MOVING CHARGED PARTICLES

An electron traveling at constant speed in a uniform magnetic field traverses a circular path with a radius of curvature that is inversely proportional to field intensity and directly proportional to speed. This radius is called a *MAGNETRON RADIUS;* it is calculated by Eq. 23.1 in Chap. 23.

As previously discussed, a charged particle moving in a magnetic field is deflected from a linear path by its interaction with that field. The faster a charged particle moves, the greater the deflecting force it experiences. Since momentum also increases with speed, however, faster moving par-

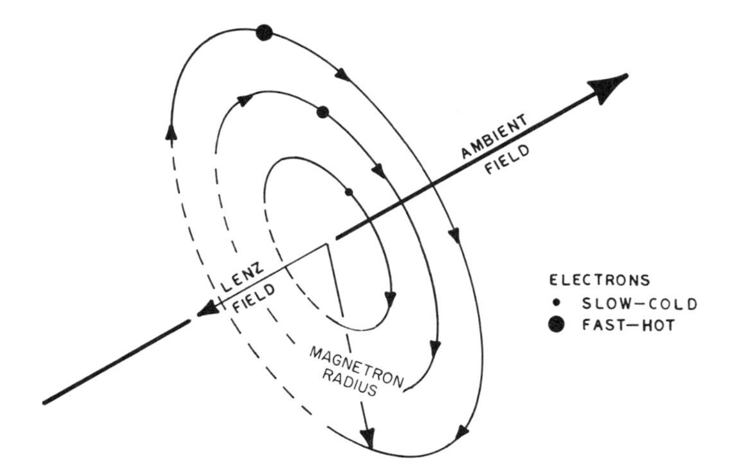


Fig. 2.3. Magnetron radius: The radius of curvature of the path of a charged particle moving in a magnetic field depends on how fast that particle is moving.

ticles are deflected less, as shown in Fig. 2.3. Since the velocity of an electron is determined by the voltage through which it falls, the strength of a magnetic field is directly related to measurements of both the voltage applied and the radius observed (see Fig. 23.3).

As shown in Fig. 2.3, an electron follows a clockwise path if it passes through an ambient field facing into the page. Because of its opposite electric charge, a proton follows a counterclockwise path if it passes through this same ambient field. In either circumstance, the energy of a charged particle is not affected because of its interaction with a magnetic field that causes a change in trajectory. The energy of the system in which the charged particle moves is minimized by the circular path, but the energy of the particle itself continues unchanged by such deflection.

A change in trajectory caused by Lorentz forces is superimposed on any other trajectory upon which a charged particle may be embarked. As shown by Fig. 2.4, the paths of charged particles are circling flux lines of a spiral nature if components of motion are present both in the direction of an ambient field and at right angles to that field.

In following traverses through "real" environments, an electron might travel its full magnetron circumference, or it might be deflected from this path by interaction either with other particles or with other fields. In large evacuated spaces, an electron can complete a full circle without interference. In the more general circumstances of travel, as within solids, electrons experience repeated deflections.

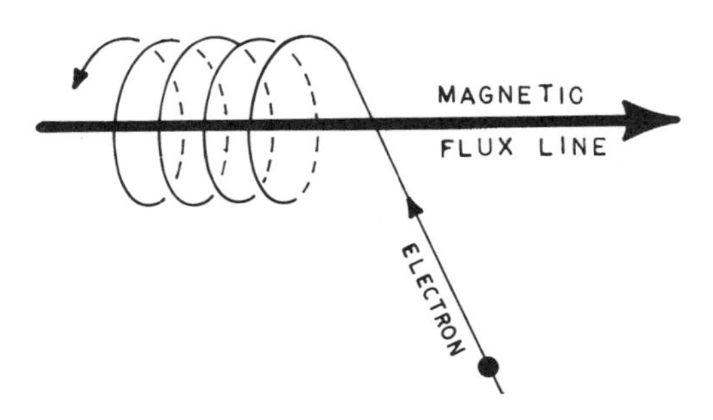


Fig. 2.4. Charged particles moving in magnetic fields in some direction other than at right angles to those fields travel in spiral paths around flux lines.

## 2.9 TOROIDAL MAGNETIC FIELDS

When electrically charged particles traverse circular paths, the cylindrical Biot-Savart field of Fig. 2.1 is bent into a circle.The resultant magnetic field has a toroidal shape.

A *SOLENOID* is a series of current-carrying loops of wire located about a common center line. Each loop adds its field to the toroid field of every other loop. As shown in Fig. 2.5, the result is a stronger field that retains the basic shape. The toroidal form of one loop's field is stretched out along the coil's axis a bit because of the spacings between the separate turns, but the summed field is strengthened at every point in a manner that follows the general toroidal pattern. Those lines of flux that are off a specific center line leave one end of the structure, curve around through space, and enter the other end. Flux lines are always continuous and complete a given path; no "loose ends" are left dangling without explanation. Even the center line is considered to leave one end and enter the other after completing a path of infinite curvature.

As shown by Fig. 3.1 in Chap. 3, an electron (or ion) continually spinning on its axis reduces the "solenoid" (or circular-path) concept to its minimum form. In this case, some number of units of electric charge continuously circulate in a loop of zero radius. Such circulation results in microminiature toroidal magnetic fields surrounding many (but not all) atomic particles. These small toroidal fields are significant contributors to the cohesive forces bonding all matter. In fact, a toroid-shaped magnetic field is a pervasive concept bearing on many of the subjects outlined in this text. Figure 12.4 illustrates variations in magnitude inherent in this field configuration.

A bar magnet repeats the same general shape. An electromagnet derives its strength from an association of current-carrying loops. A permanent

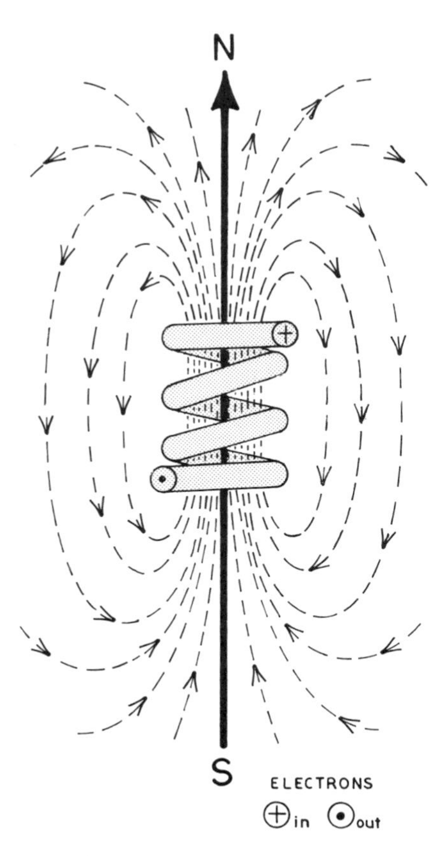


Fig. 2.5. Toroidal magnetic field: The magnetic field surrounding a current-carrying coil of wire assumes the shape of a doughnut or toroid.

magnet derives its strength from an association of spinning atomic-sized, electric-charged particles that generates many microminiature toroid fields.

## 2.10 MAGNETIC FIELD DIRECTION

The direction of a magnetic field (see Fig. 2.1) is determined according to the *RIGHT HAND RULE:* "If the right hand is placed around a conductor with the thumb pointing in the direction of current flow, the fingers point in the direction of the associated magnetic field." Similarly, if the right hand is placed around the solenoid of Fig. 2.5 with the fingers pointing in the direction of current flow, the thumb points in the direction of the magnetic field.

Since an electron, or other negatively charged particle, moves in the opposite direction from current flow (or positive particle movement), the left hand is substituted for the right in determining field directions. On the other hand, since spinning charged particles represent electric charges in a circular path, the Right Hand Rule applys to them (see Fig. 3.1).

Returning to the concept of Fig. 2.1, although the Biot-Savart field has a direction at any point in space, the net effect of all directions at all points in space does not concentrate flux lines at any particular location or in any one direction. On the other hand, a toroidal field has the same direction for all points along its axis, and the effects of many flux lines are concentrated in this same general direction. As a consequence, the magnetic field that results from an electric current in a circular path has a direction. This direction is determined by the plane of the circular path and by the direction of circulation.

Any body of material that produces a toroid field is said to be *POLARIZED* because of the direction of this field. The term "polarized" is derived from the concept of north and south magnetic poles as previously discussed. Since a polarized body produces a magnetic field with a specific direction, such a body is often depicted by a vector arrow. Although the vector arrow is only a line, it represents a complete toroid—a complete electrically charged and spinning particle. It should not be confused with the vector arrows used elsewhere to indicate field strength and direction.

## 2.11 LENZ'S LAW

According to *LENZ'S LAW,* "If the path of a moving charged particle is changed in any way by its interaction with a magnetic field, this change is always of such nature as to generate a new magnetic field which directly opposes the one which caused the change."

In Fig. 2.3, a *LENZ FIELD* is shown as a vector arrow pointing in opposition to the vector arrow that represents an ambient field.

Figure 2.2 illustrates those circumstances in which curvature is imposed on linear paths as a result of Lorentz forces acting on charged particles passing through ambient fields. Figure 2.3 extends this curve into a complete circle that can be considered equivalent to one turn of the solenoid of Fig. 2.5. As can be seen in the latter figure, the magnetic field associated with a charged particle's *circular* path is polarized, whereas the magnetic field accompanying a charged particle's *linear* path is not.

Clearly, then, the basic nature of magnetic fields surrounding charged particles is changed significantly when the paths of these particles are altered in any way.

## 2.12 MAGNETIC MOMENT

A magnetically polarized body immersed in a magnetic field is exposed to a torsional force that attempts to align that body with that field. This torsional force which is standardized in a standard magnetic field, is the *MAGNETIC MOMENT* of the magnetically charged body.

A magnetic field exerts a force on a magnetically charged body in a specific direction. This is a vector phenomenon. The charged body is accompanied by its own field that also has its own direction. This is a result of polarization. One force on a magnetically charged body, then, is attempting to change the position of the body in space at the same time that a second, torsional force is attempting to align the two fields.

Magnetic moment is a measure of the ability of the magnetically charged

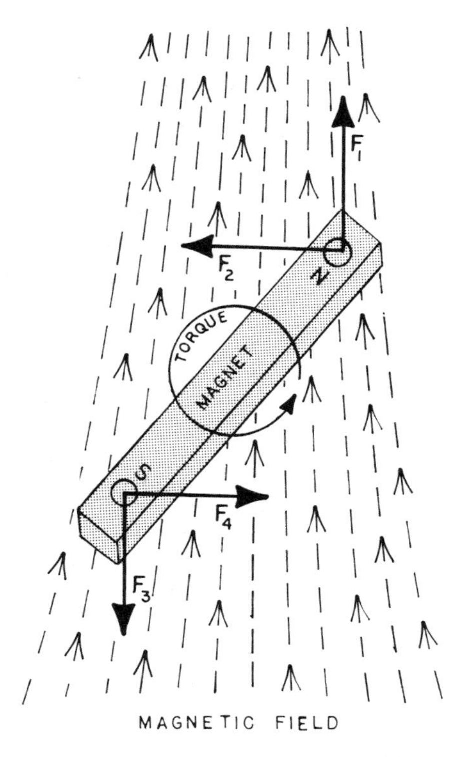


Fig. 2.6. Magnetic moment: A bar magnet is exposed to two forces when it is placed in a magnetic field, one of which attempts to align the magnet axis with the field, and the other, to move the magnet in the direction of the strongest field, the result being a torsional force known as the *magnetic moment*.

body to generate a magnetic field of its own. The greater the moment, the greater the field that is represented.

Figure 2.6 shows a bar magnet in an ambient field. If the ambient field is uniform, there is no force on the bar magnet attempting to move it through space. In a uniform field, $F_1$ and $F_3$ are equal but opposed and thus cancel each other's effects. (North poles move in one direction, whereas south poles move in the other.) If the ambient field is not uniform, however, $F_1$ and $F_3$ are not equal, and there is a differential space-moving force proportional to the degree of nonuniformity. In the sketch of Fig. 2.6, $F_1$ is greater than $F_3$ because the field at the location of $F_1$ is stronger than that at the location of $F_3$, a fact indicated by the greater density of flux lines. In any event, $F_2$ and $F_4$ assist each other in aligning fields whether they are equal or not.

## 2.13 MAGNETONS

The term *MAGNETON* is used here to describe those atomic-sized, electrically charged particles that establish an associated magnetic field by spinning on their axes. These spin-derived magnetic fields have a toroidal shape, as illustrated in Fig. 3.1.

Magnetons are commonly called "magnetic dipoles," both because of the directivity of their associated toroidal fields and because of the similarity in the shape of these fields to the electrostatic fields that surround electric dipoles. There are fundamental differences, however, between these two phenomena. Electric dipoles are constructed from at least two particles of opposite polarity, whereas magnetons are a one-particle structure. Then, too, electric flux lines begin on a particle of one polarity and end either on a particle of the opposite polarity or at infinity. Magnetic flux lines, on the other hand, are continuous. Although magnetic flux lines have direction, they have no beginning and no end.

The use of the term "magneton" instead of "magnetic dipole" is felt to dramatize both the nature of, and consequences of, these microminiature "magnets." In addition, this text refers to the phenomeon repeatedly in a number of different contexts, and the phrase "magnetic dipole" is much too clumsy for this purpose.

The term *BOHR MAGNETON* is used as the unit of magnetic moment for atomic-sized magnetic dipoles. The Bohr electronic magneton is commonly used in this sense, although the Bohr nuclear magneton is sometimes referred to.

It is a fundamental law that electrons (and also protons, nuclei, etc.) always spin at a characteristic fixed angular velocity, that is, an electron always has the same electric charge, the same mass, and the same spin rate. None of these factors are influenced in any way by the environment to which an electron is exposed.

Spinning electrons have much larger magnetic moments than do spinning nuclei. This is true because the very small mass of an electron makes it spin much faster than any nucleus.

## 2.14 FLUX DRIVING FORCE

The ability of a solenoid to drive magnetic flux through a magnetic circuit is a function of both the number of turns of wire in the solenoid and the amount of electric current flowing in those turns. Together, they create a *MAGNETOMOTIVE FORCE,* or mmf. Permanent magnets are capable of providing equivalent mmf's.

A magnetomotive force drives magnetic flux through magnetic circuits just as an electromotive force drives electric current through electric circuits. Magnetic circuits are analogous to electric circuits in some respects, although charged particles actually move through an electric circuit whereas there is no particle movement through a magnetic circuit.

The result of an electromotive force (emf) driving an electric current is described by Ohm's Law. *BOSANQUET'S LAW,* the magnetic analog of Ohm's law, establishes the flux in a magnetic circuit that results from an mmf driving a circuit reluctance as follows:

BOSANQUET'S LAW $$\Phi = NI/R \qquad (2.5)$$

where $\Phi$ is the total number of lines of flux in a circuit; $NI$ (where $N$ is the total number of turns of wire in a solenoid and $I$ is the solenoid current) is the force driving those lines (the mmf); and $R$ is the resistance to the driving force expressed as a magnetic *RELUCTANCE.*

*FIELD STRENGTH* is the driving force per unit length of driven circuit. *FIELD INTENSITY,* or flux density, is the amount of flux passing through a unit area of a driven circuit. It is sometimes called *MAGNETIC INDUCTION.*

The term "field strength" is a source of some confusion in the literature, for it is sometimes used to describe $H$ (the force driving the flux) and sometimes $B$ (the amount of flux driven). These are two different, if numerically related, parameters. It is a little like confusing voltage with current in an electric circuit of 1-ohm resistance.

## 2.15 FLUX RESISTING FORCE

*RELUCTANCE* is a characteristic of a particular magnetic circuit that defines the ability of that circuit to carry magnetic flux in response to a flux-driving force.

Ohm's resistance is proportional to the length of path traversed by an electron-stream, inversely proportional to the cross section of that path, and further inversely proportional to "conductivity." Conductivity is a term referring to the electric characteristics of the material from which a current-carrying path is fabricated.

Bosanquet's reluctance is proportional to the length of path traversed by magnetic flux, inversely proportional to the cross section of that path, and further inversely proportional to "permeability." Permeability is a term referring to the magnetic characteristics of the material from which a flux-carrying path is fabricated.

Reluctance is determined as follows:

RELUCTANCE $$R = K(L/A)\,(1/m) \qquad (2.6)$$

where $R$ is the reluctance of a path of length $L$, area $A$, and permeability $m$, and $k$ is a constant.

## 2.16 PERMEABILITY

PERMEABILITY is a per-unit characteristic of a material that describes the ability of that material to support a magnetic flux density. It is determined as follows:

PERMEABILITY $$m = (\Phi/A)/(NI/L) \qquad (2.7)$$

or

$$m = B/H$$

where $H$ (or $NI/L$ ) is the force it takes to drive flux over a unit length of path; $B$ (or $\Phi/A$ ) is the resulting amount of flux driven over a unit area of path (the flux density); $m$ is a unit-volumetric characteristic of the material from which the path is fabricated; $\Phi$ is the flux through the material; $A$ the cross-sectional unit area; $N$ the number of solenoid turns; $I$ the electric current flowing through the solenoid; and $L$ the unit length.

The permeability of free space is commonly referenced as $m_0$. Although flux does pass through a vacuum, it takes a magnetomotive force to make it do so.

## 2.17 TYPES OF MOVING CHARGED PARTICLES

All magnetic effects are derived from the movements of particles that have both mass and carry electric charge. The possibilities include the following:

### Electrons

1. The electric charge of an electron is one unit-negative, and its mass is very small.
2. The mass of all electrons is always the same, although the apparent mass is subject to variation depending on the environment.
3. All electrons always spin at exactly the same angular velocity.

### Holes

1. A hole is the positional absence of an electron in a lattice where one might otherwise be expected. As a result, a hole has an electric charge of one unit-positive and a very small mass.
2. As shown in Fig. 17.1, the movement of a hole involves the movement of an electron in the opposite direction. Therefore, the mass and the spin of a hole are exactly the same as those of the oppositely moving electron.

### Protons

1. Protons are positive ions with unit-positive charge and unit molecular mass.
2. A proton's positive-unit charge exactly matches the negative-unit charge of an electron, but the proton's mass is a great deal larger than that of the electron.
3. All protons spin on their axes at exactly the same speed, which is much slower than that of the electron.

### Positive Ions

1. Positive ions have various charges, which are integral multiples of the proton's charge, and various masses, which are integral multiples of the proton's mass plus some extra mass supplied by subatomic particles.

2. Only ions with an odd number of nucleons appear to spin on their axes.
3. Different ions with different masses appear to spin at different angular velocities.

## Negative Ions

1. Negative ions are subject to the same variations as are positive ions, but they carry negative rather than positive charges.

Any of these particles, in any combination, can move in various linear, or curvilinear, paths at various speeds. A number of similar particles, moving more or less together, is called a *BEAM*. Each particle in a beam can have a mass, a direction, and a velocity that is more or less the same as that of all its neighbors. In the more general circumstance, however, the individual velocities of the particles in a beam cover a range of possibilities that vary according to the Maxwellian Laws of Distribution. Those particles responding to a beam's mean velocity dominate in creating magnetic effects, whereas those that do not travel at the mean velocity create second-order effects.

When velocity is a criterion of interest, particles moving at high velocity are said to be *hot*, whereas those moving at low velocity are called *cold*. These are relative terms without absolute meaning.

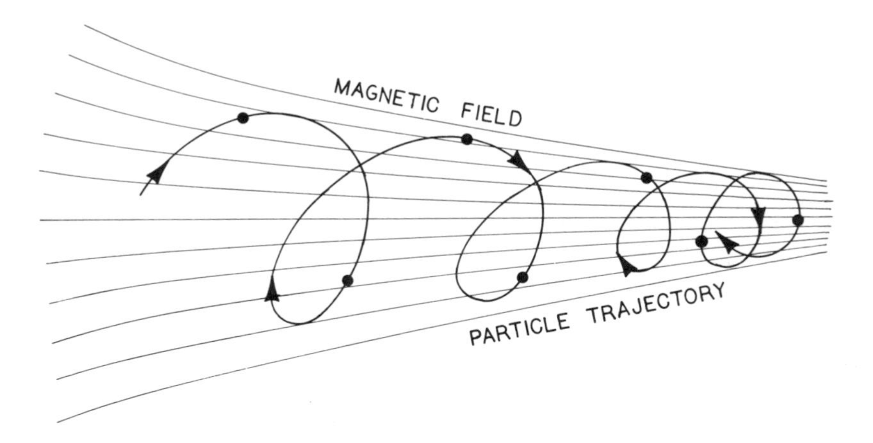


Fig. 2.7. Magnetic mirror: An electrically charged particle, embarked on a spiral path leading into a magnetic field of increasing strength, can be so configured as to reach a point of reversal ultimately and return toward its source.

## 2.18 MAGNETIC MIRROR

A moving charged particle circles in a spiral traverse if its velocity has components in the directions both of the magnetic field and at an angle to that field. In Fig. 2.7, a charged particle spirals into a magnetic field of increasing field strength. In this case, the diameter of the spiral decreases as the field strength increases. Given these circumstances, it is possible to configure a system in which the particle reverses its axial movement at some point of maximum field strength and returns along a path of decreasing field strength.

# 3
# CHEMICAL EFFECTS

Material/energy interactions depend on material structures. The component periodicity of structures, the population of conduction electrons, the forces binding orbit electrons, molecular compositions, and so on, all represent energy states that can be modified by the addition, or the abstraction, of energy in any one of several possible forms.

The cohesive forces that bind all particles together into bodies of material are an underlying theme of this entire transductive discussion. They start off with the simple electrostatic forces of attraction and repulsion, pass through a large number of complex bonding mechanisms, and finally develop into the beautifully ordered crystals and biological assemblages of the material world. In this hierarchy of association, each of the more complicated relationships is based on the force patterns of its simpler constituents.

Microminiature magnetic fields are a significant part of these bonding forces. Both magneton directions and electron trajectories contribute to structural formations. In some circumstances, an imposed magnetic field can affect a structure by influencing either magneton directions or electron trajectories. Phenomena based on material-structure/energy interactions are then modified by imposed magnetic fields. These phenomena include some chemical processes.

## 3.1 FORCE PATTERNS

All matter is constructed of discrete particles, each with as many as three different kinds of field possibilities. An association of two or more particles does not completely cancel the effects of the individual particles, that is, some force-creating field is left over in some direction and is therefore not completely expended in any particular association. Two or more of these "left-over" forces combine to form the next hierarchical level.

For instance, two electrons that spin in opposite directions are attracted to each other by their "opposite" magnetic fields.At the same time, they are repelled by their "same" electrostatic fields. Although the two magnetic fields may cancel in a close relationship, the electrostatic fields are strengthened by this same relationship. An electrostatic-force potential is then established that can be used for some other association.

Such force patterns continue to expand as larger numbers of particles combine until they include the macromanifestations of gravity, ambient magnetic fields, and large scale concentrations of electrostatic charge.

The magnetic fields that result from charged particle movements are a part of the chemical bonds that hold all matter together. Some of these charged particles traverse a variety of paths and some spin on their axes. These two fundamentally different types of particle movements are the basis for two general classes of magnetic phenomena and several subclasses. Phenomena based on path-traverse modifications are called *diamagnetic,* and phenomena based on axes-orientation patterns are called *paramagnetic, ferromagnetic, antiferromagnetic, ferrimagnetic, antiferrimagnetic,* or *helical,* depending on which particular pattern of spin-axis orientation is stable under given environmental and chemical circumstances.

## 3.2 MINIMUM ENERGY

A body of material represents a quantity of energy as well as a quantity of mass. The total energy present in a body is subdivided into its thermal, electrical, mechanical, electromagnetic, and chemical subforms. Furthermore, the total energy present in a body is distributed among the many atom-sized particles from which that body is assembled. Each energy subform is present at the minimum value it can achieve because this minimum is consistent with the minimum values of all other subforms.

For instance, an electron is attracted to a nucleus by an electrostatic force. The closer the electron comes to the nucleus, the less the electrostatic energy involved. At the same time, the electron is orbiting the nucleus, and the closer it comes, the faster it revolves and the higher its orbit speed. As high-speed orbits involve more mechanical energy than do slow-speed orbits, the electrical and mechanical forces are in opposition. The electrostatic force is centripetal, whereas the mechanical orbital force is centrifugal. These opposing forces finally reach a balance, with the electron orbiting the nucleus at a distance that represents the minimum of the sum of both the mechanical- and electrical-energy subforms.

If the total amount of energy present in a system is changed, the various ratios of energy distributions among both subforms and individual particles may also change. These changes can be rather dramatic. As an example, by adding a very small increment of heat at a particular temperature, a solid melts into a liquid without a change in that temperature!

One reason for such sudden changes lies in the mechanisms of quantum mechanics. According to quantum mechanics, energy changes are not continuous. All such changes occur in discontinuous, quantized steps. Although these individual steps are very small, they are still discrete. If many of these small steps occur under the same circumstances at the same instant, the accumulated effect can be very significant indeed. They can be explosive!

As a part of these quantized, minimum-energy relationships, a magnetic field is supported by some form of potential energy. If the field is very strong, the supporting energy must be greater than if the field were weak. In addition, more energy is required to fill large volumes with magnetic fields than small volumes. The minimum of this energy manifestation is achieved when the total field, in terms of both strength and volume, is minimized. Therefore, the various magnetic-field-producing particles within a body of material will always be found in those relationships that minimize magnetic fields because this minimum is consistent with the minimums of other energy subforms.

## 3.3 ATOMIC PARTICLES IN PATH TRAVERSE

Every atom consists of a certain number of electrons continually orbiting a nucleus. At any particular instant, an orbit-electron is traveling in a more or less circular path around its nucleus. The axis of this circular path is continually changing so that, over a period of time, an electron covers every point on the surface of a sphere.

This electric-charge circulation generates an instantaneous toroidal magnetic field. As the axis of the toroid is continually changing its direction, however, the field from one direction cancels the field from some other direction. As a result, an electron in spherical orbit does not generate a net magnetic field. In fact, the failure to create a magnetic field is what forces the electron into a spherical orbit!

If an externally supported magnetic field is imposed on this circulation, the situation will be analogous to that of the moving electrons in Fig. 2.2. Although an orbit is not a straight line, an electron in orbit will still respond to Lorentz forces when its moving charge interacts with an imposed field. The shape of the orbit is then modified by the interaction. An instantaneous circular orbit becomes elliptical whereas a continuing spherical orbit becomes ellipsoidal. This change of orbit from spherical to ellipsoidal generates a Lenz field in opposition to an imposed field.

In some complex molecules, bonding electrons travel paths that are more extensive and more intricate than those of electrons in nuclear orbit. Responding to the same Lorentz deflecting principles, however, an imposed field modifies these molecular orbits and Lenz fields result. Such Lenz reactions are not restricted to the orderly movements of electrons in orbit.

Both ions and electrons in plasmas and conduction electrons in solids are charged particles that follow generally random paths. The apparent disorder of these random activities, however, is not without purpose or even order. One consequence is a complete cancellation of all locally generated magnetic fields that might otherwise be established either by the path traverse or the oriented spins of these charged particles. Since it takes energy of some kind to support a magnetic field in space, any cancellation mechanism is a minimum-energy manifestation. The imposition of a magnetic field on these disordered paths modifies the disorder enough to form Lenz fields in opposition to imposed fields. It is of interest to note that random activities and spherical orbits are two quite different mechanisms designed to achieve the same end—local field cancellation.

These basic principles apply to the paths of all charged particles in all circumstances. The magnitude of the effect depends on how closely a charged particle is held to its original path. In this respect, the diamagnetism of the Meissner Effect of Sec. 24.1 is much larger than the diamagnetism of orbit electrons because the restrictions on the movements of conduction electrons near absolute zero are much less than those on orbit electrons.

## 3.4 LENZ OPPOSITION

*DIAMAGNETISM* is a direct result of those Lenz responses that are created when materials are immersed in magnetic fields. Diamagnetic materials act to reduce any magnetic field in which they are placed. As shown by Fig. 2.3, the vector of a Lenz field is always established in direct opposition to the vector of an imposed field. This is true for every direction, without regard for the orientation of a diamagnetic material in an imposed field.

Not only are field strengths reduced by Lenz oppositions, but a force is experienced by any body constructed from a diamagnetic material when that body is exposed to a magnetic field gradient. This force, which is a function of gradient direction and not of field polarity, seeks to move the body from a region of relatively strong magnetic field to a region where the field is weaker, toward a volume of space where electron-orbit modifications will be minimal.

The mechanical force on a diamagnetic material in a magnetic field is a measure of the atomic forces that attempt to maintain orbit-electrons in spherical orbits.

All materials are diamagnetic because, at the very least, all are constructed from atoms with orbiting electrons. Some materials support spin fields as well as Lenz fields. As spin fields tend to be much stronger than Lenz fields, various effects associated with spin fields are likely to dominate wherever the two occur together.

The diamagnetism that results from an electron's orbit modification is usually weak because the local fields seen by individual electrons are much stronger than the imposed fields that seek to change the orbits of these electrons. Since the orbit change is minimal, the Lenz opposition that accompanies this change is also minimal.

On the other hand, the diamagnetism associated with the randomly moving components of a plasma is much stronger than that associated with electron orbits because the ions and electrons in plasmas are not subjected to strong bonding forces. In this case, relatively weak magnetic fields introduce significant changes in particle trajectories.

The diamagnetism of many individual microparticles in various types of circulation can be equated to a current in a coil surrounding the body of a material that contains these particles. Diamagnetism can then be quantified through measurements of such a current.

## 3.5 SPINNING PARTICLES

All matter is constructed from some combination of fundamental particles. The three particles relevant to this discussion are electrons, protons, and neutrons. Neutrons and protons have the same mass, which is much greater than that of an electron; a proton's electrostatic charge is equal, but opposite, to that of an electron; a neutron has no electrostatic charge. All three of these types of particles spin on their axes.

In this discussion, the term *magneton* is used to describe those atomic-sized particles that exhibit both magnetic and mechanical moments as a consequence of a combined spinning mass and spinning electrostatic charge. Following this definition, spinning electrons and spinning protons can function as magnetons whereas spinning neutrons cannot because the latter have no electrostatic charge and therefore no magnetic moment. (Actually, neutrons do have a small magnetic moment, as explained by second-order theory that cannot be gone into here.)

Atomic nuclei are constructed from some combination of protons and neutrons collectively called NUCLEONS. When two nucleons are brought together, they pair, with one spinning in a clockwise direction and the other in a counterclockwise direction. The clockwise spins cancel the counterclockwise spins as far as the effects of mechanical moments are concerned. If the combination has no mechanical moment, it cannot function as a magneton. When neutrons and protons associate as nucleons, those nuclei with odd numbers of neutrons plus protons function as magnetons, whereas those with even numbers do not.

In Fig. 2.5, a toroidal magnetic field is shown to result from many electrically charged particles traveling a circular path. Figure 3.1 shows this same field surrounding a spinning body that carries an electric charge. In this case, the axis of the toroidal field is directed along the spin axis. Figure 3.1 illustrates the magnetic field associated with a magneton.

It is customary to represent a magneton pictorially by a vector arrow pointing in the field direction. In Fig. 3.1, this vector arrow is shown directed along the axis of rotation. When viewing such an arrow in a sketch, one should envision the entire toroid field.

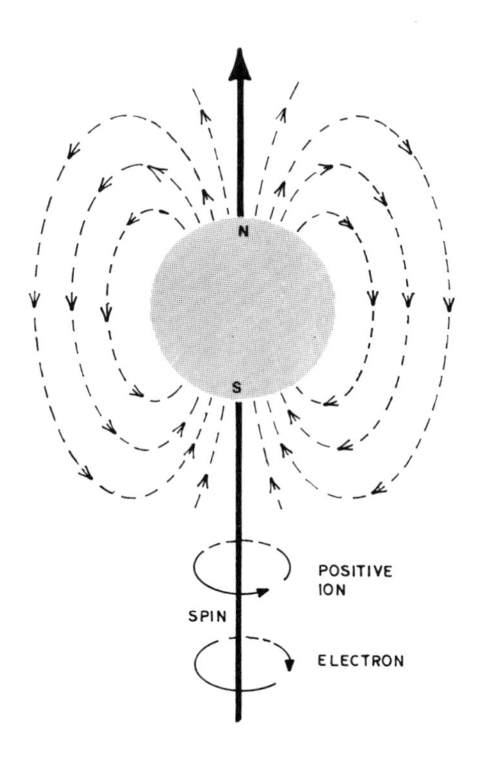


Fig. 3.1. Magneton: A spinning charged particle that generates a magnetic field of toroidal shape.

If a magnetic field is imposed on a magneton, Lorentz forces cannot act to change the shape of a circulating path because there is no such path. Since the radius of circulation is zero, it is not subject to change. Instead, a magneton reacts to an imposed field by changing its orientation, that is, by changing its direction. This orientation change is indicated by a change in the direction of the representative vector arrow.

A magneton's mechanical moment tends to maintain the spin axis in whatever direction it happens to lie, whereas its magnetic moment tends to align this same axis with any magnetic field to which it may be exposed. As long as these two aligning forces act in harmony, a magneton spins about its axis aligned with the magnetic field of interaction.

Electrons, protons, and some nuclei function as magnetons. The magnetic effects of electronic magnetons are much more pronounced than are those of other types of magnetons because the electron's smaller mass makes a faster spin possible. Since toroid field strength is proportional to spin velocity, high-speed electrons generate magnetic microfields that are a thousand times stronger than those of protons.

## 3.6 MAGNETIC FIELD EFFECT ON ORBIT ENERGY

In an isolated hydrogen atom, a single spinning electron continually orbits a spinning proton. No energy is consumed in maintaining either the spins or the orbit.

As specified by quantum mechanics, a spinning electron can occupy any one of several discrete orbits around a spinning proton. Each such orbit is equivalent to a potential energy level in which the difference between two orbits is represented by a finite quantity of energy. This quantity is sometimes called an "energy gap." At a temperature of absolute zero, a spinning electron circles a spinning proton in the lowest possible energy orbit. Under these circumstances, no energy is emitted by the atom involved.

At temperatures above absolute zero, spinning electrons move back and forth between allowed orbits. At low temperatures, this movement takes place only between the lowest energy orbits. At higher temperatures, movements between more orbits are involved. Each time a shift takes place from one orbit to another, the atom containing these orbits either emits or absorbs a burst of electromagnetic radiation. This activity occurs at a frequency determined by the quantum change in energy that exists between the two orbits. The *PASCHEN*, *BALMER*, and *LYMAN SERIES* of emitted frequencies can be used to define relative orbit energies.

When a number of independent hydrogen atoms are associated as a low-pressure gas, that gas continuously emits electromagnetic radiation at those precise frequencies determined by the various possible energy levels of single atoms. These frequencies are detected as "lines" in a refracted or diffracted spectral dispersion of the radiant energy.

In the absence of a magnetic field, the directions of the spinning-electron axes and the spinning-proton axes are completely random and continually changing. Since there are no forces in a gas that effect these directions, no energy is consumed in the process. The energy of each orbit is then determined by taking the random nature of these polar alignments into consideration.

On the other hand, some of the spinning electrons, and some of the spinning protons align their axes in an imposed magnetic field. Some align with the field, some against the field, and some continue their random activities. Consequently, there are forces present as a result of ambient fields that affect directions, and energy is consumed in movements between these directions.

By exerting a force for either parallel or antiparallel alignments, a magnetic field splits the single energy increment, represented by the difference between two orbits, into as many as three different increments. These three include the original, one increment of slightly greater energy, and a second one of slightly less. The energy of the system is not necessarily changed under these circumstances, merely redistributed.

If the magnetic field is very strong, all otherwise randomly directed axes are aligned either with the field or against it. The original energy increment, which depended on the randomness of the axes' directions, is thus eliminated. Under these circumstances, the spectral analysis of monatomic hydrogen gas shows two lines in the presence of a magnetic field for every single line displayed in that field's absence.

When atoms are more complex than hydrogen atoms, the number of orbital possibilities becomes greater, with more spectral lines indicating this complexity. The same splitting of spectral lines into pairs is also observed, however, in the presence of magnetic fields.

The splitting of spectral lines into two or more components by magnetic fields is called the *ZEEMAN EFFECT* in the presence of fairly weak magnetic fields. If an imposed field is strong enough to introduce orbital change, as well as alignment change, the emission-change phenomena observed are called the *PASCHEN-BACK EFFECT*.

Magneton alignments by imposed fields introduce directional coherence in molecular structures. As a result, the Zeeman-displaced spectral lines are constructed from polarized radiation even when an original line was not.

Interactive energies of any kind between adjacent atoms affect spectral distributions. When the pressure of a monatomic hydrogen gas is increased, the toroidal magnetic field of a spinning electron in one atom is exposed to the toroidal fields of electrons in other atoms. Energy levels in orbits are then slightly modified by the interactions between these toroid fields. When this happens, the original low-pressure, discrete spectral lines are broadened. Higher gas pressures increase the line broadening. Although imposed magnetic fields also split broadened lines, any distinction between the splitting and the broadening tends to blur.

As atomic relationships become more intimate, the number of spectral lines increases, and each line is further broadened. From molecules, to liquids, to glasses, to crystals the process continues until, in the circumstance of a solid, the spectral lines of single atoms have become broad spectral bands. When spectral lines have been sufficiently broadened, the effect of magnetic field splitting cannot be observed.

## 3.7 SPINNING PARTICLE PAIRS

Viewed from any position, individual particles can spin in either a clockwise or counterclockwise direction. There is little difference between these two directions because any spinning particle can change its spin direction simply by flipping over on its axis.

Spinning particles, charged or uncharged, tend to associate in pairs whenever the spin of one is opposite that of the other. If the two members of such a pair are physically close enough, the effects of the two spinning mechanical moments cannot be detected at any significant distance from the pair. As particles can function only as magnetons in the presence of both mechanical and magnetic moments, an association of pairs eliminates any magnetic effects the alignment of a magneton's magnetic field might otherwise have.

One reason for pair association lies in the basic principle that all matter seeks to exist in its minimum energy form. As it takes energy to support a magnetic field in space, the elimination of a magnetic field by pair association relieves the need for such support.

As previously stated, atomic nuclei are derived particles comprised of an assemblage of protons and neutrons. Together these are classified as *nucleons*. The number of protons present in a nucleus determines the type of atom, and the number of nucleons determines its atomic weight. *ISOTOPES* of the same type of atom are differentiated by their different atomic weights. They have the same number of protons but a different number of nucleons.

Nucleons are always close enough together to force their association in pairs. Therefore, a nucleus can exhibit the characteristics of a magneton only in the presence of an odd number of nucleons. As isotopes of the various atoms have different numbers of nucleons, at least one isotope of each atom exhibits magnetic characteristics and at least one does not. Nuclear magnetons are quite complex in their magnetic consequences. Although their spin can be derived only from the activities of one nucleon, their mass and electric charge are determined by a collection of nucleons.

The organization of electron orbits for all atoms follows the same general pattern. Each orbit is matched to an adjacent orbit whose quantum number is exactly the same except for the quantum number of spin. If each of two such orbits is occupied, one electron of the pair spins in one direction and the other electron in the opposite direction. As shown in Fig. 3.2, the field of one interacts with the field of the other within their own little microenvironment, and there is a complete cancellation of both fields as far as external forces are concerned.

Since any filled subshell is constructed from paired electrons, an externally detectable moment can result only from a partially filled subshell. This can happen only if there is an odd number of electrons in the atom. In fact, if the unfilled subshell is the outer subshell, two unpaired electrons of two adjacent atoms in a molecule can pair to cancel moment reactions. Since free electrons pair with ease, they have no detectable moments unless forced apart by thermal or other violent activities.

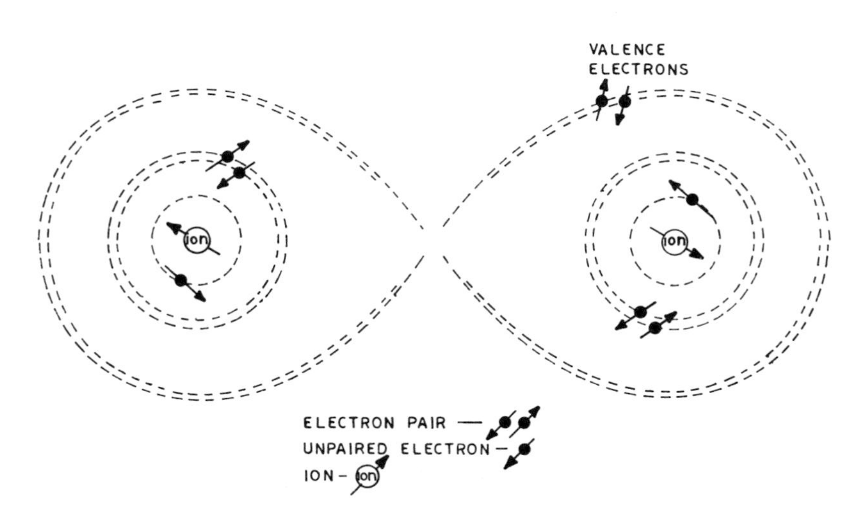


Fig. 3.2. Electron pairs: Electrons orbiting either a single nucleus or several nuclei tend to pair in field cancellation attempts. Only unpaired electrons are available to create magnetic phenomena.

Pair association eliminates the formation of detectable magnetons in most molecular associations, but if detectable magnetons are present, either nucleonic or electronic, they are free to interact with other magnetons and other magnetic fields.

## 3.8 UNPAIRED SPINNING PARTICLES

There is a force of attraction between two magnetons that is transmitted through an interaction of their two magnetic fields. Although this force is a complicated function of the distance of separation, its strength decreases at a rate that is inversely proportional to some high power of this distance.

Two magnetons pair whenever their two fields interact at close range since close-range fields are strong enough to completely overpower weaker fields coming from more distance sources. At the same time, the combination of the two strong fields at close range cancels whatever effects either field might otherwise have at a farther range. As a result, once magnetons pair in close association, they neither respond to, nor contribute to, other magnetic fields.

On the other hand, if forces are present that prevent a close approach, pair formation is not enforced. One magneton's field can still interact with that of another, but now the relationships are transmitted through much weaker fields. These weaker fields are of the same order of magnitude, or even less, than fields arising from other sources. Under such circumstances, each magneton functions as a single entity rather than as one of a pair. A magneton that is not a member of a pair can both react to, and contribute to, other fields.

Nuclei that contain an odd number of nucleons act as *NUCLEAR MAGNETONS*. Of the many types possible, single protons produce the most pronounced magnetic effects.

Unpaired electrons in inner unfilled subshells of an atom exhibit detectable moments as far as magnetic forces external to the molecular environments are concerned. In this case, the individual electrons are separated from each other by enough atomic distance to be out of range of each other's strong localized fields. Whenever the magnetic fields within a molecule are weak because the electrons are not close enough to each other, these same electrons are open to influence by magnetic fields that originate beyond the confines of the molecule in question. Because of their relatively high-speed spins, such electrons function as the strongest magnetons. The term *LANGEVIN MAGNETON* is an appropriate one for these electrons in recognition of Langevin paramagnetism, of which they are the constituent parts.

Although most conduction electrons pair, a few do not. (Conduction electrons are those whose movement represents electric current; that is, they "carry" the current.) Since the energies of conduction electrons follow Fermi-Dirac distribution patterns, the greater energy of these few separate their influences from the influences of the many having lesser energy. In such circumstances, the few electrons that have extricated themselves from pairing by their high-energy movements can function as *PAULI MAGNETONS*. The effects of such magnetons are not very pronounced since only a few electrons are involved. In fact, since conduction electrons also participate in diamagnetism, the negative diamagnetic effect can be stronger than the positive Pauli effect. In some materials, one effect is stronger, whereas in other materials, the reverse is true.

Electrons in atomic orbits follow spherical paths that do not support magnetic fields. Electrons in molecular orbits follow paths that are not spherical. The asymmetries of molecular orbits support magnetic fields that, for most purposes, resemble those of magnetons. As a result, the term *VAN VLECK MAGNETON* can be applied to this phenomenon even if, strictly speaking, the polarized magnetic field is not derived from a spinning charged particle.

## 3.9 DIRECTIONAL COHERENCE

Electrons and nuclei combine into various orderly structures partially as a consequence of their electrostatic charges. These structures include atoms, molecules, radicals, gases, liquids, glasses, crystals, and biological assemblages. The electrostatic charges are the main factor that determines the position of each particle in these structures.

Electrons and nuclei are also spinning charged particles functioning as magnetons. The magnetic nature of these magnetons is one of the bonding forces that act, in coordination with the electrostatic charges, to establish a particular structure. Although the electrostatic and magnetic patterns always act in coordination, they can be considered to be two different phenomena. The position of a magneton is established by the electrostatic forces, whereas the direction of its axis is controlled by magnetic forces.

In many circumstances, magneton directions can be manipulated without inducing a major effect on the electrostatic structure. In most circumstances, however, a particular pattern of magneton direction does have a minor effect on the electrostatic structure. Magnetotransduction is one of the subjects concerned with these minor effects.

In the absence of an aligning force, Langevin magneton directions are incoherent; that is, the direction assumed by a particular magneton is random and is constantly changing because of the thermal jostling of the various charged particles within their chemical bonds. Figure 3.3 illustrates a random alignment, providing one imagines a dynamic condition in which the magneton axes are constantly changing their directions in response to thermal forces.

Although random jostling is often considered to be a thermal phenomenon, it also achieves a minimum energy objective by minimizing the strength of the composite magnetic field that results from summing individual magnetic fields. The field of one magneton is cancelled both by the random alignments of its neighbors and by its own random changes in direction.

If a common magnetic field is imposed on all of the magnetons contained within a body of material, some of them will align themselves with that field. Two alignments are possible. One of them is parallel with the imposed field, and the other is antiparallel and opposite to the imposed field. The relative densities of the parallel and antiparallel populations depend on the strength of the imposed field. Stronger fields have larger parallel and smaller antiparallel populations.

Figure 3.4 shows a condition in which most of the magnetons are aligned but a few are still responding to thermally inspired random activities. In this sketch, those magnetons that achieve parallel alignment are balanced by those in antiparallel alignment except for a very small surplus parallel population. This surplus population is indicated by the larger arrows.

Thermal jostling causes individual Langevin magnetons both to pass back and forth between coherence and incoherence and to flip back and forth between parallel and antiparallel alignments. Although individual magnetons are in a continuous state of directional change, however, the two alignment populations remain constant for a given set of circumstances. This pattern rather neatly underlines one facet of the quantum theory: The magnetization of a material changes in intensity each time one magneton flips over, and this "flip" changes the relative population densities by two magnetic quanta.

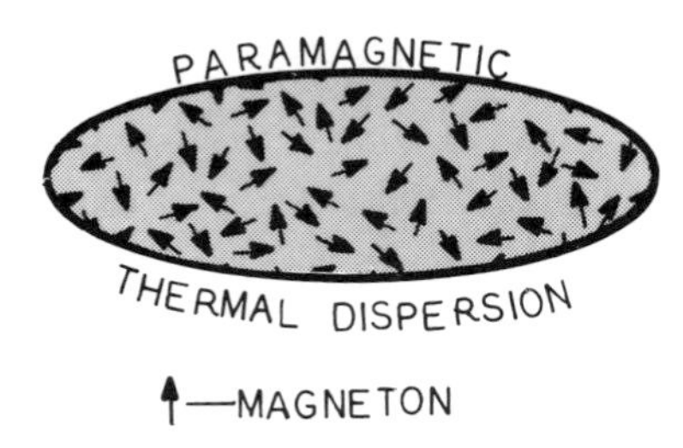


Fig. 3.3. Random magneton alignments: The thermal jostling of paramagnetic magnetons in what is otherwise considered to be a random pattern is in actuality a field cancellation phenomenon that minimizes magnetic effects.

The relation between parallel and antiparallel populations can be viewed as an expansion of the pairing principle since the parallel and antiparallel magnetons form in-transit pairs in the presence of relatively weak local fields in contrast to the locked-in pairs that result from strong local fields.

## 3.10 MAGNETIZATION

*MAGNETIZATION* is the term used to describe the magnetic field established within a material as a result of polarization. It is established by an imposed field because of two effects. One is the effect of the polarizability of the atom or molecule. This is the Lenz response. The other is the polarizing effect of magneton alignments.

Magnetization has the following properties:

1. In the absence of an imposed field or other aligning force, the magnetization of a material is zero.
2. In an imposed field, magnetization is a function of the strength of that field.
3. In diamagnetic materials, magnetization is negative; in other materials, it is positive.
4. In both diamagnetic and paramagnetic materials, magnetization is proportional to the imposed magnetizing force.
5. In other materials, magnetization is a function of an imposed force acting in concert with localized aligning forces. The magnetization of a ferromagnetic material is a complex function best described by the hysteresis loop of Fig. 4.2.
6. The magnetization of any material can be described as a magnetic moment per unit volume.

## 3.11 IMPOSED FIELD ALIGNMENTS

*PARAMAGNETISM* is the tendency for individual magnetons, the directions of whose axes are not determined by other forces, to strengthen an imposed field by aligning themselves with it. The individual magnetons can be influenced by external fields both because their atomic-scale influences are not cancelled at the local level and because their alignment is not enforced by local fields.

In most materials, magnetons are aligned by local fields below some critical temperature. In such a state, materials are not paramagnetic. Above this critical temperature, magnetons have a "normally random" orientation. This is illustrated in Fig. 3.3 where magneton axes point in every possible direction throughout a unit volume of material without any ap-

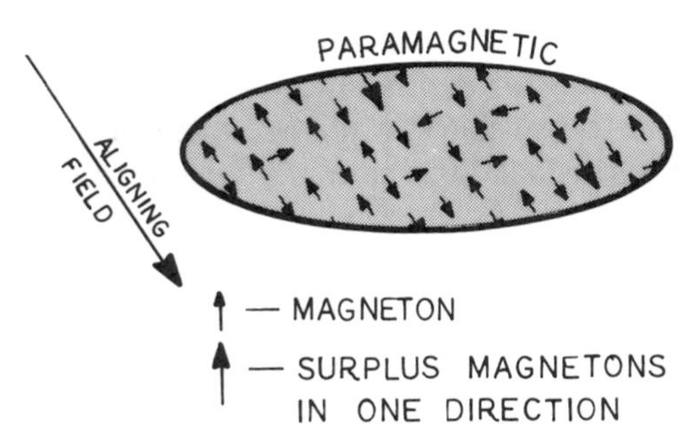


Fig. 3.4. Paramagnetic magnetons: When a paramagnetic material is placed in a magnetic field, some of the magnetons align themselves with the field in pairs opposition, some continue to respond to thermal jostling, and a few surplus unpaired magnetons align with, and strengthen, the imposed field.

parent regularity in their pattern. Thermal agitations tend to maintain this random condition on a dynamic basis in the absence of any other forces. In this condition magnetons are susceptible to paramagnetism. As shown by Fig. 3.4, imposed fields influence some of these incoherently directed magnetons to line up in a direction coincident with the imposed field direction.

Because paramagnetic magnetons align themselves with an imposed field, paramagnetic materials act to increase the strength of magnetic fields into which they are placed. Such alignment occurs whatever the orientation of the material in the imposed field.

Not only are field strengths increased by paramagnetic alignments, but a force is experienced by any body constructed from a paramagetic material when that body is exposed to a field gradient. As shown by Fig. 2.6, this force seeks to move the body from a region where the magnetic field is relatively weak to a region where it is stronger. This force is a function of gradient direction but not of field polarity.

All materials are diamagnetic, but some are also paramagnetic. Orbital diamagnetism has a small, negative effect, whereas Langevin paramagnetism has a larger, positive effect. Orbital diamagnetism is changed little by temperature changes because electron paths are rigidly held by quantized rules, and thermal forces do not change these rules.

Paramagnetism based on either Pauli or Van Vleck magnetons is also little changed by temperature because the populations of these magnetons tend to be very small and their alignments are saturated at low field intensities. In addition, these populations do not change significantly over wide temperature ranges. Although there is some variation in paramagnetism, it is very small.

On the other hand, Langevin paramagnetism is temperature-sensitive. The populations of Langevin magnetons are very large, and saturation can be achieved only with very strong fields not normally encountered in nature. In this case, the thermal agitation that tends to disperse magneton orientations is of the same order of magnitude as the imposed field forces that tend to regularize these orientations; that is, the stronger the imposed field, the more the magnetons tend to line up, but the higher the temperature, the more the magneton directions disperse. This aligning/disaligning relationship is described by the *CURIE-WEISS LAW*, as follows:

CURIE-WEISS LAW $$X = C/(T - T_C) \qquad (3.1)$$

where $X$ is the susceptibility of that material; $T$, the absolute temperature; $T_C$, the Curie-point temperature (the critical temperature below which magneton alignments are established by local fields); and $C$, the *CURIE CONSTANT* of the material. The relationship of the Curie-Weiss Law to other magnetic phenomena is shown in Fig. 3.12.

In paramagnetic materials, high magnetic fields at low temperatures can give paramagnetic saturation, in which case the Curie-Weiss Law no longer applies. Materials of various kinds exist in which the electrons alone, the nuclei alone, or both nuclei and electrons are paramagnetic; in which the nuclei are paramagnetic and the electrons diamagnetic; in which the electrons alone or both the nuclei and the electrons are diamagnetic (plasmas); and in which some electrons are diamagnetic and others paramagnetic.

## 3.12 FIELD INTENSITY

The flux density (field intensity) supported by a material that is exposed to a particular per unit-length of magnetizing force (magnetomotive force) is a characteristic of that material. It is mathematically stated as follows:

FIELD INTENSITY $$B = mH \qquad (3.2)$$

where $H$ is the magnetomotive force per unit-length of circuit; $B$, the resulting flux density per unit area; and $m$, the permeability of the material.

When a field is imposed on a material, the field that is established within that material is different than it would have been in a vacuum occupying the same space. The material either assists, or opposes, the efforts of the magnetizing force to drive flux through the space occupied by the material. Diamagnetic materials act to reduce the flux density below that which would have existed in a vacuum, whereas paramagnetic and other materials manage to increase the flux density.

A material's *PERMEABILITY* is the factor that determines flux density. Permeability, in turn, is a quantity that results from the magnetic moment per unit volume. The permeability of a vacuum is a convenient standard against which other permeabilities are compared. Considering the permeability of a material in relation to the permeability of a vacuum, we can determine a measure of the relative ease by which flux is established in a material, or *PER-UNIT PERMEABILITY,* as follows:

$$m_m/m_o = 1 + M/H \tag{3.3}$$

where $H$ is the forcing field strength; $M$, the magnetization of the material, that is, the magnetic moment per unit volume; $m_o$, the permeability of a vacuum; and $m_m$, the permeability of the material that has a magnetic moment per unit volume of $M$.

The ratio of magnetization divided by magnetizing force ($M/H$) is called the *SUSCEPTANCE* of a material. Susceptance is a measure of the ease by which paramagnetic magnetons align themselves with an imposed field. The susceptance of various materials can vary over a wide range from zero to over one hundred thousand. For isotropic media, susceptance is scalar; for anisotropic materials, it is tensor.

## 3.13 MICROENVIRONMENT

Each magneton responds to whatever field it sees within its own microenvironment. This is the vector sum of local fields and externally imposed fields. If the local fields are very strong, as they may well be, an imposed field does not add much to the local field, and magneton alignments are little affected by the imposed field.

In a close association of magneton pairs, the local field seen by each member of a pair is strong enough to exclude the effects of other magnetic fields completely. Even when magnetons are prevented from pairing by their position in a crystal lattice, or other molecular structure, field interactions between adjacent magnetons can still be much stronger than imposed fields. On the other hand, if a large number of magnetons can respond to an imposed field and align themselves with that field, their alignment will increase the field's strength profoundly. The *MAGNETIC MOMENT PER UNIT VOLUME,* or the ability of a material to add to an imposed field, is a function of the effect an imposed field has at the local level in relation to the effect of the local field at this same level.

## 3.14 SELF-ALIGNMENT

In some materials, magnetons, which would otherwise be paramagnetic, align themselves with local atomic fields of one kind or another. Within a group of magnetons, the orientation of a particular magneton can be established by the interactions of its field with the fields of its neighbors. At the same time, the orientation of that particular magneton will also influence the direction of its neighbors. The result of these mutual interactions can be the imposition of a regularity of some kind on the pattern of magneton orientations.

A number of stable magneton configurations are theoretically possible within a given volume of different types of materials. Figure 3.5 illustrates the possibilities with two simplified sketches. Here the direction of magneton 0 is determined by adjacent magnetons 1 and 3 if the "columns" are closer together than the "rows", or by 2 and 4 if the rows are closer together than the columns.

The particular orientation pattern assumed by a group of magnetons depends on a number of different atomic circumstances. Magnetic inter-

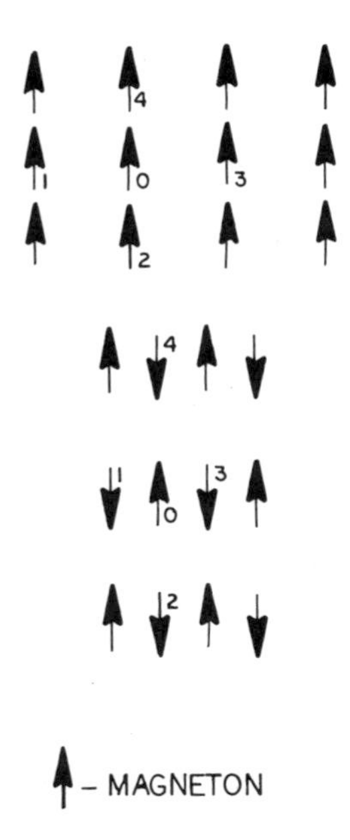


Fig. 3.5. Magneton alignments: Alignments are influenced by the directions of a magneton's closest neighbors.

actions are only a part of all bonding forces. Since bonding forces differ with different materials, each type of material has its own pattern of bonding forces and therefore its own preferred pattern of magneton orientations.

The magnetic forces that maintain a particular pattern are called *EXCHANGE FORCES*, whereas the orientation of many magnetons into some regular pattern is called an *EXCHANGE EFFECT*.

There are three general classes of exchange effects. Each is enforced by the overriding need to minimize external magnetic fields. Figure 3.3 illustrates random opposition (paramagnetism); Fig. 3.6 illustrates individual opposition (antiferromagnetism and helical magnetism); and Fig. 3.7 illustrates group opposition (ferromagnetism).

In all circumstances, the bonding forces formed are those that have the lowest possible energy content compatible with a given set of microenvironmental conditions. These conditions change with changes in the energy distribution. Since the exchange forces are relatively weak, an exchange effect will be modified by a significant change in the distribution of energy. For instance, thermal energy tends to bounce magnetons around to the point of their assuming a random orientation, whereas exchange forces tend to maintain a stable pattern of some kind. Since the exchange forces tend toward constant strength whereas the thermal forces increase with temperature, there is a temperature below which exchange forces dominate and above which thermal forces dominate. As a result, all materials with unpaired spinning electrons are subject to some kind of exchange effect below a critical temperature and are paramagnetic above that temperature. This critical temperature is analogous to the melting point of a solid.

## 3.15 INDIVIDUAL OPPOSITION

In *ANTIFERROMAGNETIC* materials, exchange forces tend to encourage magnetons to associate mutually in a pattern of polar opposition. As shown by Fig. 3.6, the magnetons take on a "normally opposite" arrangement in contrast to the "normally random" arrangement they assume in paramagnetic materials, as shown by Fig. 3.3.

The permeability of an antiferromagnetic material is positive, but it is less positive than that of a paramagnetic material under the same circumstances. The reason is that there are no exchange forces to restrict paramagnetic magnetons, whereas antiferromagnetic exchange forces tend to prevent magnetons from responding to or assisting an imposed field. Some few antiferromagnetic magnetons respond to an imposed field, however, if the populations between the two directions are altered by the

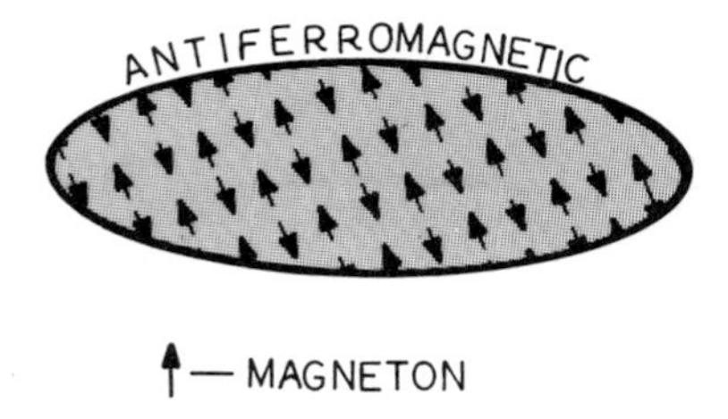


Fig. 3.6. Antiferromagnetic magnetons: In antiferromagnetic materials, magnetic effects are minimized when magnetons are aligned in pairs opposition.

imposed field. As in the circumstances shown in Fig. 3.4, the stronger the imposed field, the larger the number of magnetons that align with it and the fewer the number of magnetons that maintain their opposite, antiferromagnetic directions.

The "normally opposite" conditions break down with an increase in temperature until the point is approached at which the thermal energy equals the exchange energy. Thereafter breakdowns occur rapidly with increasing temperature until there are no more exchange forces to break. This state occurs at a finite temperature called the *NEEL TEMPERATURE.*

The permeability of an antiferromagnetic material increases with temperature until it reaches a maximum at the Neel temperature. Above this temperature, the material becomes paramagnetic, and the permeability decreases with further increases in temperature, as it does with all paramagnetic materials.

## 3.16 INDIVIDUALLY SKEWED OPPOSITION

In antiferromagnetic materials, adjacent magnetons are in polar opposition; that is, each magneton is oriented at an angle of 180 degrees from each of its closest neighbors. The net result is a complete cancellation of magnetic fields and a zero magnetic field for each unit volume.

In three-dimensional crystalline space, it is possible to have stable orientations between adjacent magnetons that are other than 180 degrees and still yield a zero magnetic field for each unit volume. Such arrangements are called *HELICAL MAGNETISM.*

The exchange susceptance of helical-magnetic materials is analogous to that of antiferromagnetic materials in the presence of either changing magnetic fields or changing temperatures. In other words, helical magnetism also breaks down at a Neel temperature.

## 3.17 GROUP OPPOSITION

In *FERROMAGNETIC* materials, magnetons are associated in groups. A *WEISS DOMAIN* consists of a volume of material in which exchange forces have aligned all magnetons in the same direction. This self-reinforcement is in contrast to a random orientation of magnetons all in total opposition, as is true of paramagnetic materials or an orientation of magnetons in individual opposition, as is true of antiferromagnetic materials.

A *DOMAIN* is a volume of material supporting a net magnetomotive force. All of the magnetomotive forces derived from the individual magnetons within a single domain add their effects since all of the magnetons are aligned in the same direction.

If one domain could exist by itself, it would be required to support a magnetic field that extended beyond its own confines. This would take quite a bit of energy, which exchange forces are generally not strong enough to supply. If a number of domains form within a single crystal, however, they can mutually direct themselves in a manner that minimizes external fields. In Fig. 3.7, although magnetons in one domain, or one group, are uniformly directed, several groups are organized in polar opposition.

In ferromagnetic crystals, a number of domains are found whose magnetic fields are oriented in directions of mutual opposition. Such an arrangement is temperature-sensitive because of the thermal effects on the relatively weak exchange forces. At a temperature at which the thermal energy equals the exchange energy, the group action is destroyed, and the material becomes paramagnetic. For ferromagnetic materials, this happens at a temperature called the *CURIE TEMPERATURE* or the *CURIE POINT.*

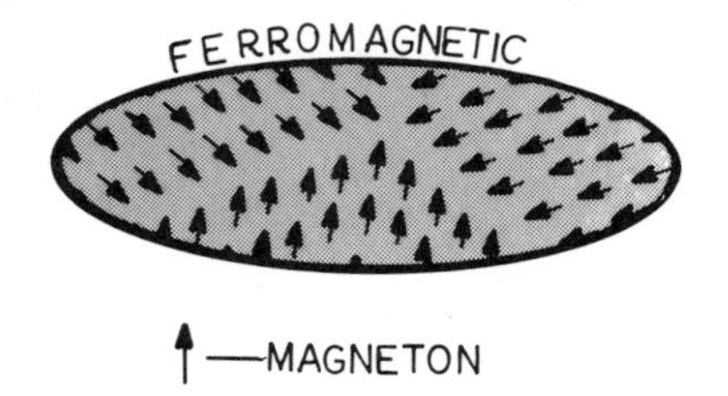


Fig. 3.7. Ferromagnetic magnetons: In ferromagnetic materials, magnetons are aligned together in groups where the groups are organized in field opposition in order to minimize magnetic effects.

## 3.18 DOMAIN WALLS

The boundaries separating domains are called *BLOCK WALLS*. These are not imaginary lines. On the contrary, they are physical entities several hundred atomic distances wide. All magnetons inside a domain are ordered in the same direction, whereas each within a wall has its own direction, one that differs slightly from those of adjacent magnetons. As shown by Fig. 3.8, a gradual transition takes place in the direction of these wall magnetons, from the direction of one domain to that of an adjacent domain.

The greater the angle between adjacent wall magnetons, the greater the force needed to maintain their wall against the self-aligning exchange forces. Smaller domains require larger angles, whereas larger domains have more difficulty in cancelling each other's magnetic fields. The wall forces tend to make each domain as large as possible, whereas the field cancellation forces tend to minimize its size. The actual size as well as the actual direction of a domain is a compromise, or balance, between the exchange forces, the forces that are required to maintain the angle between wall magnetons and the forces that minimize external fields.

A given ferromagnetic crystal is in a state of equilibrium in which all forces are in balance. Under these circumstances, there is no lateral force on a wall; that is, a wall can be easily moved if it is exposed to a force

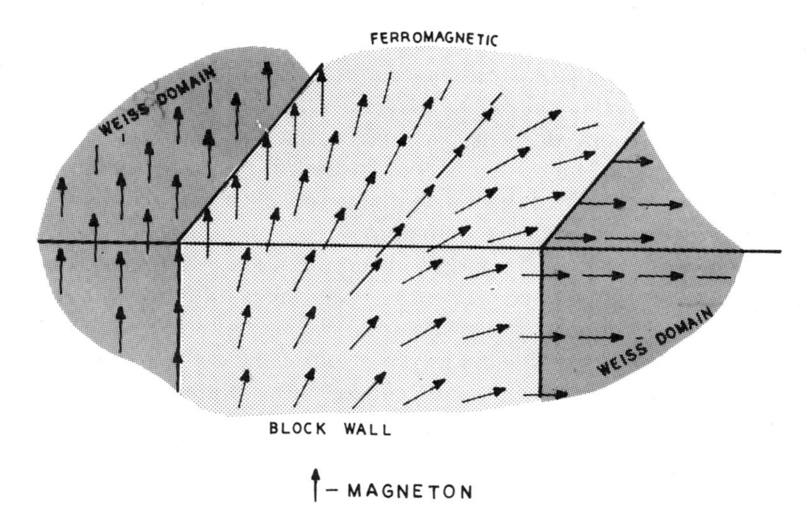


Fig. 3.8. Block wall: A "wall" exists between two groups of ferromagnetically aligned magnetons. The alignment of each wall magneton changes gradually from that of one group to that of its neighboring group.

directed at right angles to its own plane. For instance, if an externally imposed force on a domain attempts to change its size, location, or direction, walls will readily move to accommodate that change.

Wall movement comes about when one domain magneton becomes a wall magneton on one side of a wall at the same time that a wall magneton becomes a domain magneton on the other side of the wall. After any such movement, the wall-angle/exchange-force balance is exactly the same as it had been before the movement. There is, then, no force on a wall that attempts to locate its position in space other than the extenal field minimization forces.

Although walls can be easily moved, the *RAYLEIGH-EWING EFFECT* provides a force that opposes such motion. Since this force is a function of movement rate, it is also called *MAGNETIC VISCOSITY*.

## 3.19 CRYSTAL RESTRICTIONS

There are forces within a crystal that attempt to align magneton axes with one of the crystal axes. In the absence of other forces, each domain field assumes one, out of some finite number, of the preferred crystalline directions. For instance, a cubic crystal has six preferred directions. This phenomenon is called *MAGNETIC ANISOTROPY*. It is a direct result of crystal lattice anisotropy.

These crystal-axis aligning forces can be countered by other forces and the magnetons consequently aligned in directions other than crystal-axis directions. *MAGNETOCRYSTALLINE ANISOTROPIC ENERGY* is the force needed to rotate domain directions against crystal anisotropy.

In the absence of an external force, several domains form in ferromagnetic crystals, each of them oriented in one of the preferred directions so that the net crystal field is minimum.

In addition, fields are created by discontinuities in a material such as those found between adjacent crystals, at surfaces, at empty lattice sites, at inclusions, and the like. These *MAGNETOSTATIC FIELDS* oppose magneton self-alignments. This interruption of the crystal symmetry unbalances the exchange forces and reduces the spontaneous tendency to create domains.

Magnetostatic fields disrupt the equality of magneton angular difference that is found in domain walls and reduce the energy required to maintain wall-angle differences. A relatively large force is then required to move a wall from any position that includes a magnetostatic field.

## 3.20 MULTIPLE DOMAINS

It is the domain concept that distinguishes ferromagnetism from paramagnetism and diamagnetism. The latter two are characteristics of individual atoms or molecules whereas ferromagnetism is characteristic of groups of certain types of atoms.

The size, shape, location, and direction of domains are determined by all of the forces present in the crystalline system since these forces participate in a joint minimum-energy configuration. Domains vary in their maximum dimension from microns to centimeters depending on the geometry, history, and environment of the material involved. The exchange forces tend to give the domain magnetons the same direction; the magnetocrystalline energy determines what this direction will be; and the requirement for a minimum external field forces a number of domains to form in a field-cancellation effort. (See Fig. 3.9.)

Magnetocrystalline energy makes it easier to magnetize crystals in the direction of a crystal axis. If crystals are joined together in a completely random fashion, the effect of crystal directivity is lost, and bulk materials exhibit more or less equivalent characteristics in all directions. If crystals are aligned in a pattern of some axis regularity, however, it is easier to magnetize bulk material in one direction than in another because of bulk anisotropy.

Ferromagnetic materials show remarkable differences in their magnetic properties. These include their Curie Temperatures, saturation points,

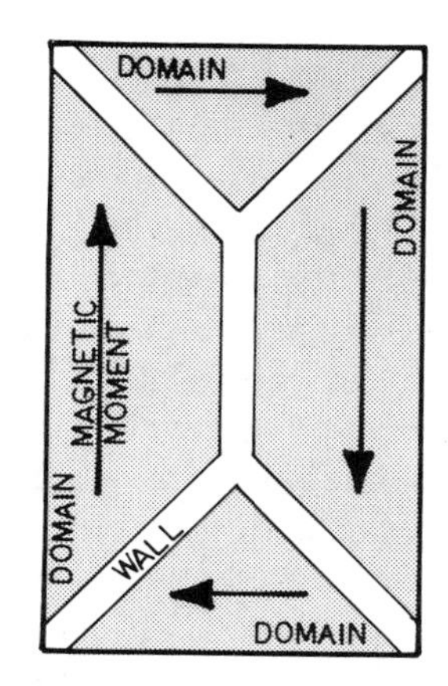


Fig. 3.9. Multiple domains: In ferromagnetic materials, each group of magnetons is normally aligned according to one of the crystallographic electrostatic axes.

permeabilities, coercive forces, hysteresis losses, and, in general, the whole size and shape of their hysteresis loop. These properties are affected by gross composition, impurities, mechanical strain, heat treatments, crystal structure, and crystal orientation.

Alloying and forming techniques are used to develop the most desirable characteristics for specific applications.

## 3.21 PHASE TRANSITIONS

The cooperative behavior of atomic particles in a unique association is commonly called a *PHASE*. Solids, liquids, and gases are well-recognized phases. One phase manifestation is a characteristic pattern of energy distribution established at the particle level.

Materials assume the configuration of a particular phase under specific environmental conditions. One phase may persist over one range of environmental variations but change to some other phase for some other environmental range. Between the two ranges there is a critical region of transition.

The characteristics of the many types of phase transitions encountered vary widely. Some, as in the case of an explosion, take place rapidly after being triggered by a small energy increment. Nevertheless, all phase transitions are continuous processes, not discontinuous jumps from one state to another. In fact, some phase transitions occur over a stable, reproducible, and reversible environmental range.

Both ferromagnetic and antiferromagnetic materials are paramagnetic above a critical temperature. Just below this temperature, there is a short range of temperature variations in which materials change from one phase to another and partake of the attributes of both.

In both ferromagnetic and antiferromagnetic materials, exchange forces attempt to align magnetons at the same time that thermal forces attempt to disperse these alignments. The temperature of a body of material is the mean of the temperature of each of its individual particles. The temperature of a particular particle, however, follows a Fermi-Dirac distribution, which differs for each temperature gradient. In addition, each particle's temperature undergoes a continous change within this distribution. In this seething mass of activity, the hotter magnetons shed their exchange forces at a lower body temperature than do the cold magnetons. Each magneton reaches the critical temperature at a slightly different body temperature.

Figure 3.4 illustrates the phase transition of a material from antiferromagnetic to paramagnetic almost as well as it does the alignment of a paramagnetic material in an imposed field. Figure 5.3 shows the magneton alignments for a phase change from paramagnetic to ferromagnetic.

Other forms of energy redistribution can also affect the exchange forces. For instance, an imposed field can force ferromagnetism upon some antiferromagnetic materials. *METAMAGNETISM* is the phenomenon of transition between antiferromagnetism and ferromagnetism in which some magnetons are oriented in individual opposition and others in domain opposition.

In all circumstances, when the environment of one phase is changed suddenly to the environment of another, a time delay is experienced in the change of phase.

## 3.22 DOUBLE LATTICE

Crystals of a pure substance have well-defined lattice structures extending in three-dimensional space. If two materials are combined by dissolving one in the other, the result can be a new crystal in which each constituent type of molecule assumes a unique position in a more complicated lattice pattern. This lattice can take on a symmetry of novel significance since the resulting structure can be considered to be two crystals of two different substances that simultaneously occupy the same space. Although these "two" crystals are really only one, each substructure can assume characteristics that differ from those of the other substructure in certain circumstances.

For instance, the ferromagnetic exchange forces can function separately as if there were two crystals instead of just one. In this case, two different sets of magnetons are individually influenced by their own lattice forces

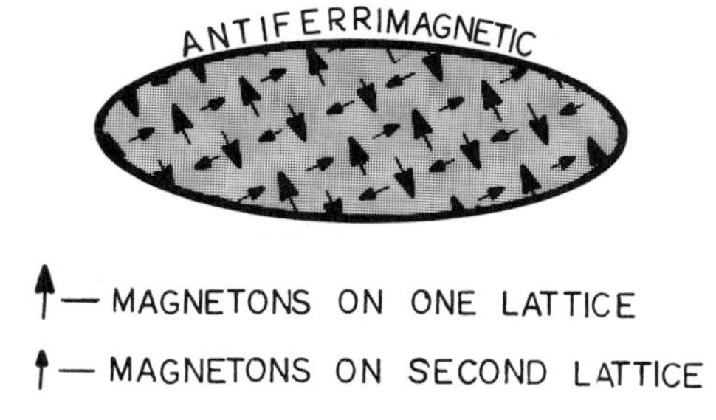


Fig. 3.10. Antiferrimagnetic magneton alignment: In contrast to antiferromagnetic materials, antiferrimagnetic materials function as if there were two sets of magnetons occupying the same space, where each set is independently organized in pairs opposition.

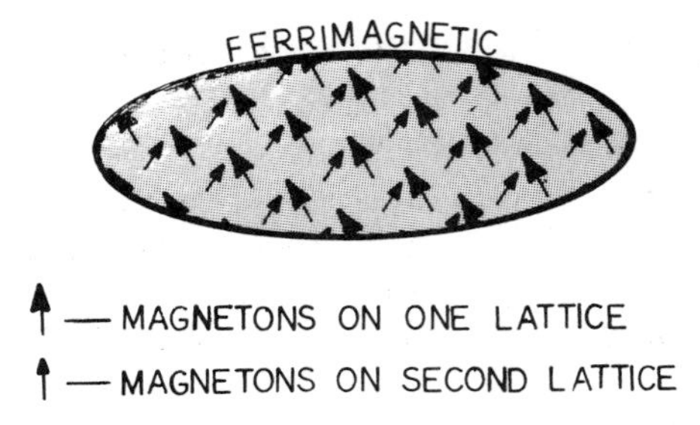


Fig. 3.11. Ferrimagnetic magneton alignment: Ferrimagnetic materials contain two sets of ferromagnetic magnetons simultaneously occupying the same space but not in pairs opposition. In ferrimagnetic materials each group forms in group opposition as is true for ferromagnetic materials.

as well as by those of their diluent. When this occurs, the material is said to be *FERRIMAGNETIC* (or *ANTIFERRIMAGNETIC*). In Fig. 3.10, the larger arrows represent the magnetons on one lattice; the smaller arrows, the magnetons on the other.

The phenomenon of the double lattice leads to a wide variety of theoretical possibilities; for example, lattices in which both crystals are ferromagnetic in the same, or in opposite, directions; in which the two are individually antiferromagnetic; in which they form around different crystal axes; and so on. In Fig. 3.10, magnetons on different axes form an antiferrimagnetic pattern, whereas those in Fig. 3.11 are ferrimagnetic. Ferrimagnetic as well as ferromagnetic materials form domains, whereas antiferrimagnetic materials can have two Neel temperatures.

As commonly used, the term *FERRITE* applies to an insulating ferrimagnetic material that has enough resultant magnetic moment with the appropriate exchange forces to form domains in a manner similar to that of ferromagnetic materials.

## 3.23 MAGNETIC ANISOTROPISM

An *ISOTROPIC MATERIAL* possesses the same properties in like degree in all directions. *ANISOTROPIC MATERIALS*, on the other hand, have characteristics that differ in different directions.

It is possible for a property to be constant no matter where it happens to be measured in a bulk quantity of material, or under what orientation it happens to be viewed. This type of property is a scalar quantity characterized by magnitude only. Other properties can be vector quantities that have both magnitude and direction.

It is also possible, however, for a property to be constant throughout a volume of material and still assume a different value depending on the direction from which it is viewed. This type of property is called *TENSOR*. For instance, a magnetizing force that is not directed parallel to a crystal axis might produce a flux that is parallel to that axis. Under these circumstances, the flux is not in the same direction as the driving force. The ratio $B/H$ is then tensor, as is the susceptance $M/H$.

In a *GYROMAGNETIC* material, the dielectric constant is scalar, whereas the permeability is made tensor by an applied magnetic field. In a *GYROELECTRIC* material, the dielectric constant is tensor, whereas the permeability is scalar.

## 3.24 SUSCEPTANCE VARIATIONS

As indicated by Eq.3.3, susceptance represents the quantity of magnetic moment that is established in a unit volume of material as a result of exposure to a magnetic field intensity of unit magnitude. Different materials have significantly different values of susceptance. Figure 3.12 illustrates some general susceptibility characteristics exhibited by various materials under changing temperatures. No attempt has been made to draw this sketch to relative scale. These curves give an indication of the relationships between thermal forces and the forces required to align magneton axes as these relationships change with temperature. As shown in Fig. 3.13 some materials exhibit hysteresis in the temperature–susceptance relationship.

## 3.25 FLUX REFRACTION

When a ferromagnetic body composed of two materials of different permeabilities is placed in a magnetic field, the resulting flux lines change direction at the surface interface between the two materials. In passing from one ferromagnetic medium to another, the ratio of the tangents of the angles of incidence and refraction is constant. This phenomenon of *FLUX REFRACTION* is in contrast to the refraction of light, for which the ratio of sines is constant.

## 3.26 SPIN WAVES

All of the magneton axes throughout a volume of material can be aligned either by exchange forces or by imposed fields. If one aligned magneton is momentarily forced from its orientation in the common alignment, this deviation does not remain localized but propagates as a spin wave throughout the lattice structure of the material.

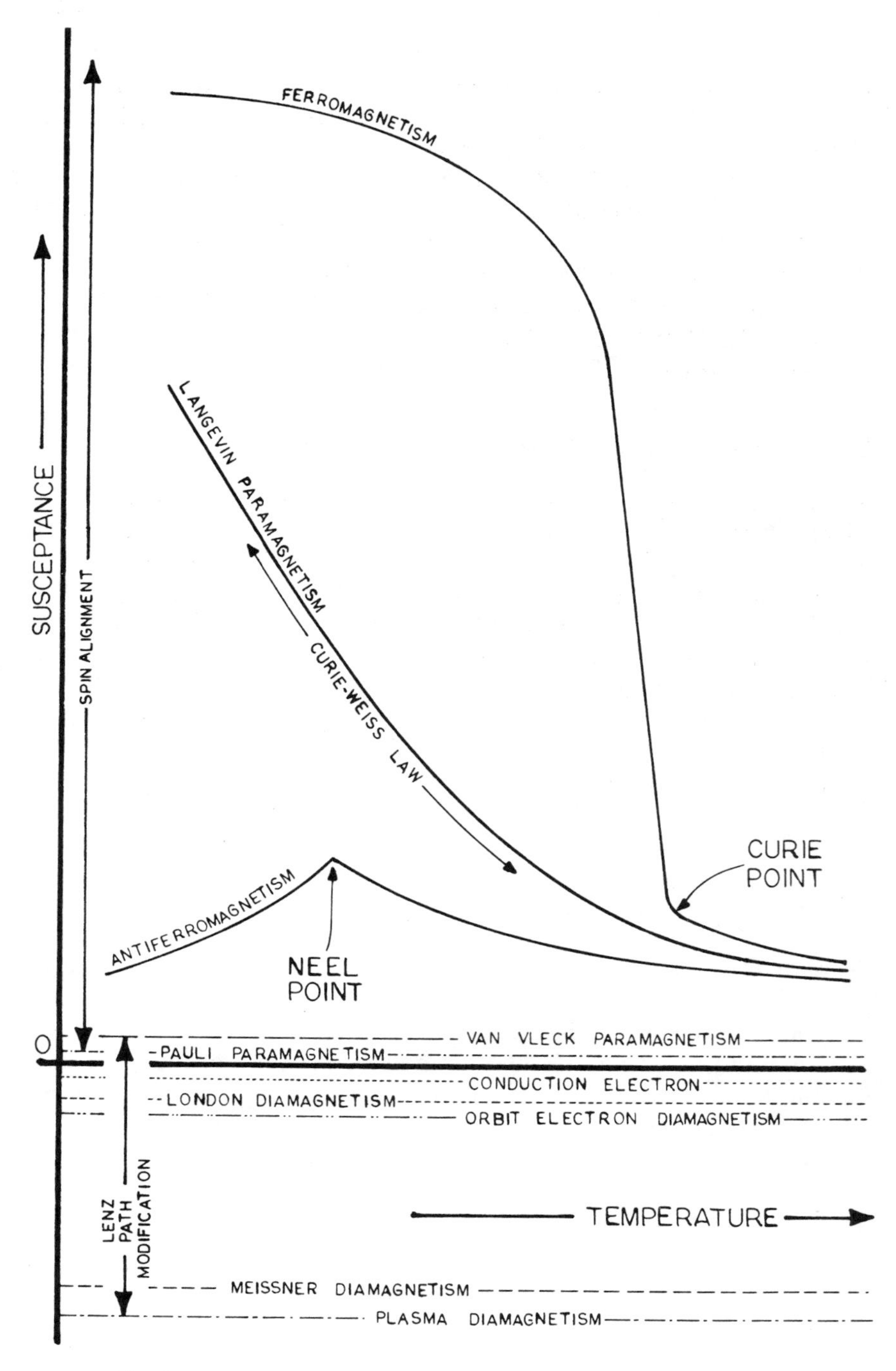


Fig. 3.12. The effects of temperature on various magnetic phenomena.

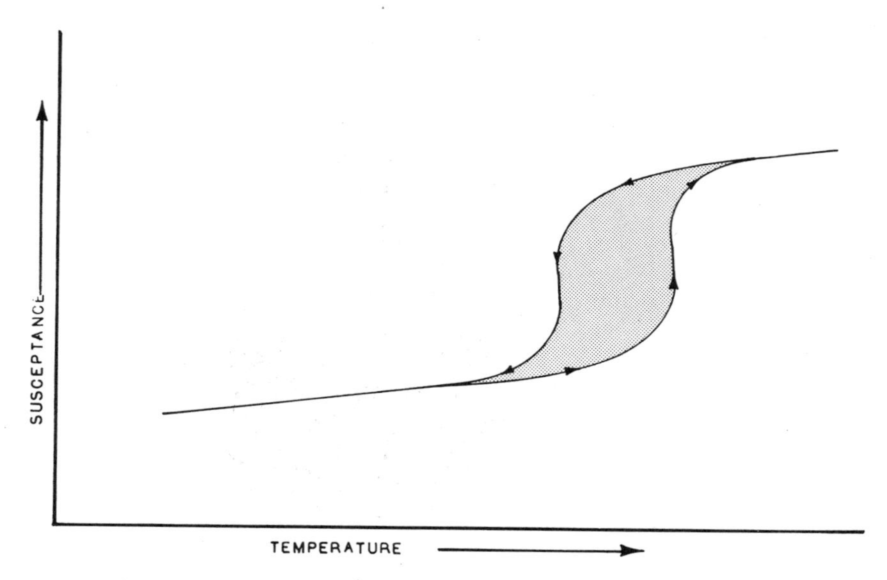


Fig. 3.13. Susceptance/temperature hysteresis: In some magnetic phenomena, the value of magnetism achieved at a particular temperature depends on whether the material has recently been exposed to either a cooling or heating cycle.

*SPIN WAVES,* which are oscillations in the relative orientations of magneton axes on a lattice, are in contrast to lattice vibrations, which are oscillations in the relative position of atoms on a lattice. Figures 8.7 and 8.9 depict this phenomenon. When lattice vibrations are quantized, they are called *PHONONS*. When spin waves are quantized, they are called *MAGNONS*. Magnons affect the flow of heat, electric current, and other energy transport phenomena.

## 3.27 MAGNETO-ELECTRIC EFFECT*

Magneto-electric materials are magnetized by electric fields and electrically polarized by magnetic fields. They contain molecules that are asymmetrical, the asymmetry including a nonuniform distribution of electric charge. In other words, each asymmetrical molecule represents an electric dipole.

As a further contribution to asymmetry, the spins of the various spinning particles of each participating molecule are not completely balanced by oppositely directed spins of adjacent particles. As a result, a net spin is present in each molecule in addition to the net distribution of electric

*Note that areas of magnetotransductive interest have been indicated by underscoring.

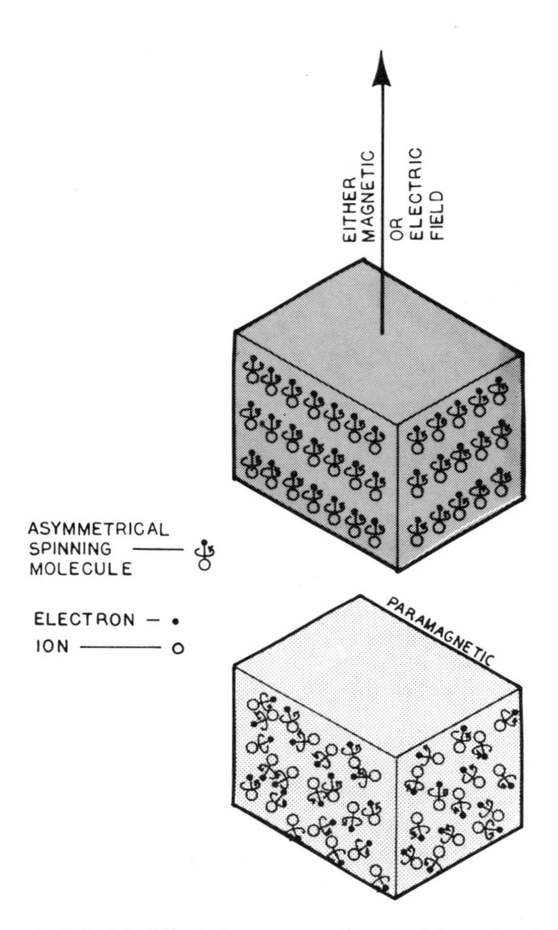


Fig. 3.14. Magneto-Electric Effect: A paramagnetic material constructed from asymmetrical nuclear magnetons is electrified in a magnetic field and magnetized in an electric field.

charge. In other words, magneto-electric materials contain spinning electric dipoles. In the absence of aligning forces, these spinning dipoles are randomly oriented in such a manner that both their electric fields and their magnetic spins are cancelled by neighboring alignments.

Those materials that support random alignments might be called *PARAMAGNETO-ELECTRIC* materials. FERROMAGNETO-ELECTRIC phenomena have been observed, and it is probable that *FERRIMAGNETO-ELECTRIC* materials also exist.

Since both electric fields and magnetic fields are created uniaxially by these spinning, asymmetrical molecules, any force capable of aligning the

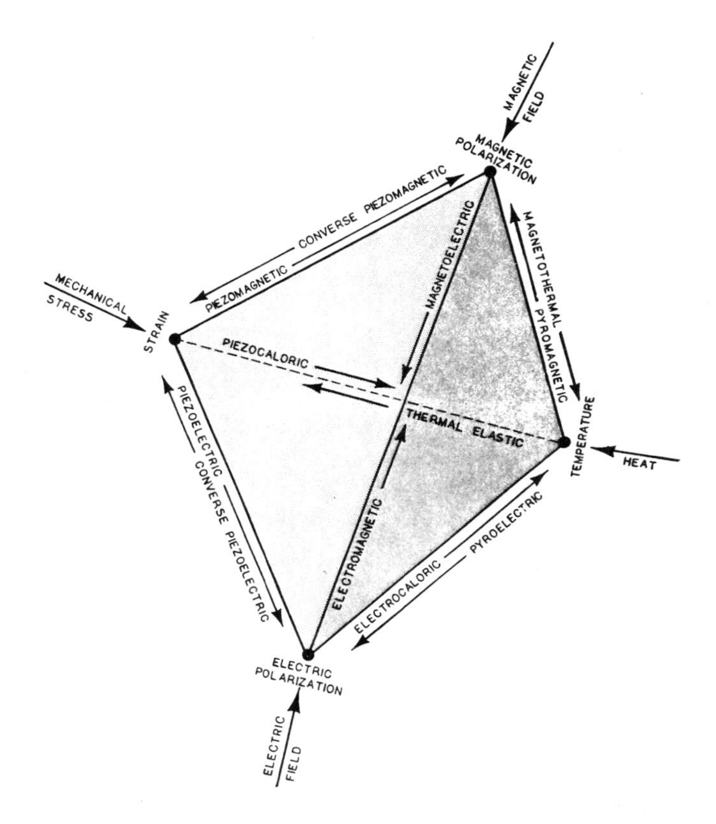


Fig. 3.15. Magneto-electric Voight diagram: In a material exposed simultaneously to mechanical force, heat, and magnetic and/or electric fields, any one of these factors will influence the effects that would otherwise be imposed by another.

latter will create both electric and magnetic fields throughout magneto-electric materials. (See Fig. 3.14.)

A magneto-electric Voight Diagram, as shown in Fig. 3.15, indicates the presence of both a *PYROMAGNETIC EFFECT* and a *PIEZOMAGNETIC EFFECT*, respectively associated with the pyroelectric and piezoelectric effects inherent in all crystalline materials.

# 4 MAGNETIC HYSTERESIS

Magnetization resulting from either diamagnetism or paramagnetism is proportional to magnetizing force. Only one level of such magnetization is possible for each value of magnetizing force. With ferromagnetism, on the other hand, a range of possible magnetization levels exists for each value of magnetizing force. The particular level of magnetization achieved depends on a material's magnetization history. Any phenomenon whose value depends on this history is called *hysteretic*.

Magnetic hysteresis is derived from the characteristics of domains. A domain is represented by a volume of material in which all of the available magnetons are oriented in the same direction. It is said to be *saturated* when no more magnetons are present within its bounds that can be used to increase the material's magnetization.

Domains form spontaneously in the crystals of ferromagnetic materials as long as their temperatures are below their Curie points. Above the Curie point, random thermal activities dominate, domains are lost, paramagnetism takes over, and hysteresis does not exist.

## 4.1 DOMAIN MATRIX

Domains form spontaneously as members of related groups. They are oriented in various directions, the field of any one of them being practically neutralized by the fields of its neighbors. In fact, all of the domains within a single crystal are organized to minimize the magnetic field external to that crystal. As shown in Fig. 4.1, the domains of a crystal in polycrystalline materials are oriented in a field-opposition mode that also helps neutralize the fields of neighboring crystals. The lighter shading shown indicates the presence of one crystal as a part of a matrix of many crystals.

Domains tend to be very small, usually smaller, in fact, than a single crystal since several of them should be present in each crystal if the latter is to have a minimum external field. If a particle of magnetic material is small enough, however, it cannot support the forces required of a domain

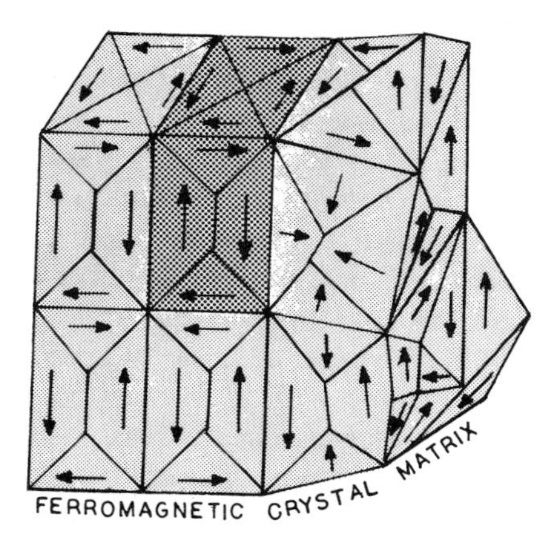

Fig. 4.1. In both ferromagnetic and ferrimagnetic materials, magnetons form in group opposition within individual crystals, whereas orientations of groups within adjacent crystals are influenced by their neighbor's orientations.

wall and therefore exists as a single domain unto itself. Figure 20.3 illustrates two single-domain crystals.

## 4.2 WEAK MAGNETIC FIELDS

The configuration of domains within ferromagnetic crystals represents a minimum energy condition. These domains are organized to minimize external magnetic fields. If a ferromagnetic material is immersed in a weak magnetic field, those domains that are oriented in the general direction of the imposed field become larger in volume and those that are not so oriented shrink. Such volumetric changes occur because the imposed field adds force to the crystalline fields, which then establish a different minimum-energy configuration.

The elasticity of domain walls makes domain volumetric changes easily possible. All of the magnetons within a domain share the same direction. Within the walls, however, the directions of adjacent magnetons progressively change from the direction of saturation in one domain to the direction of saturation in a neighboring domain (see Fig. 3.8). The angular difference is very small between adjacent magneton directions and practically constant from one wall magneton position to the next. A slight change in the same direction for each one of these small angular differences has the net effect of moving the Block wall.

As long as all adjacent magnetons are moved very slightly, there is no chemical force present that attempts to maintain them at any particular angle. After a wall movement, each magneton sees the same directional difference in angle between itself and its neighbors regardless of where the wall happens to be. At least this happens to be the case for all except the magnetons directly neighboring the adjacent domains. The latter merely interchange their relationships with their neighbors. On one side, wall magnetons become domain magnetons, whereas on the other side, domain magnetons become wall magnetons. The field external to a crystal increases, however, whenever a domain wall shifts from a minimum external field position. If this field exceeds a certain magnitude, and if it is not supported by an imposed field, domains are forced back toward their original configuration.

The range of possible stable "minimal" energy configurations is illustrated by Fig. 4.3. Here, $0_1$, $0_2$, and $0_4$ represent three different combinations of four domains that are reasonably stable in themselves but permit smooth changes to other stable positions that lie between the positions shown. (The Block wall in $0_3$ is hung up on an imperfection, and smooth movement from one position to another becomes impossible.) The net result is a range of domain configurations that are each individually stable in that they do not represent enough external field to force a return to a single zero pattern. This broad range of possible stable positions for each magneton is the basis both for hysteresis loops and for permanent magnetic effects. The differences in external magnetic fields that are detectable between these various stable positions are supported by wall energy.

Slight changes in magnetic condition vary smoothly with changes in the applied field as long as this field is fairly weak. These changes are reversible in that a reversed field can return them smoothly to their original position. A *WEAK MAGNETIC FIELD* is here defined as one whose effect on a magnetic material is restricted to wall movements.

## 4.3 MEDIUM MAGNETIC FIELDS

When an imposed field is increased beyond some nominal value, additional effects can take place. One of these effects is caused by crystal imperfections. The magnetostatic fields that are associated with all crystalline irregularities hold a wall in a fixed location until a field is imposed that is strong enough to break the wall loose from that imperfection. In another effect, domains continue to grow with increasing field strength. Now, however, the favored domains take over those less favorably oriented until each crystal in a material becomes a single domain unto itself.

These two effects are commonly associated whenever a field that is strong enough to tear a wall loose from a magnetostatic field is also strong enough to force a crystal into a single domain. Under these circumstances, the axis of magnetization for each crystal is that one of its finite number of possible crystallographic axes that is most nearly coincident with the imposed field direction.

After the most favored domain has taken over all of the others, a crystal is saturated in one direction. If a material is polycrystalline, each such domain has a slightly different orientation in relation to every other domain. These differences are maintained by magneto-crystalline energy.

## 4.4 STRONG MAGNETIC FIELDS

When an imposed field is further increased, each domain axis of magnetization rotates from its preferred crystal axis until it is parallel to the imposed field direction. When all domains reach this direction, the bulk material is said to be saturated. *STRONG MAGNETIC FIELDS* are those that rotate the axes of crystal magnetization after crystals have been saturated by weak or medium magnetic fields.

Magnetization of ferromagnetic materials is accomplished by different mechanisms. These include wall shifts, magnetostatic ruptures, domain captures, and axis rotations. The concepts of weak, medium, and strong magnetic fields derive from the mechanisms each stimulates. These mechanisms occur at different levels in different materials. In fact, since there are no imperfections in a single perfect crystal there are no medium-field effects either. In this case, the weak-field phenomena merge smoothly into the strong-field phenomena without passing through the discontinuities of magnetostatic field ruptures.

The combination of all these mechanisms establishes the concept of the hysteresis loop.

## 4.5 HYSTERESIS LOOP

A plot of an imposed field of strength $H$ against the resulting flux density $B$ pictorially presents the phenomenon of magnetic *HYSTERESIS*. In homogeneous materials, these curves are always symmetrical about the plot center, although their shapes differ widely among the various ferromagnetic materials. A particular curve includes all of the possible stable conditions

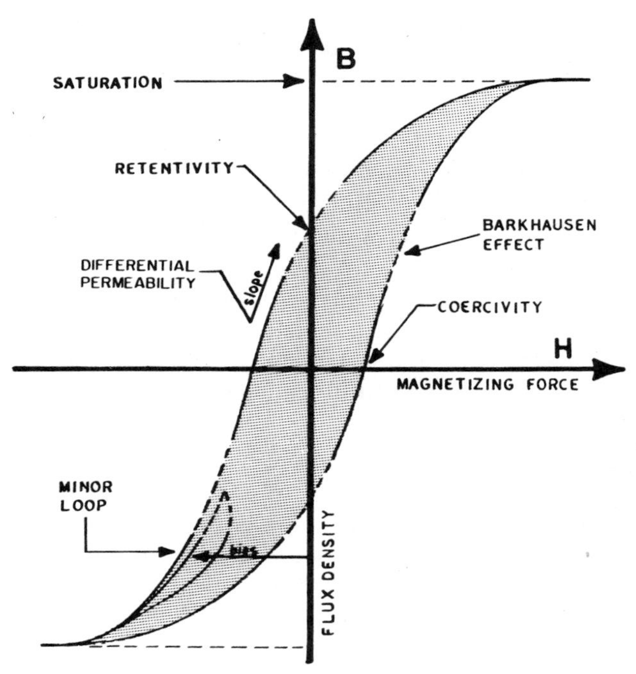


Fig. 4.2. Hysteresis loop: The magnetization of both ferromagnetic and ferrimagnetic materials depends on their magnetic histories.

that can be assumed by the magnetons of a particular material in either the presence or the absence of an imposed field. Figure 4.2 illustrates the various characteristics of a hysteresis loop, these being defined as follows:

*RETENTIVITY* is the magnetic force required to return domains to their original zero-balance condition once this balance has been upset by an imposed saturating field. It is represented by the intercept of the $B$ axis when $H$ is zero.

*COERCIVITY* is the residual magnetism left in a material after an imposed saturating magnetic field has been removed. It is represented by the intercept of the $H$ axis when $B$ is zero.

*SATURATION* is the maximum amount of flux density $B$ that can be forced through a material regardless of the magnetizing force $H$. In actuality, the flux density continues to increase beyond saturation, but the increase is inconsequential for most purposes. In this region, since magnetization of the material no longer strengthens the resulting field, the permeability drops to a very low value.

*DIFFERENTIAL PERMEABILITY* is a measure of the slope of a hysteresis curve at any point on that curve.

The circumference of a hysteresis loop is a trace of the magnetic flux density in a material that results from a cyclical magnetic field imposed on that material. If an imposed field carries the flux density to both plus and minus saturation, it is called a *MAJOR LOOP*. If the flux is not driven to both extremes, it is called a *MINOR LOOP*. The shape of a minor loop depends both on the strength of the cyclical field and on the particular location of the minor loop in respect to the major loop. If the center of a minor loop differs in some respect from the center of the major loop, the amount of difference in magnetizing force is said to be a *MAGNETIC BIAS* in the operating point.

*RECOIL PERMEABILITY* is the slope of a minor loop in the vicinity of an operating point.

## 4.6 BARKHAUSEN EFFECT

The *BARKHAUSEN EFFECT* is a series of minute "jumps" in magnetization that occurs when a magnetizing force is changed continuously. It is observed only over the middle range of a hysteresis loop.

If there is perfect symmetry in a crystalline lattice, wall magnetons can be moved smoothly in either one direction or the other. Conversely, if there is a lack of symmetry, the resulting magnetostatic fields adversely affect this ability to shift directions. Magnetostatic field differences assist wall magnetons to accomplish their required angular differences, and a domain wall minimizes its own energy content by drawing on magnetostatic energies. If a wall has less energy because it includes a nonhomogeneity, more energy will be required to move it away from such a nonhomogeneity.

Under these circumstances, domain walls tend to be *PINNED* by, or to "hang up" on, lattice discontinuities; that is, a domain wall can be moved smoothly until it encounters a "break" in the lattice regularity. The wall tends to stick to such an irregularity until it is supplied enough energy by an imposed field to cause it to break away. Once the wall does break away, it moves on well beyond the point in space where it had been pinned, jumping suddenly from one position to another and, in the process, causing a sudden jump in its material's magnetization.

In fact, the breakaway energy is usually strong enough to involve two domains. As the force of the imposed field increases, two domains suddenly combine to form a single larger domain each time a wall breaks away from a nonhomogeneity. These sudden combinations in a rising field (or decombinations in a falling field) give rise to the Barkhausen Effect. This is a noise observed in electric circuits that make use of ferromagnetic

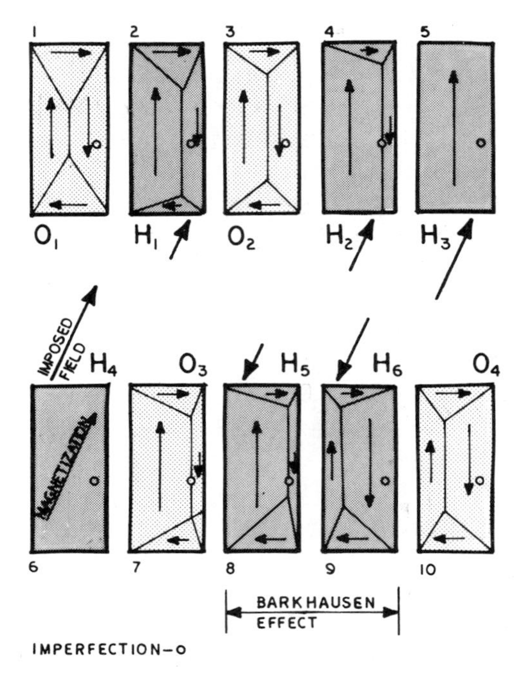


Fig. 4.3. Barkhausen Effect: Domain walls tend to "hang up" on crystal imperfections (electrostatic discontinuities), and there is a sudden jump in magnetization when a wall is forced to break loose from such a fixed position.

materials. The increase in flux density caused by one of these sudden jumps is on the order of magnitude of one millionth of a gauss.

Regularly patterned crystals have no Barkhausen Effect. If the Barkhausen Effect is present in a "defective" crystal, it can only occur over the center portion of the hysteresis loop. Below the center, domains grow smoothly as a result of wall shifting, pinned walls being left in place if necessary. Above the center, no wall movement takes place, and domain axes rotate smoothly from the preferred crystal axes.

Figure 4.3 shows a domain wall pinned both by a magnetic field $H_2$ and under zero field conditions, $0_3$. Furthermore, it shows $H_5$ as not strong enough to break the wall away from its pinned position, the breakaway (the Barkhausen Effect) not occurring until the imposed field increases to $H_6$.

## 4.7 ENERGY LOSS

The area enclosed by the hysteresis loop of a material is an indication of the energy loss that will be experienced if the material is cycled in an alternating field. This field-extracted energy causes a rise in the temperature of the cycled material.

Three different types of losses are involved in a field-reversal process. First are the losses incurred when the loop is driven as mentioned above; second are the eddy-current resistive losses; and third are the losses resulting from various mechanical deformations of the material. The latter can be caused either by Lenz forces or by magnetostriction.

When domain walls are pinned, they "snap away" at relatively high velocities. The result is high eddy-current losses since these are proportional to wall velocity.

## 4.8 ROTATIONAL HYSTERESIS

A ferromagnetic disk placed in a magnetic field that is aligned with the plane of the disk resists rotation. Rotating such a disk against this torque therefore consumes energy. In such circumstances, the magnetization of the disk material is directed at an angle that is not parallel to the applied magnetic field but rather depends on the shape of the material's hysteresis loop and on the speed of rotation.

## 4.9 ANCILLARY EFFECTS

A change in field can cause a change in energy distribution at the atomic level. Such changes include redistributions of thermal energies. If redistributions cause some lattice sites to become hotter than others, a period of time will elapse before the bulk material reaches thermal equilibrium.

The *MAGNETIC THERMAL AFTEREFFECT*—a delayed change in magnetization accompanying a change in an imposed field—is caused by various thermal-magnetic effects redistributing their local temperature differences.

In some circumstances, a given ion can occupy either of two different lattice sites separated by a small energy increment. The redistribution of energies caused by changes in magnetic field can supply the difference in energy between these two sites. An ion can then move from one site to the other when it experiences the appropriate field change. This movement is not necessarily instantaneous since it can also depend on the random thermal activities of the ions in the lattice.

The *MAGNETIC CHEMICAL AFTEREFFECT*—a delayed change in magnetization accompanying a change in imposed field—is caused by a diffusion of ions from one set of lattice sites to another set.

The *MAGNETIC ANNEALING EFFECT* is a change in crystal size caused by extrapolations of the preceding nuclear motions. In this case, ions move in the presence of changes in magnetic field, but, in the process of repeated field reversals, continue to move in the same general direction; that is, they move when they sense a field change but do not return when the field reverses. This unidirectional net motion in the presence of bi-directional field changes continues until the most stable size of crystal has been achieved.

# 5
# THERMAL EFFECTS

An electron in molecular orbit represents a quantity of potential energy. For that matter, an orbit represents a potential energy level in its own right.

The potential energy levels of "basic" orbits are defined for those circumstances in which the directions of all electronic magnetons are kept incoherent by thermal activities. Because of this directional incoherence, there are no exchanges of magneton-derived forces between orbiting electrons. If orbiting electronic magnetons are aligned by imposed magnetic fields, however, interactions are possible between those that circulate in orbits of adjacent molecules. These interactions add potenial energy to some orbits and subtract energy from others. Figures 8.3 and 8.4 illustrate the mechanisms involved. Although these particular sketches show protonic magnetons, the same relationships of either attraction or repulsion affect electronic magnetons as well. In short, a magneton is exposed to a different magnetic field when its neighbors are directionally oriented than it does when their orientations are random. These interactive energy-level alterations of magneton orbits are not very large in terms of basic orbit energy, but they are detectable.

In the *ZEEMAN EFFECT*, the frequency of emitted radiation is shifted slightly in the presence of magnetic fields. This shift is derived from the slight changes in orbital potential energy that accompany electronic-magneton interactions. In addition to the Zeeman Effect, changes in bonding relationships may be experienced in the presence of magnetic fields. These, too, are brought about by those changes in orbit energy that occur in the presence of imposed magnetic fields. Since thermal activities also cause changes in bonding relationships, various interactions between magnetic fields and temperature are encountered that have combined effects on molecular structures.

As a direct consequence of structural variations, a material's resistivity, permittivity, susceptibility, and the shape of its Fermi surface are affected by patterns of magneton alignment. Just as magneton alignment patterns respond to both thermal and magnetic forces, so are a material's electronic characteristics subject to modification by these factors.

Since magnetization is a contributor to chemical bonding, a change in a material's temperature can change that material's magnetization. Conversely, a change in magnetization can affect temperature. The *MAGNETOCALORIC EFFECT* is a temperature change caused by spontaneous magnetization, whereas the *MAGNETOTHERMAL EFFECT* is a temperature change caused by an imposed magnetization. These two phenomena are related and both result from energy redistributions among particles and bonds in molecular structures. Several different mechanisms of energy manifestation can cause these energy redistributions.

*MAGNETIC COOLING*, on the other hand, is the result of only one mechanism. In this case, a system's entropy is reduced by the imposition of order on the thermal disorder of paramagnetism. In general, an ordered system represents less energy than an equivalent disordered system. In magnetic cooling, the temperature of a material is slightly reduced by the presence of an imposed field. If the material's temperature is already very low, a slight change can be significant, and the technique is useful for reaching temperatures very close to absolute zero.

## 5.1 CHEMICAL POTENTIAL

According to *VOLTA'S LAW*, a potential difference can be detected between any two different materials in contact. This potential difference is a function of the chemical natures of the two materials.

*FERMI ENERGY* is the average energy of free electrons moving through a body of material. Under given enviromental conditions, each material has a characteristic Fermi energy. When two different materials with different Fermi energies are brought into contact, electrons flow from one material to the other until the two Fermi levels are adjusted to the same potential. This migration of electrons in response to Fermi-level differences creates Volta's potential difference.

When enviromental conditions change, the potential difference between two materials that is characteristic of a particular environment also changes. Variations in this difference are the bases for thermocouple effects since the potential difference between two materials has different magnitudes at different temperatures. This "difference" in potential difference gives rise to voltages capable of driving currents through electric circuits.

In addition, the magnitude of Fermi energy is a function of crystal anisotropy. As a result, the Fermi energy has different values when viewed in different directions as the latter relate to crystal axes. A *FERMI SURFACE* is a spherical plot of the magnitude variation of this Fermi energy.

Any phenomenon that depends on Fermi energy is affected by crystal organization, by the direction in which this organization is viewed, and by changes to this organization. A variation of thermocouple output under mechanical strain is one electronic consequence of Fermi surface manipulation.

## 5.2 NERST EFFECT

The magnetic coefficient of the Seebeck Effect is called the *NERST EFFECT* in which the output of a thermocouple is a function of both the strength and the direction of a magnetic field as well as a differential temperature.

Thermoelectric effects are consequences of Fermi energy differences encountered under different temperature conditions. The imposition of a magnetic field on an anisotropic crystal can change the Fermi surface of that crystal from one anisotropic form to another. This can result in a change in the Fermi energy differences experienced at two different temperatures. Because magnetization modifies Fermi surfaces, imposed magnetization modifies both the Thomson Effect and the Peltier Effect, with resulting variations in the Seebeck Effect.

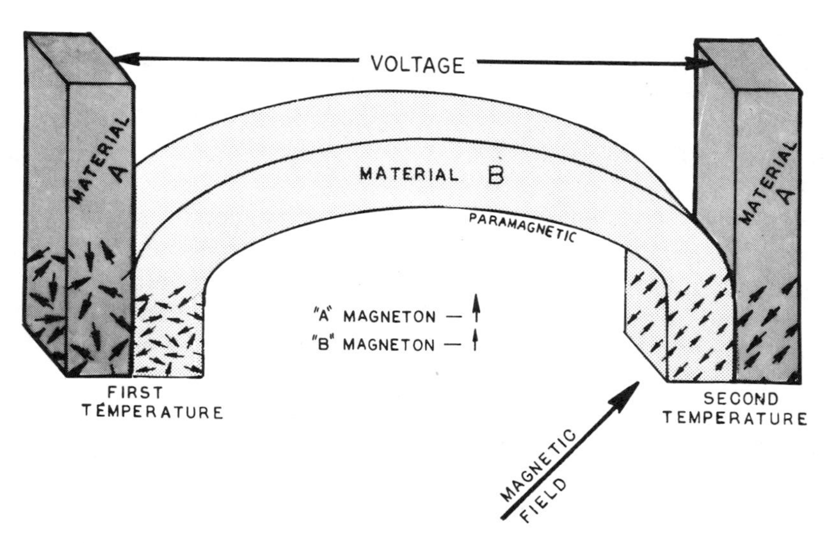


Fig. 5.1. Nerst Effect: The voltage generated by a thermocouple is altered when one (or both) junctions is exposed to a magnetic field.

In Fig. 5.1, one junction is shown exposed to a magnetic field that imposes order on its magneton alignments. The other junction shows disorder in the absence of a magnetic field. Both junctions could be exposed either to the same or different magnetic fields, and these fields are subject to both directional and magnitude variations. Although the possibilities are endless the mechanisms are always the same; that is, the Seebeck Effect is dependent on a relationship of two Fermi surfaces, whereas the Nerst Effect is a result of changes to these surfaces as these are induced by imposed magnetic fields.

## 5.3 MAGNETON ALIGNMENTS

The magnetons in ferromagnetic and antiferromagnetic materials are aligned in regular patterns by spontaneous exchange forces. Antiferromagnetic magnetons are aligned in patterns of individual opposition, whereas ferromagnetic magnetons are aligned in grouped opposition. Paramagnetic magnetons, on the other hand, are dynamically dispersed by thermal activities into continuously changing, randomly directed alignments. These relatively free magnetons can then be aligned by externally imposed fields.

Ferromagnetic materials become paramagnetic above their Curie points, whereas antiferromagnetic materials become paramagnetic above their Neel points.

The dispersion of magneton alignments from some regular pattern, at critical temperatures, is analogous to solids melting into liquids. There is a transition between the molecular positional regularity of solids and the positional dispersion of liquids, just as there is between magneton alignment regularities and alignment dispersions. In this analogy, paramagnetism may be thought of as the liquid element.

Magneton directional dispersions at critical temperatures are less precipitous functions of temperature, however, than the melting points of crystalline solids. The transition from self-imposed regularity to paramagnetism extends over a range that is both reversible and reproducible. In addition, this transitional range can be varied both in slope and in temperature by alloying and other techniques.

In these variations, the consequences of changing ferromagnetic dispersions tend to be much more significant than those of changing antiferromagnetic dispersions.

## 5.4 THERMOFERROMAGNETIC EFFECT

SUSCEPTANCE is a measure of the ability of magnetic fields to align magnetons in a manner that strengthens those fields. Since ferromagnetic

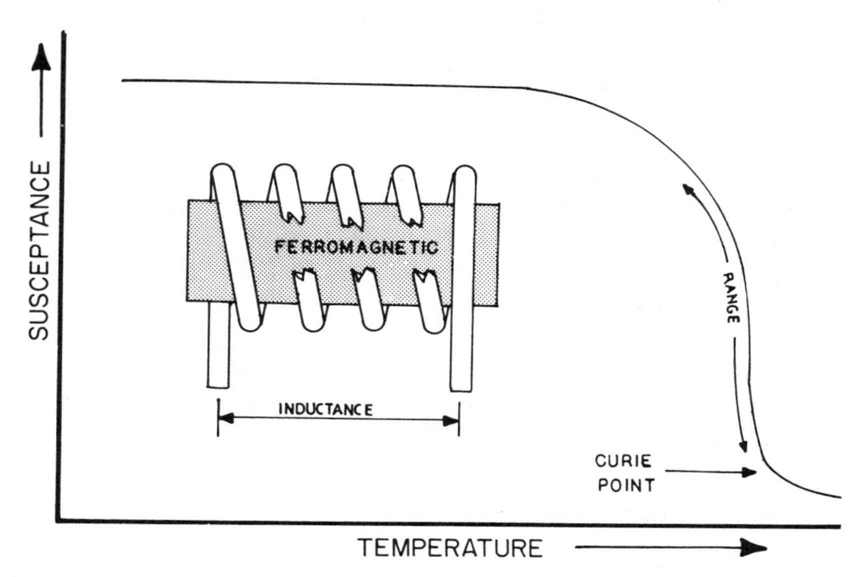


Fig. 5.2. Thermoferromagnetic Effect: The temperature of a ferromagnetic material over the range of its transition from ferromagnetic to paramagnetic may be deduced by measuring the inductance of a coil surrounding a sample of that material.

materials provide self-help in the aligning processes, the susceptance of ferromagnetic materials tends to be much greater than the susceptance of paramagnetic materials. Ferromagnetic materials, however, change to paramagnetic above their Curie point. As a result, susceptance magnitudes change sharply during Curie transitions (see Fig. 5.2). Susceptance changes experienced by ferromagnetic materials near their Curie points are called the *THERMOFERROMAGNETIC EFFECT*.

As long as the exchange force of ferromagnetism can hold domains together, magnetons remain more or less aligned, and there is little variation in their mutual relationships with changes in temperature. Although the susceptance of ferromagnetic materials changes with temperature, it does not change very much. Above the Curie temperature, however, the exchange forces are lost. Thermal agitations disperse magneton directions, and the susceptance of ferromagnetism is reduced to that of paramagnetism. This can be a significant change. Figure 5.3 gives some idea of the magneton alignment mechanisms at work in the Thermoferromagnetic Effect as well as it does in the Cabrera-Torroja Effect.

The random nature of crystals in metals and the finite nature of both crystals and domains blunt the effect of a Curie transition. The domains do not all break down at the same temperature or even in the same magnetic field. There is, then, a range of both temperature and magnetic field strengths over which some of the domains break up and others do not. The slope of this susceptibility curve in transition from complete ferromagnetism to complete paramagnetism is a trace of incremental susceptibility. It is this curve that makes a useful electrotransductive phenomenon possible.

A device based on this curve is useful only for that temperature range in which a material changes from ferromagnetism to paramagnetism. It is possible to adjust the Curie point, however, by altering the chemical composition of a material. For instance, if various proportions of $NiFe_2 0_4$ and $ZnFe_2 0_4$ are combined, the Curie temperature of the mixture can be made to lie anywhere in the range of 300° to 800°K.

Obviously, there is also a *THERMOANTIFERROMAGNETIC EFFECT*. This effect is always accompanied either by a Cabrera-Torroja Effect or by a Thermoferrimagnetic Effect.

## 5.5 CABRERA-TORROJA EFFECT

Resistivity changes experienced by ferromagnetic materials near their Curie points are here called the *CABRERA-TORROJA EFFECT*.

Resistivity is a function of the scatter mechanisms experienced by conduction electrons passing through materials. Magneton directional align-

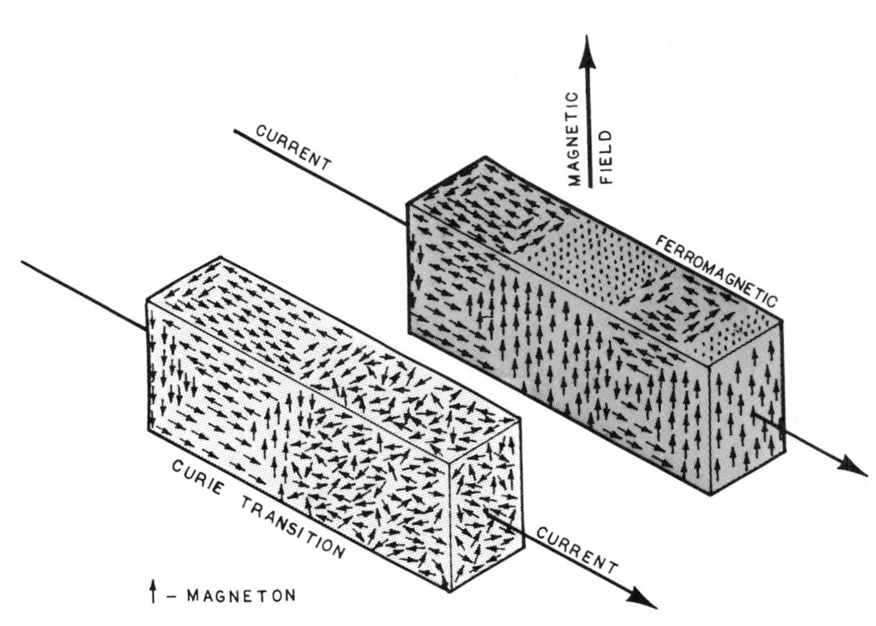


Fig. 5.3. Cabrera-Torroja Effect: Because of changes in magneton alignments, the resistivity of a ferromagnetic material changes significantly near its Curie temperature.

ment patterns provide one scatter mechanism. If alignments are orderly, the scatter action is different than it is when they are random. Furthermore, different scatter actions are associated with different order patterns.

Just as magneton alignments change at a Curie point from the self-imposed order of ferromagnetism to the disorder/imposed-order of paramagnetism, so will a ferromagnetic material's resistivity change with changing magneton order.

The Cabrera-Torroja Effect of Fig.5.3 and the Gauss Effect of Fig. 18.3 are related in that the Gauss Effect functions strictly under ferromagnetism whereas the Cabrera-Torroja Effect occurs during a transition between ferromagnetism and paramagnetism.

As magneton alignment patterns depend both on the strength and the direction of magnetic fields, there is a different Cabrera-Torroja Effect for each imposed-field circumstance.

A Cabrera-Torroja Effect is always accompanied by a Thermoferromagnetic Effect.

## 5.6 THERMOFERRIMAGNETIC EFFECT

The significant permittivity changes experienced by ferrites near their Curie points are here called the *THERMOFERRIMAGNETIC EFFECT*

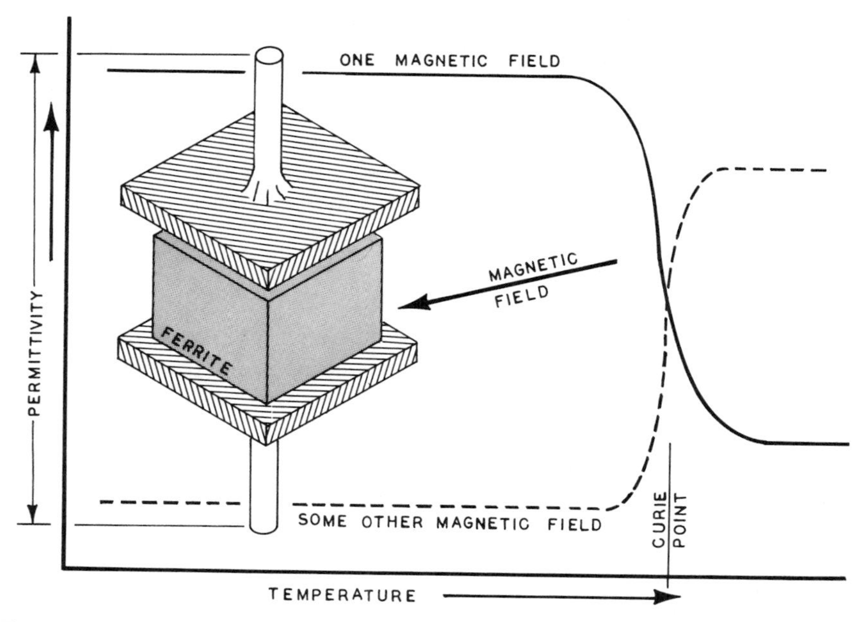


Fig. 5.4. Thermoferrimagnetic Effect: Because of magneton alignments, the ability of a ferrite capacitor to store electric charge depends both on temperature and on the magnetic field to which the ferrite sample is exposed.

(See Fig. 5.4). As a result of these changes, the capacity of a condenser that utilizes a ferrite as a dielectric changes significantly in the vicinity of that ferrite's Curie point.

Magnetons (magnetic dipoles) and electric dipoles interact in ferrites. The alignment or dispersion of one influences the alignment of the other. Since permittivity is a function of the ease with which electric dipoles are aligned by an electric field, anything that influences this ease affects permittivity.

Since magneton alignment patterns depend both on the strength and the direction of imposed fields, there is a different Thermoferrimagnetic Effect for each imposed-field circumstance.

A Thermoferrimagnetic Effect is always accompanied by a Thermoferromagnetic Effect.

## 5.7 FERROMAGNETIC-THERMOELECTRIC EFFECT

Domain structures of ferromagnetic materials influence Fermi surface configurations. Since these domains are lost during a Curie transition, Fermi surfaces are subject to more radical changes during a Curie transition than over any other similar temperature range covering the states of either ferromagnetism or paramagnetism.

The *FERROMAGNETIC-THERMOELECTRIC EFFECT* is the change in thermoelectric energy experienced during a Curie phase transition. The *ANTIFERROMAGNETIC-THERMOELECTRIC EFFECT* is a related phenomenon that may be observed during a Neel phase transition. Both of these effects are related to the Nerst Effect in a manner similar to that between the Cabrera-Torroja and Gauss Effects.

## 5.8 DECALESENCE

A heated ferromagnetic material passes through a Curie temperature at which the magnetic state changes from one of order to relative disorder—from ferromagnetism to paramagnetism. Since disorder represents a higher energy state than order, the transition from order to disorder absorbs heat, whereas the reverse transition releases heat.

The absorption of heat experienced in passing through a Curie temperature is of sufficient magnitude to lower the temperature of a heated specimen for a short period.

*DECALESENCE* is the phenomenon in which a specimen of ferromagnetic material that is absorbing heat is cooled as it passes through a Curie point, and *RECALESENCE* is the inverse phenomenon in which a specimen releasing heat has its temperature increased as it passes through a Curie point.

## 5.9 MAGNETIC SEMICONDUCTOR EFFECT

In any semiconducting material, the carrier population increases with temperature, and, as a result, the resistivity of that material decreases.

A *ferrite* is an insulating ferrimagnetic material that experiences enough resultant magnetic moment with the appropriate exchange forces to form domains in a manner similar to that of ferromagnetic materials.

For a magnetic semiconductor constructed from a "doped" ferrite, the environment observed by a carrier changes from one of relative order below the Curie point to increasing disorder above that point. Since thermal disorder tends to scatter carriers, the latter experience restrictions on their orderly movements, and the material's resistivity increases as the thermal

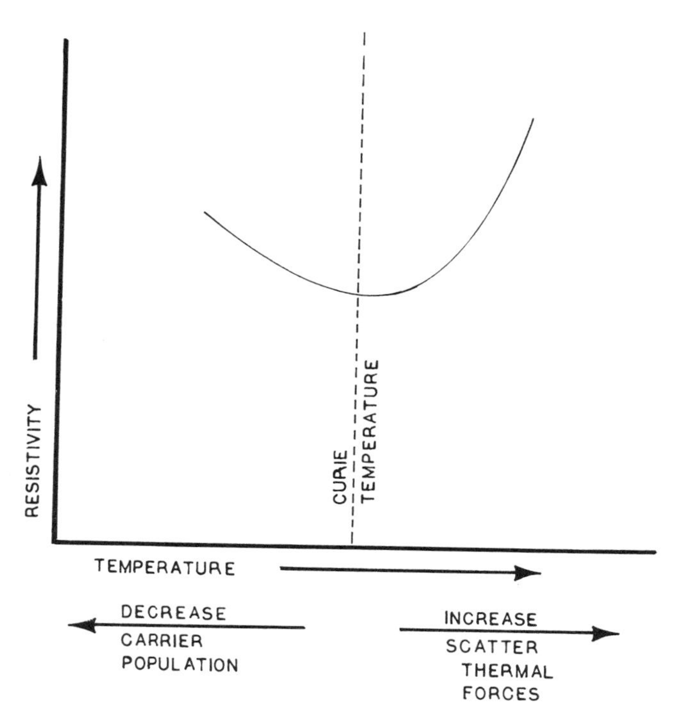


Fig. 5.5. Magnetic semiconductor: The resistance of a magnetic semiconductor experiences a minimum near its Curie temperature.

disorder increases. The relative order thus changes from the composure of ferrimagnetism to the rage of paramagnetism.

The *MAGNETIC SEMICONDUCTOR EFFECT* combines the effects of an increase in carrier population with increases in temperature and of the change of magneton order to disorder at the Curie point to form a resistivity response whose mimimum is somewhere near the Curie point (see Fig. 5.5).

# 6
# MECHANICAL EFFECTS

Every system is subject to restrictive forces that are directed to the task of minimizing its energy content. The Coulomb forces of magnetic systems are a part of these restrictive forces. If the movement of two magnetically charged bodies toward each other tends to reduce the total system energy, there is a force of attraction between these two bodies. Conversely, if such motion increases the total system energy, the two bodies repel each other.

Magnetic field strength is one manifestation of system energy. If the movement of a body in a magnetic field affects the strength of that field in any way, the force on that body attempting to move it in a particular direction will minimize the strength of that field.

## 6.1 AMPERE'S FORCES

According to Biot-Savart's Law, an electric current flowing through a wire creates a magnetic field around that wire. This field has a direction identified by the Right Hand Rule.

Ampere discovered that two wires carrying currents in opposite directions repel each other, whereas two wires carrying currents in the same direction attract each other. In the first circumstance, the Biot-Savart fields that exist between the two wires, are oriented in the same direction so that they reinforce each other and the total field between the two wires is thus strengthened. Since strengthening a field requires an increase in supporting energy, the wires repel each other. In the second circumstance, since the two fields are in opposition, the resulting field in between the wires is reduced, and the energy required to support this reduced field is also reduced; the wires therefore attract each other.

A loop of wire assumes a circular shape when it is carrying an electric current unless it is prevented from doing so by mechanical constraints. This happens because the elements of current on opposite sides of the loop are travelling in opposite directions so that they repell each other. In contrast, the current elements flowing in the adjacent loops of a solenoid travel in the same direction and therefore attract each other. As a result, the coils of a solenoid experience two sets of Ampere's forces. One is a radial force directed toward expanding the loops to their maximum possible size; the other is an axial force that acts to compress the loops.

## 6.2 PINCH EFFECT

The *PINCH EFFECT* is a phenomenon in which fluid conductors respond to compressive forces caused by the parallel flow of their constituent carriers.

Following the dictates of the Ampere force concept, attractive forces are experienced between individual carriers undergoing unidirectional, parallel flow in single conductors. These same carriers also experience repulsive forces derived from their common electrostatic charges. Various circumstances exist in which either one or the other of these two force patterns is dominant.

Attractive forces between carriers provide radially directed compressive forces on conductors. Solid conductors easily withstand such radial forces, but liquids and gases cannot support shear. As a result, Ampere's compressive force on liquid or gasseous conductors can cause them to contract radially even to the point of being reduced to zero radius, at which point they are said to be "pinched" off!

Oscillating systems based on the Pinch Effect can be constructed. In these, the flow of current through a fluid conductor causes mechanical pinch-off. The pinch-off breaks the conductor, thereby interrupting the electric continuity and stopping the current flow. As soon as the current flow is interrupted, the Pinch Effect forces are eliminated, and the fluid conductor reestablishes its continuity. The process repeats itself, continually making and breaking contact.

## 6.3 MAGNETIC RIGIDITY

The motion of conducting fluids tends to follow flux lines. According to Lorentz's Law a carrier experiences a force when it moves across flux lines but does not do so as long as it does not cross flux lines.

As conducting fluids move through magnetic fields, Lorentz forces tend to restore the carrier movements to field axis flow whenever there is any tendency for the carriers to deviate and move across flux lines.

*MAGNETIC RIGIDITY* is the longitudinal tension and lateral compression exerted by a magnetic field on a moving conducting fluid that opposes any lateral displacement of that fluid's movement in relation to the magnetic field direction.

## 6.4 DIMENSIONAL CHANGES

Materials experiencing polarization of any kind undergo slight dimensional changes. These changes result from the attraction or repulsion that individual dipoles of either an electric or a magnetic persuasion have for each other in different configurations. Figure 3.5 illustrates two different configurations in which opposite dimensional changes might be experienced in the presence of aligning fields.

In ferromagnetic materials, the magnetostriction of Chap. 16 is a domain phenomenon in which magnetic field strength and mechanical strain are reciprocally related. In nonferrous materials, magnetic fields might produce strain, but reciprocity does not exist. Figure 6.1 shows the strain created by a magnetic field in a paramagnetic material. Here the magnetic field forces the magnetons into an alignment pattern that produces strain. On the other hand, strain does not compel randomly directed dipoles into any kind of regular pattern that would create a magnetic field.

Any force that changes the dimensions of a body—whether thermal, mechanical, magnetic, or electric—will change both the electric and heat

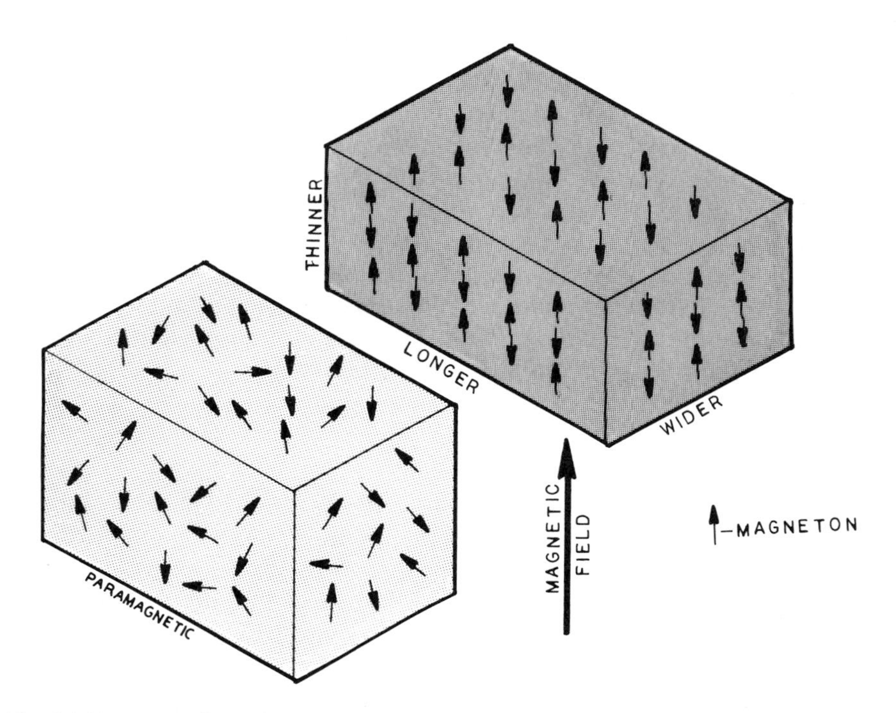


Fig. 6.1. Paramagnetic strain: The dimensions of paramagnetic materials change slightly in the presence of magnetic fields.

conductivities of the material from which that body is fashioned. The following sections will discuss these effects in some detail as individual phenomena.

## 6.5 FERROMAGNETIC STIFFENING

Ferromagnetic materials become stiffer in the presence of magnetic fields. If one of these fields affects the atomic structure of a material in any way, it changes those characteristics of the material that depend on its atomic structure, including its mechanical nature as well as a wide variety of electrotransductive phenomena. Chapter 16 deals with some of these effects. In addition, the elastic strain coefficient is altered depending on the degree to which structural rigidity is affected.

## 6.6 MECHANICAL FORCING FIELDS

Spinning magnetons have both magnetic and mechanical moments. The combination of these moments in one body provides opportunities for interactions between either of them and forcing fields of various types.

In the *EINSTEIN-DE HASS EFFECT,* a paramagnetic body tends to rotate when it is suddenly magnetized. The magnetic field tries to align the revolving electronic systems and, in the process, makes then all precess in the same direction until they settle down in their new alignment direction. The transient torsional force observed on the body is a net reaction of all the individual magneton precessions.

The Einstein-De Hass, Barnett, and Maxwell Effects are all examples of *GYROMAGNETIC* phenomena. A body constructed from the magneto-electric material of Fig. 3.14 will tend to rotate in an electric field for reasons similar to those of the Einsten-De Hass Effect.

## 6.7 BARNETT EFFECT

As a converse to the Einstein-De Hass Effect, a paramagnetic cylinder is magnetized when it is rotated at high speed. The magnitude of this magnetization is proportional to the speed of rotation.

In the *BARNETT EFFECT,* magnetons are aligned by the centrifugal forces on the spinning cylinder (see Fig. 6.2). In an analogy to the Barnett Effect, a magneto-electric material fashioned into a cylinder will become electrically polarized if that cylinder is rotated at high speed.

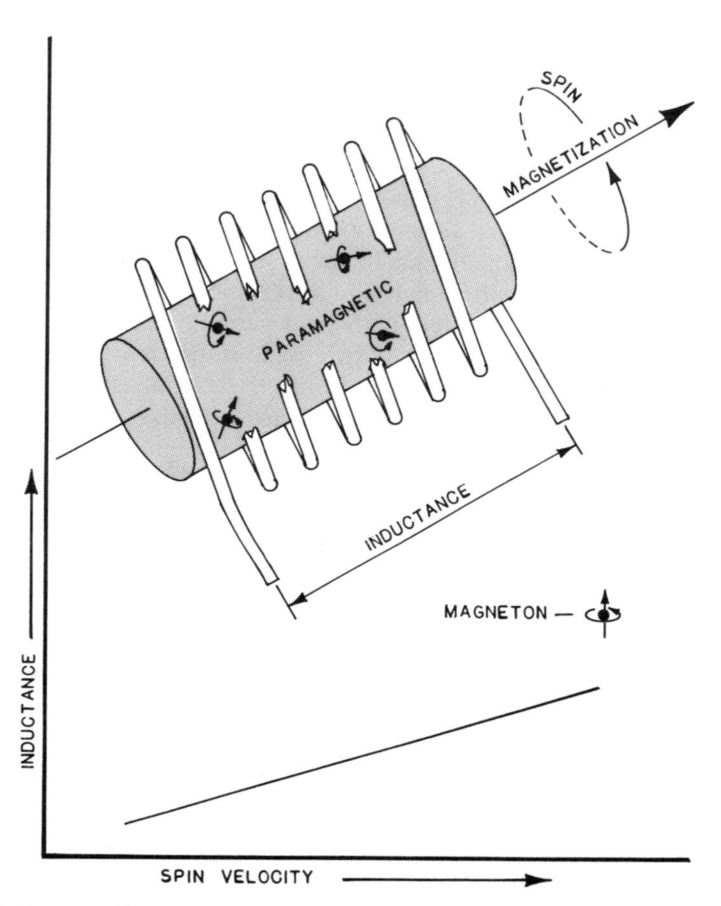


Fig. 6.2. Barnett Effect: A spinning sample of paramagnetic material is magnetized along its spin axis.

## 6.8 MAXWELL EFFECT

In some materials and under certain conditions, the summed gyroscopic effect of all spinning magnetons can be observed. Consider, for instance, a disk fashioned from a material containing magnetons aligned in the same direction as the disk axis. When such a disk is exposed to torsional acceleration, the magnetons will precess, and the result of this precession is a force that attempts to tip the disk out of its plane of rotation. Such a "tipping" force is here called the *MAXWELL EFFECT.*

## 6.9 PARAMAGNETIC TENSION

A container filled with a nonmagnetic gas immersed in a paramagnetic gas experiences a force when exposed to a magnetic field gradient. This force is here called *PARAMAGNETIC TENSION.*

When a paramagnetic gas is exposed to a magnetic field, its magnetons align themselves along flux lines. Some of these magnetons are directed with the field and some against, as happens with any other paramagnetic response. If the magnetic field is not uniform, the flux lines follow paths of variable curvature, variable length, and variable density. Each such flux line can be envisioned as an elastic band under tension that repels every other band. The tension on the shorter bands is always greater than it is on the longer bands. In paramagnetic gases, this band tension is proportional both to the magnetic field intensity and the number of magnetons present. Since the number of magnetons present is proportional to gas pressure, band tension is consequentially proportional to gas pressure.

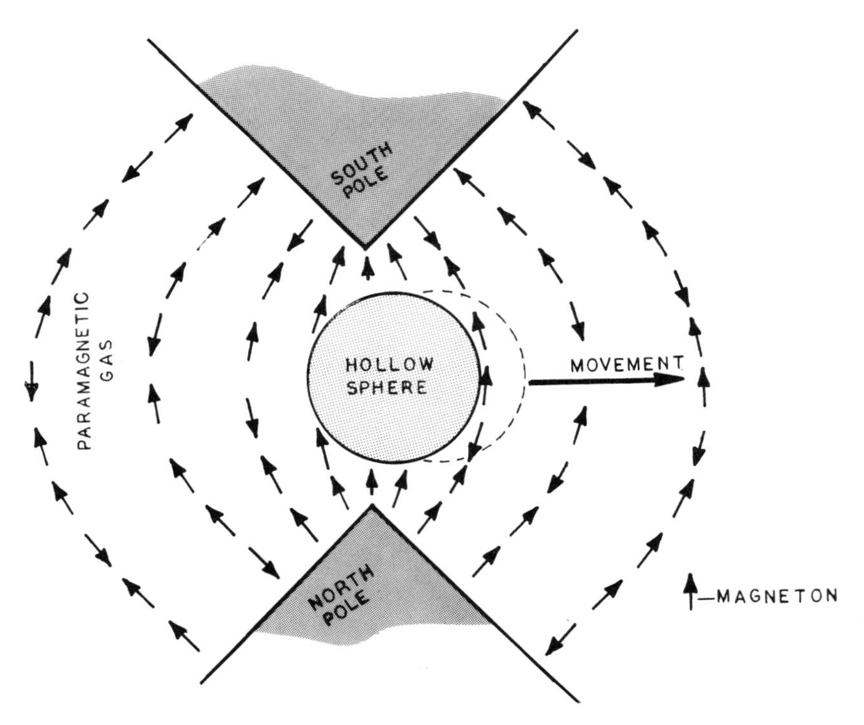


Fig. 6.3. Paramagnetic tension: A hollow sphere placed in a space containing a magnetized paramagnetic gas experiences a force that attempts to move it toward fields of decreasing magnetic field strength.

Figure 6.3 shows a nonlinear field symmetrical about a center line. If a paramagnetically hollow sphere is placed on the center line of this nonlinear field, it interrupts the symmetrical alignment of magnetons along the flux lines on both sides of center. The presence of such a sphere eliminates magnetons from those flux lines that pass through it. Those flux lines without magnetons are weaker than equally long flux lines with magnetons.

If this sphere is displaced slightly off the centerline, it is bounded by relatively short, strong flux lines on one side and longer, weaker flux lines on the other. This nonsymmetrical, interruptive pattern produces a force that acts to move the sphere further off the centerline because the motion of the sphere away from the centerline reestablishes stronger flux lines in its wake even as it interrupts weaker flux lines in its ongoing path. The sphere continues to experience a force that attempts to move it toward that volume of space in which the flux density is minimal.

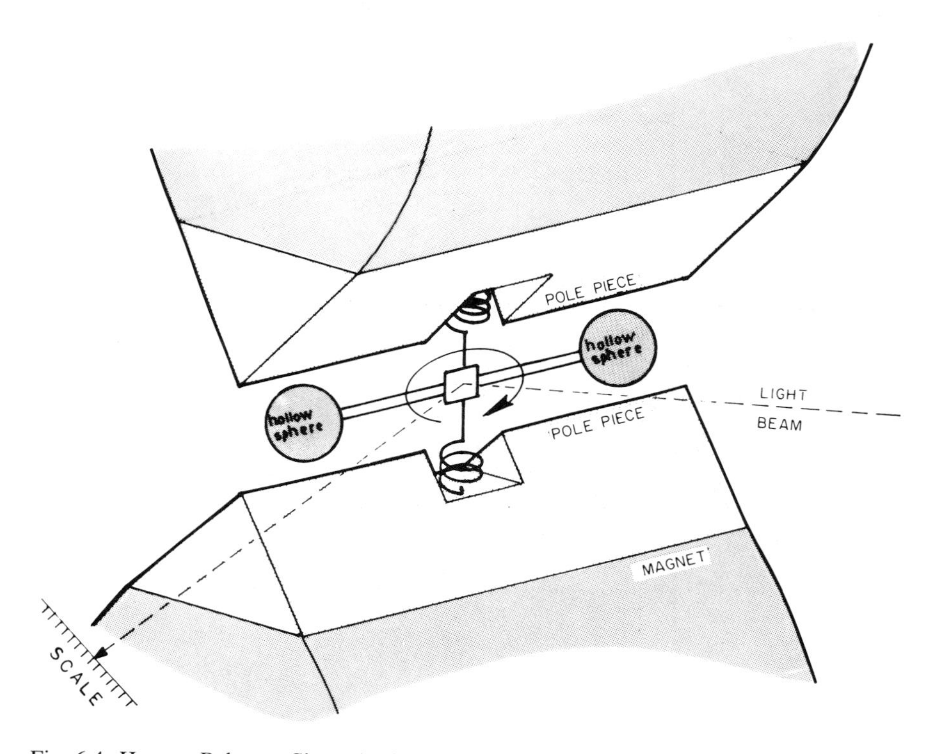


Fig. 6.4. Havens Balance: Since the force on a hollow sphere immersed in a magnetized paramagnetic gas depends on the concentration of that gas, it may be balanced against restrictive spring forces to measure gas concentration.

## 6.10 HAVENS BALANCE

As shown in Fig. 6.4, two hollow spheres can be associated in a spring-loaded balance. In this configuration, paramagnetic tension attempts to rotate the balance in one direction, whereas the springs oppose this rotation. The strength of the rotating force is proportional to the number of paramagnetic magnetons present in a paramagnetic gas.

When a paramagnetic gas is mixed with a nonmagnetic gas and held at constant pressure, the degree of rotation of a HAVENS BALANCE can be calibrated in terms of the percentage of paramagnetic gas in the mixture.

## 6.11 PARAMAGNETIC LEVITATION

The principles illustrated by Fig. 6.3 can be used to separate particles of various nonmagnetic materials of differing densities. In the configuration of Fig. 6.5, in which particles of varying density are carried by a paramagnetic liquid, paramagnetic tension operates in opposition to gravity. Since the lighter particles respond more to the paramagnetic tensile forces

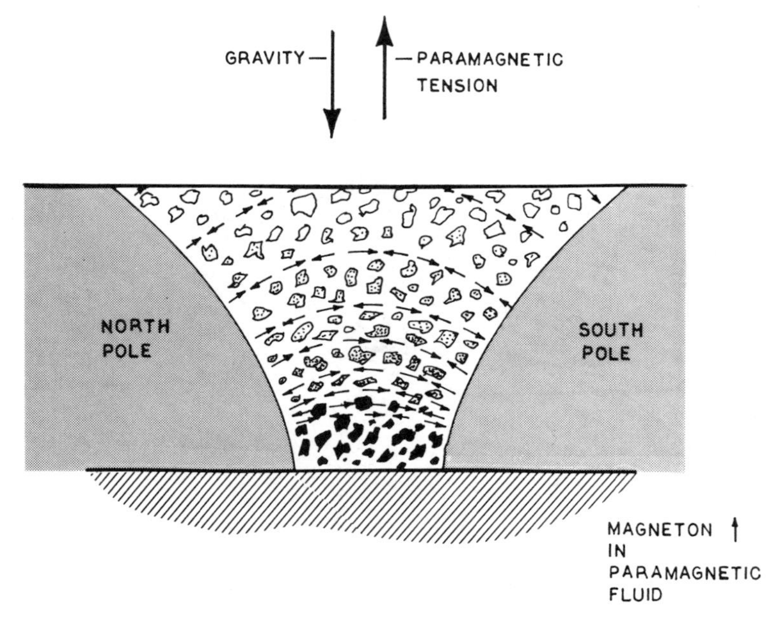


Fig. 6.5. Paramagnetic levitation: Particles immersed in a magnetized paramagnetic fluid may be separated according to their relative densities by balancing paramagnetic tension against the forces of gravity.

and the heavier particles more to gravity, the lighter particles tend to rise and the heavier particles to sink to the bottom.

## 6.12 MAGNO-THERM EFFECT

A combination of magnetic and thermal gradients imposed on a quantity of paramagnetic gas in which the field maximum and temperature maximum are displaced slightly one from the other induces a mechanical force on that gas. This force, as shown in Fig. 6.6, is here called the *MAGNO-THERM EFFECT*. The Magno-therm Effect functions as a pump!

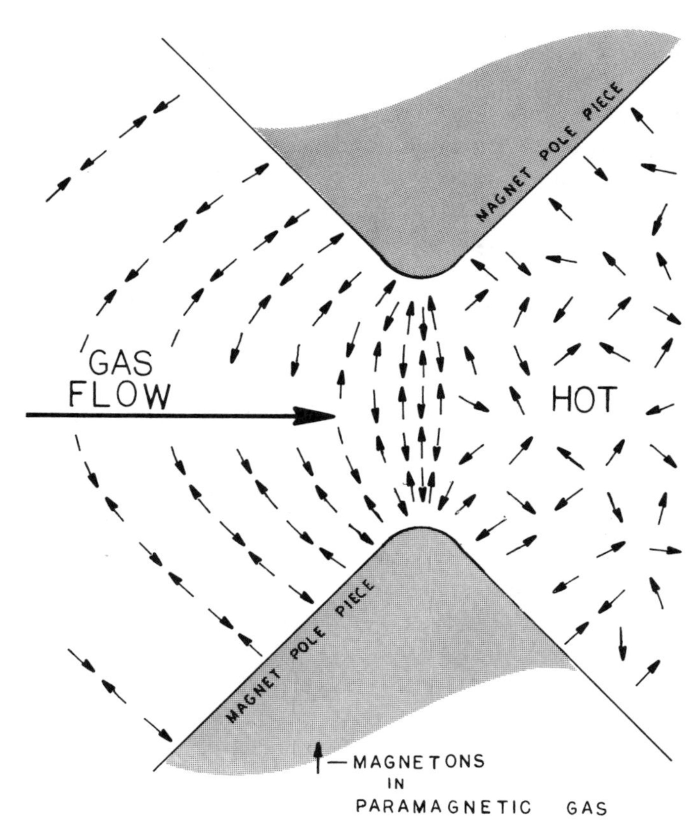


Fig. 6.6. Magno-therm Effect: Paramagnetic tension tends to pull a paramagnetic gas toward a magnetic field of increasing field strength. If such a gas is heated to the point at which magneton alignments become random, paramagnetic tension is destroyed. In the configuration shown, paramagnetic tension impels the gas to flow toward the heated zone and on beyond it, thus acting as a pumping force.

Although most gases are diamagnetic, a few, such as oxygen and the oxides of nitrogen, are paramagnetic. A paramagnetic gas, as is the case with all paramagnetic materials, is characterized by magnetons whose directions are incoherent whenever they are exposed solely to thermal forces. Paramagnetic magnetons become directionally coherent to a degree, however, upon being exposed to magnetic fields. Figure 3.3 illustrates magneton directions in the absence of a magnetic field; Fig. 3.4 shows magneton directions responding to both thermal and magnetic forces; and the Langevin paramagnetism of Fig. 3.12 traces the relationship between thermal and magnetic effects in paramagnetic materials.

As is true of all paramagnetic materials, a paramagnetic gas that is exposed to magnetic field gradients experiences a force that tends to move it in the direction of increasing field strength. The force on the gas depends on the number of excess magnetons aligned in the field direction. This force increases with increasing field strength and decreases with increasing temperature, as shown by Fig. 3.12. Held in a container at constant temperature but exposed to a magnetic field gradient, the paramagnetic gas experiences a pressure gradient whose pressure maximum coincides with the point of maximum field strength. This pressure maximum decreases with increasing temperature.

Consider the circumstances in which a contained gas is exposed to both magnetic and thermal gradients and the temperature maximum is offset slightly from the magnetic field maximum. The pressure on the low-temperature side of the point of maximum magnetic field is greater than the pressure on the high-temperature side. A gas current then circulates in response to this pressure differential. This circulation can be put to useful task.

When any gas is passed through two identical channels, each containing identical gas heaters, the flow of gas cools both heaters to the same degree. If the gas is paramagnetic, however, and one channel is exposed to combined magnetic/thermal gradients as just described, the flow of gas through the magnetized channel will be greater than the flow through the non-magnetized channel. With different flow rates in the two channels, the two heaters are cooled at different rates. If the two heaters form two legs of a Wheatstone bridge, the bridge imbalance is an indication of the quantity of the paramagnetic constituent in the gas stream.

## 6.13 FERROMAGNETIC FLUID

*FERROMAGNETIC FLUIDS* are fluids that possess magnetic properties and whose viscosity is significantly affected by magnetic fields. Consider those circumstances in which very small particles of ferromagnetic material

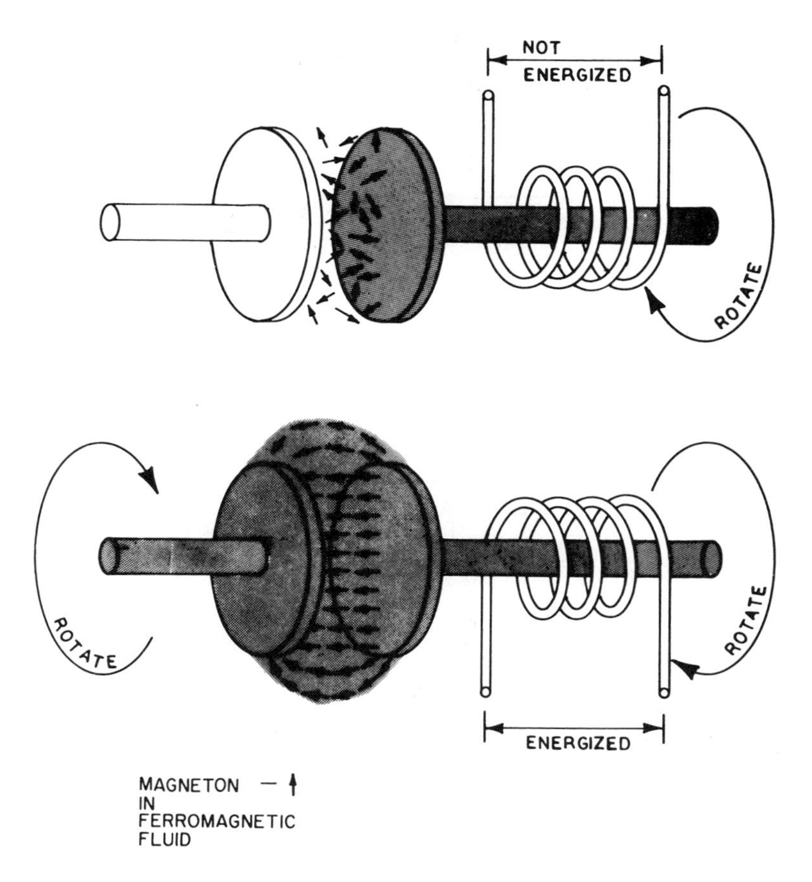


Fig. 6.7. Magnetic clutch: Mechanical force is transmitted through magnetized ferromagnetic fluids.

are suspended in a nonmagnetic fluid to form a "slurry." The viscosity of a slurry of this kind will be a function of the strength of an imposed magnetic field. As shown by Fig. 6.7, the slurry can be interposed between the plates of a "clutch." Since the viscosity of the clutch fluid varies with field strength, a torsional force of varying magnitude can be transferred from the drive shaft to the driven shaft.

## 6.14 RELAY

It takes more energy to support a magnetic field in space than in a magnetic material. As a result, a force of attraction exists between bodies fabricated from magnetic materials whenever these bodies are exposed to magnetizing

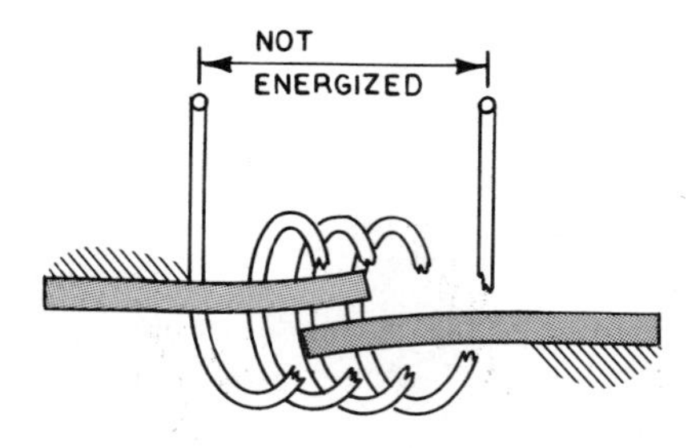


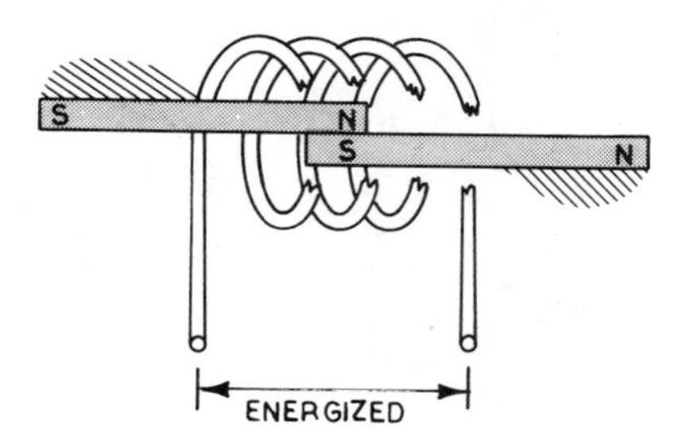


Fig. 6.8. Relay: Two ferromagnetic members are attracted toward each other when they are both magnetized.

fields. As shown by Fig. 6.8, the gap between ferromagnetic "springs" is closed when the magnetizing force is greater than the spring force and opened if the spring force is greater than the magnetizing force. The opening and closing of such a gap can be used to maintain or interrupt the continuity of an electric circuit.

## 6.15 FLUID ACTUATOR

Consider a mercury-containing capsule mounted in a magnetic field, with electrodes arranged as shown in Fig. 6.9. An electric current passing through the mercury will position the mass of mercury at one end of the capsule or the other, depending on the current's direction. If an electrode pair is mounted at one end of the capsule, the electric continuity of that pair can be made or broken depending on the driving current's direction.

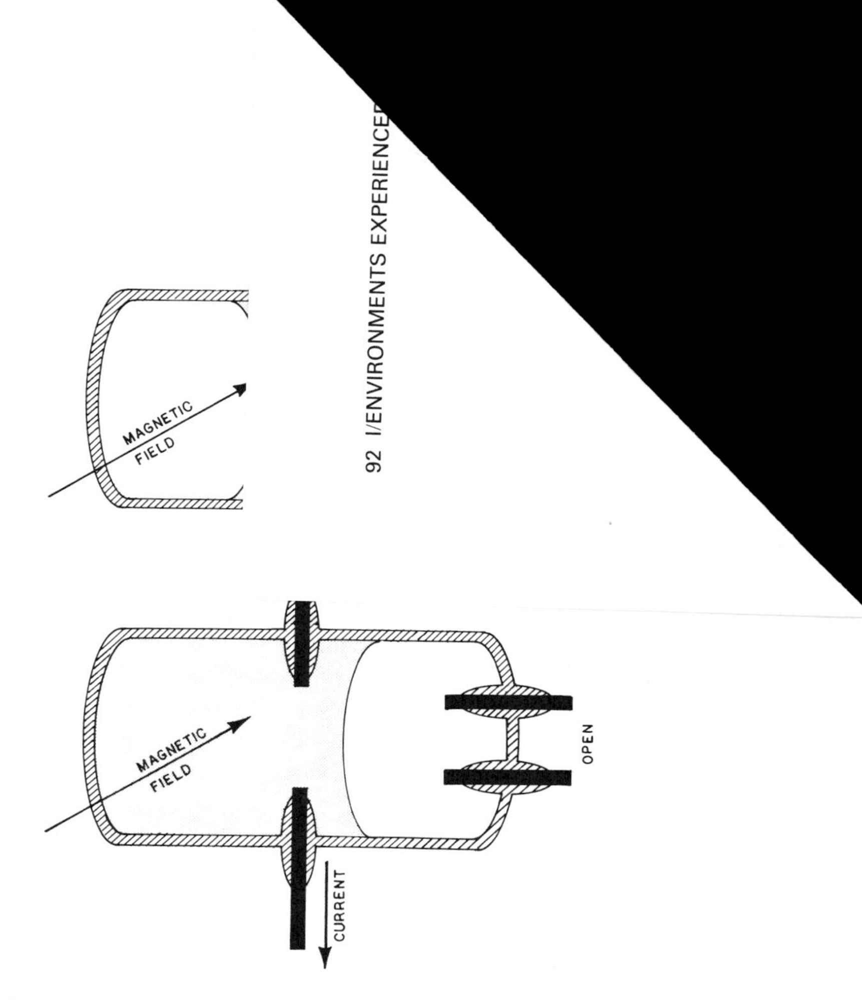


Fig. 6.9. Fluid actuator: A mechanical force is experienced by a current-carrying fluid immersed in a magnetic field.

## 6.16 FERROFLUID

*FERROFLUIDS* are fluids that possess magnetic properties but whose fluidal characteristics are not significantly affected by magnetic fields. Magnetic fields are rather used to control a ferrofluid's position undisturbed by gravitational, centrifugal, or other extraneous forces. Furthermore, objects that are otherwise denser than a ferrofluid can be levitated in the latter by imposing magnetic fields.

Ferrofluids owe their magnetic properties to submicroscopic magnetic particles colloidally suspended in a carrier liquid. Each such particle is smaller than a magnetic domain, which is typically 10 nanometers in di-

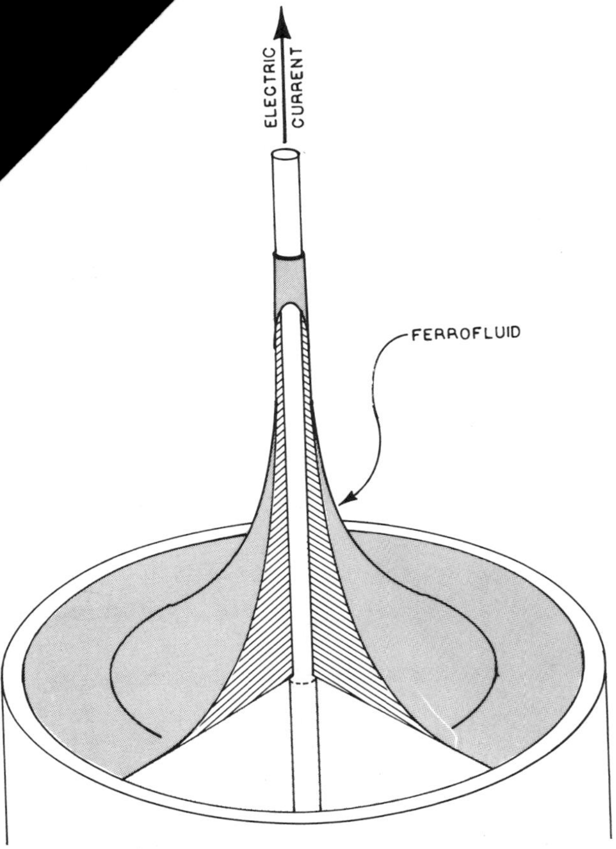

Fig. 6.10. Ferrofluid in a Biot-Savart Field: The surface shape assumed by a ferrofluid when gravity forces are offset by the pull of a Biot-Savart field.

ameter. The liquid properties are maintained even when the fluid is magnetically saturated because these particles are treated with a surfactant that keeps them dispersed in the carrier fluid. The fluid becomes magnetized when the particles are aligned by an applied magnetic field.

Figure 6.10 involves a current-conducting rod that passes vertically through a dish of ferrofluid. As the fluid is drawn toward the rod in the direction of the field gradient, the fluid that is already near the rod surface is squeezed up the rod by the fluid that is drawn in from the dish edge and forms a conically shaped meniscus.

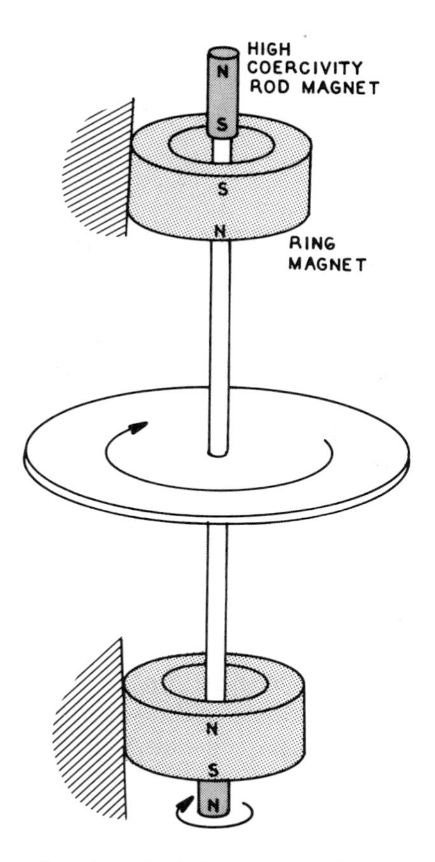


Fig. 6.11. Magnetic suspension: A mechanical structure can be supported entirely by magnetic forces.

## 6.17 MAGNETIC SUSPENSION

Using permanent magnets of appropriate shapes and coercivities, it is possible to construct a mechanical assemblage one of whose members is supported entirely by magnetic fields; that is, it can be made to "float" in such fields. As shown by Fig. 6.11, the field surrounding a ring magnet has a "cup" shape. Although it is usually toroidal in form, the field is weaker along the centerline than in all directions off that line. As long as the floating magnet is strong enough to overcome the demagnetizing force of the ring's field and is constrained from flipping over, the repulsion between the ring's field and the floating magnet's field supplies the suspending force.

# 7
# MAGNETIC MEASUREMENTS

Magnetic measurements are concerned both with evaluating magnetic fields and with determining the magnetization of materials.

Magnetic fields are defined in terms of strength, intensity, direction, and shape. Field "intensity" and field "strength" are related by the factor of permeability as this term is described by Eq. 3.2. In a nonmagnetic medium such as air, permeability is unity, and the terms "intensity" and "strength" are commonly used interchangeably. This usage is unfortunate since it causes a certain amount of confusion in the literature that leads to misuse of units, misstatements, and the like.

Almost any phenomenon that depends on a magnetic field for its performance can be used to measure field strength in a vacuum, in air, or in various other nonmagnetic media. If all other variables to which a phenomenon might respond are held constant, the transductive principle in question responds primarily to differences in field strength. If one field is used as a reference, the strength of a second field can be evaluated in terms of the first. An oscillating magnet (see Fig. 7.2) provides one very crude method of evaluating relative field strengths.

Since magnetic fields have direction, most field evaluation devices are sensitive to orientation. They respond to the field vector that extends in the particular direction determined by the device axis. For instance, the orientation of a freely suspended bar magnet is dictated by field direction. Such a device tends to align itself with an immersion field. Both a compass needle and the dip needle of Fig. 7.1 are useful devices that are based on this response.

Certain transductive principles such as magnetoresistance or magnetoinductance operate in any field direction as far as theory is concerned. Because of the anisotropic nature of the material from which such devices are fabricated, however, they are useful only when calibrated in the direction of their application. Devices that depend on a measurement of Larmor frequencies, on the other hand, are relatively insensitive to direction because all of the participating magnetons are automatically aligned in the field direction and there is no practical way of telling what that

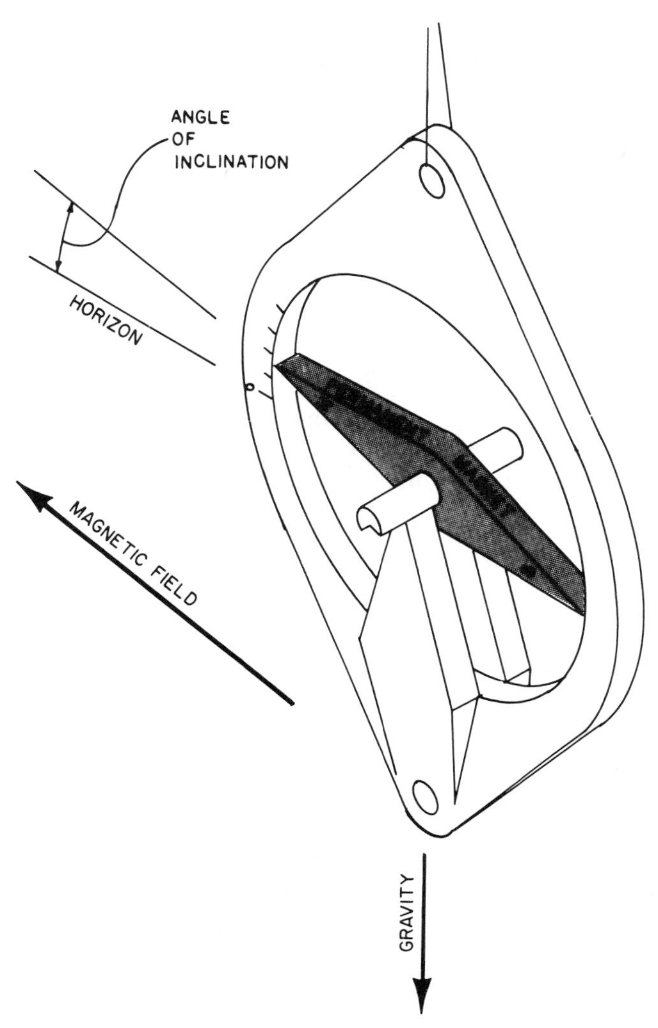


Fig. 7.1. Dip needle: A magnetized body tends to align itself with the earth's magnetic field.

direction is. The Larmor frequency is determined only by field strength, regardless of field direction, but the magnitude of the detected signal is a function of direction. In fact, the Larmor frequency cannot be detected at all if the Larmor precession-stimulating alternating field is applied in line with the immersion field.

No single phenomenon can sense field shape without mechanical manipulation. This parameter is deduced from magnitude measurements made in different directions and in different locations. The term *MAGNETO-*

*METER* is commonly applied to instruments that are designed to make almost any type of magnetic measurement, and there is no other narrower definition. The term *GAUSS METER* is often applied to instruments that measure field strength (Fig. 7.3) rather than field intensity, as the terminology might otherwise imply.

The magnetization of a material can be judged either by its ability to distort a standard magnetic field or by the magnitude of the force it experiences when immersed in a standard field. The vibrating sample of Fig. 7.4 is based on field distortion, whereas the Gouy method of Fig. 7.6 is based on force measurement.

The strength of a magnetic field in a nonmagnetic medium is commonly measured in terms of either "gamma" or "oersteds" (Oe). One gamma ($\gamma$) represents $10^{-5}$ oersted. The gamma has been chosen as a convenient magnitude for earth field measurements. A nail in a shoe about 5 feet away will affect the earth's magnetic field by about 1 gamma, as will an automobile at 100 feet.

## 7.1 FIELD DIRECTION

The body of a permanent magnet mounted so that it is free to move in the vertical direction will align itself with the earth's magnetic field. If a scale is maintained in an orientation that is fixed in relation to the earth's gravitational field, the movement of this magnet in relation to that scale can be used to indicate the direction of the earth's magnetic field in relation to the horizon, as shown in Fig. 7.1.

The presence of an otherwise undetected (buried) magnetic body can be deduced by recording the inclination angles encountered during an area traverse and noting any anomaly encountered in the otherwise uniform pattern of the earth's magnetic field.

## 7.2 OSCILLATING MAGNET

Figure 7.2 shows a body constructed from some magnetized material and suspended by means that allow free rotation in any direction defined by its plane of rotation. If this body is placed in a magnetic field whose direction lies in the body's plane of rotation, the body will align itself with that field. If it is then displaced from the field direction and released, it will respond with a damped oscillation whose frequency is a function of field strength.

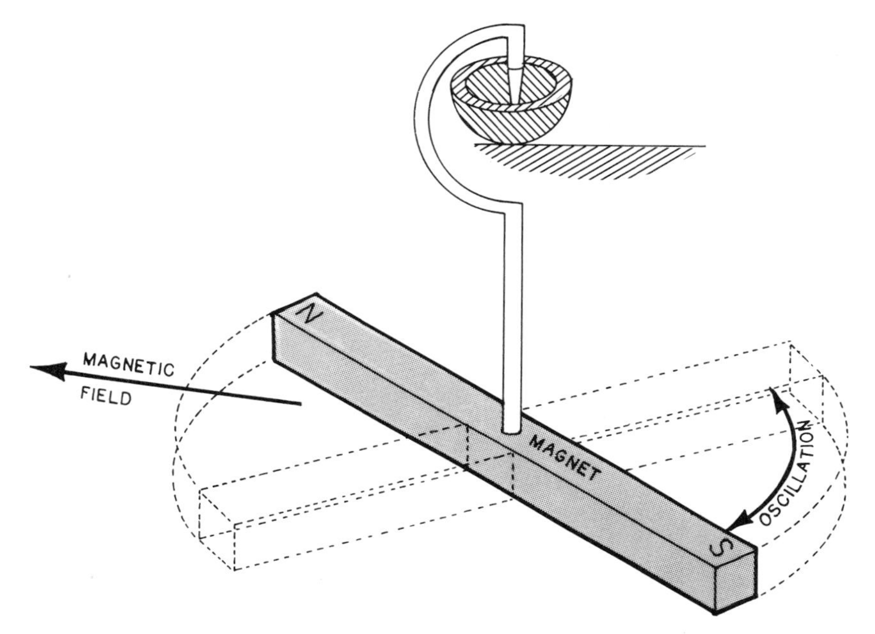


Fig. 7.2. Oscillating magnet: A magnetized body freely suspended in a magnetic field can be made to oscillate at a frequency that is a function of field strength.

## 7.3 PERMANENT MAGNET "GAUSS METER"

Magnetic bodies constructed from permanent magnetic materials tend to align themselves with any magnetic field in which they are immersed. The aligning force, standardized in a standard magnetic field, is the *MAGNETIC MOMENT* of the body. Once a body's magnetic moment is established in a standard magnetic field, the same magnetic moment can be used to evaluate unknown fields.

Consider, for instance, the configuration shown in Fig. 7.3. Here a permanent magnet is aligned by an immersion field against the tension of a coil spring. If such a device, commonly called a "gauss meter," is rotated through 360 degrees, the spring force at some point will be strong enough to overcome the aligning force; that is, as the assembly is twisted, it reaches a maximum and then "snaps" away from that maximum if twisted further. (If this does not occur, the immersion field is too strong for the gauss meter in question and is likely to damage the device by changing its magnetism.) This point is indicated by the point of maximum deflection of

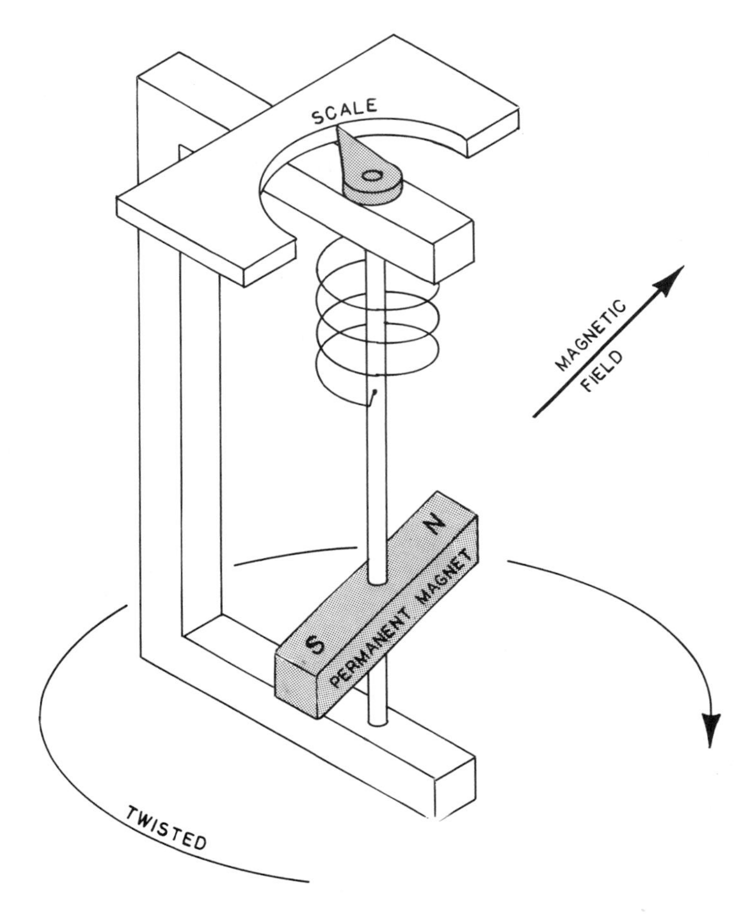


Fig. 7.3. Permanent magnet gaussmeter: The force aligning a magnetized body in a magnetic field may be balanced against a restraining spring force to indicate field strength.

the pointer on the scale, the maximum deflection being calibrated in terms of the immersion field's vector strength.

## 7.4 GRADIOMETER

Two physically separated magnetometers can be used to detect the difference in magnetic field strength between their two positions. If the output signals of both are connected in series opposition and balanced to zero, the result will ignore the general ambient field in which they are immersed and produce only the difference signal. This dual arrangement is called a *GRADIOMETER*. If a local field difference exists in the vicinity of one of these two instruments, their summed output will register either a positive or negative difference increment.

These devices can be used to detect buried bodies of ore (or the like) that provide very small anomalies over very long distances. To do so, the gradiometer is merely moved along a line whose length is adequate to reveal the anomaly of interest. The fixed distance between the two magnetometers and the distance over which the combination is moved both affect the resolution and sensitivity of the measurement.

## 7.5 VIBRATING SAMPLE MAGNETOMETER

A magnetic moment is induced in a material exposed to a magnetic field. Its magnitude is proportional to the product of susceptibility multiplied by the field strength.

If a sample of material is made to undergo sinusoidal motion in a magnetic field, an electric signal will be induced in a strategically placed pick-up coil (see Fig. 7.4). The signal detected in such a coil is proportional to the magnitude of the magnetic moment and to both the amplitude and the frequency of vibration. If both the amplitude and frequency are held constant, the magnitude of the detected signal will be proportional to the magnetic moment. If the magnetic field strength is held constant, this technique may be used to compare the susceptibilities of various materials.

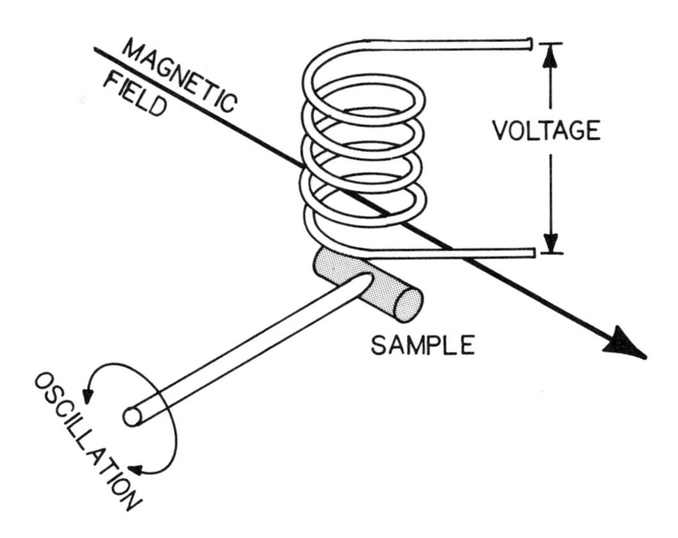


Fig. 7.4. Vibrating sample magnetometer: The relative magnetization of a sample immersed in a standard magnetic field may be deduced by noting the voltage generated in a nearby coil when the sample is oscillated uniformly.

## 7.6 BOZORTH CONFIGURATION

When a *BOZORTH CONFIGURATION* is used to determine the susceptibility of a material, two identical serially driven bridge solenoids are mounted side-by-side, with a means of measuring field intensity located centrally between the two. This assembly is first tested to be sure the magnetic effects of the two bridge coils cancel as far as the detection apparatus is concerned.

The material of interest is fabricated into rod form, the dimensions of the rod coinciding with the dimensions of yet a third, test solenoid. This sample is then placed in one of the bridge solenoids, and the test solenoid is similarly placed in the other bridge solenoid (see Fig. 7.5). The two bridge solenoids are energized at some particular level of magnetomotive force. The current through the test solenoid is then adjusted to eliminate the effects of the tested sample as far as the detection system is concerned.

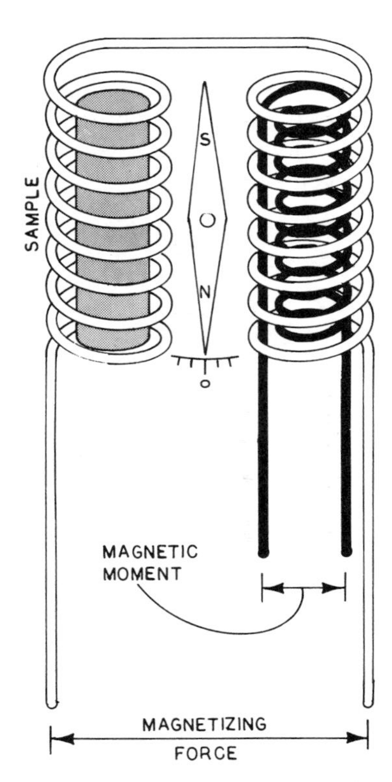


Fig. 7.5. Bozorth Configuration: The relative magnetization of a sample of material may be balanced against an equivalent magnetic field supplied by a current-carrying coil in which magnetization is a function of coil current.

Under nulled conditions, the magnetic moment of the tested sample will be the same as the magnetic moment of the test coil. This magnetic moment can be calculated, given the physical characteristics of the test coil and the current through that coil.

## 7.7 GOUY METHOD

As shown in Fig. 7.6, a sample of material is prepared in rod form and then suspended with one of its ends in a magnetic field. If the rod is long

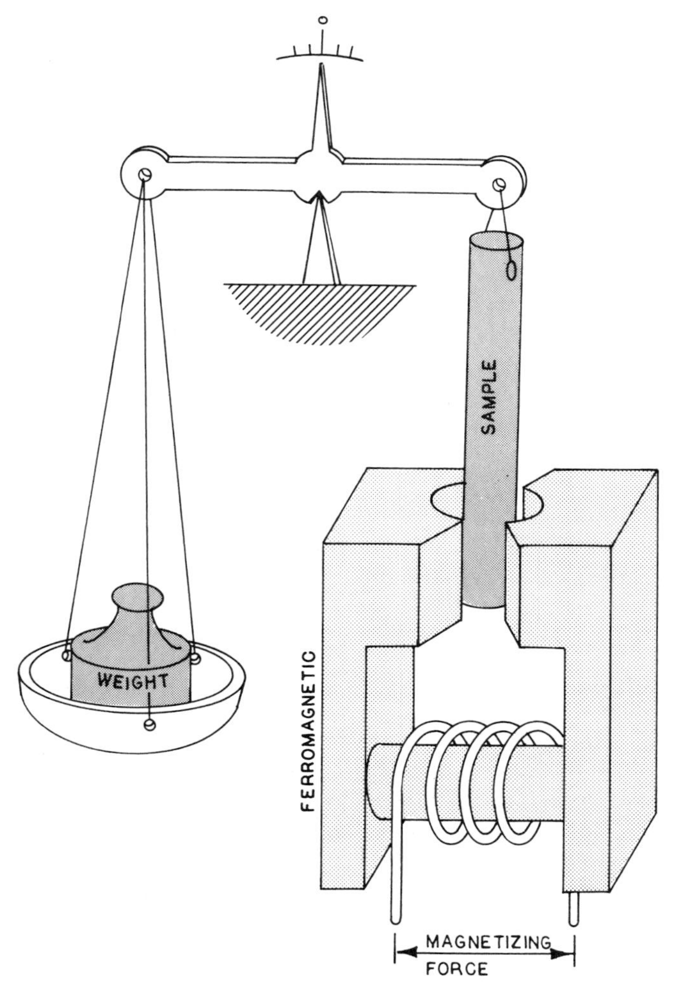


Fig. 7.6. Gouy Method: The relative magnetization of a sample of material may be balanced against a weight.

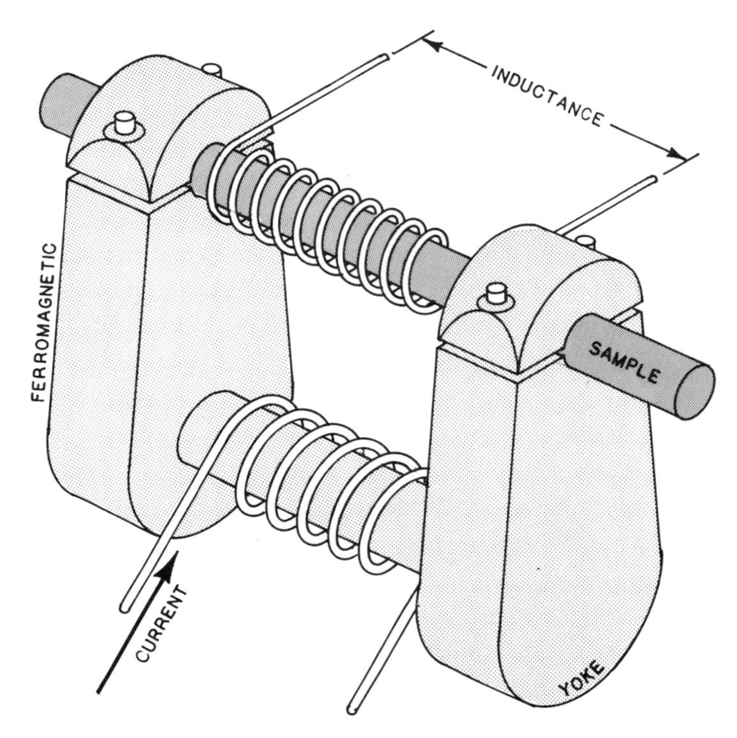


Fig. 7.7. Permeameter: The relative permeability of a sample of material may be deduced at different field strengths by measuring the inductance of a coil surrounding that sample.

enough for its other end not to be magnetized, the susceptibility of the rod material can be calculated from a knowledge of the field strength, the rod cross-sectional area, and the force attracting the rod towards the magnetic field.

## 7.8 PERMEAMETER

Figure 7.7 presents a two-coil configuration capable of measuring the permeability of a sample material under varying degrees of magnetizing force. In this *PERMEAMETER,* the inductance of the sample coil can be calibrated in terms of permeability, whereas the current through the driving coil is calibrated in terms of magnetizing force. When the inductance of the sample coil is measured with minimum current, under such circumstances, the permeability measured is that of the differential permeability of Fig. 4.2.

An indication of absolute permeability can be obtained by wrapping the sample coil tightly around the specimen and connecting it to a ballistic galvanometer. When the current in the major coil is reversed, the throw of the galvanometer is an indication of absolute permeability.

## 7.9 OPTIC FIBER

A magnetostrictive film can be coated onto a single optic fiber. When such a combination is placed in a magnetic field, the longitudinal strain induced in the fiber by the magnetostrictive action of the film changes the length of the fiber. This change in length is measured by an optical interferometer that detects the phase change of laser light passing through the fiber.

# 8 MAGNETIC RESONANCE

Exposed materials absorb energy from changing magnetic fields. When this absorption varies with frequency and exhibits the characteristics of a resonant condition, the phenomenon is called *MAGNETIC RESONANCE.*

There is a direct analogy between mechanical structures containing subelements of spring-supported masses that resonate at various frequencies and bodies of a material containing numerous magnetons that also resonate at various frequencies. If either an increment of material or a mechanical structure acting as a unique entity is stimulated by an appropriate source of oscillatory energy, the absorption of that energy is maximized at those frequencies where subresonances occur.

## 8.1 COHERENT PRECESSION

A magneton represents both a spinning electric charge and a spinning mass. The spinning electric charge involved produces a magnetic field directed along the spin axis.

A magneton is supported in molecular space by the interactions of both its electric field and magnetic field with other electric and magnetic fields. To a certain degree such a suspension resembles that provided by mechanical springs; at least the magneton is free to bounce around a bit. In addition, the orientation of the magneton is not necessarily constrained by these "springs" but is sometimes free to respond to the direction of any additional magnetic field to which it is exposed. When a large number of similar magnetons are exposed to a common magnetic field, some of these align themselves with that field.

Now if the direction of an imposed field is suddenly changed from an "old" direction to a "new" direction, the magnetic moments of the responding magnetons attempt to pull their magneton axes to the new direction whereas the mechanical moments attempt to maintain the old direction. The residual force that extends between these two directions causes all responding magnetons to precess until they settle down in the new direction. The characteristics of this precession are derived from the magnitudes of the magneton's mechanical moment, of its magnetic moment, and of the strength of the aligning field.

Magneton alignments do not instantaneously follow the imposition of a new magnetic field but rather take place over an interval of time. This interval is variously called the *LONGITUDINAL RELAXATION TIME,* the *SPIN-LATTICE RELAXATION TIME,* or the *THERMAL RELAXATION TIME.* In addition, the time required for transient precessions to die down is called the *PARAMAGNETIC RELAXATION TIME,* or the *TRANSVERSE RELAXATION TIME.*

When the time/space conditions experienced by each one of a number of magnetons are the same, the precessions for all similar magnetons are said to be "coherent." *COHERENCE* is a term used to describe those circumstances in which the same things are happening to many systems at the same time and in the same phase.

Thermal activities are not coherent. Those taking place within a body of material keep all particles bouncing around and, in the process, undoubtedly keep all member magnetons precessing continuously. The random nature of thermal movements, however, makes it impossible to separate one small precession from all of the other activities that are in simultaneous process. As a result, thermally induced precessions are incoherent and cannot be detected by external means.

On the other hand, as illustrated by Fig. 8.1, a coil mounted with its axis at right angles to an ambient magnetic field can be used to provide the required directional transient in that field. The magnetic field that accompanies a current pulse in this coil instantaneously "pushes" all of the magnetons in the same direction at the same time. As contrasted to the incoherence of thermal energy, this push is a coherent impulse of energy delivered at the same instant to all participating magnetons. After the current pulse is completed in such an excitation coil, the force that brings all of the magnetons back to their previous orientation is also a coherent push; its results can be detected as a transient oscillation at the precessional frequency. This oscillation can be detected in the same coil that was used to give the initiating "push."

## 8.2 ENERGY/FREQUENCY EQUIVALENCY

Whenever a system can exist in either one of two different energy states, a transition between these two states can be expressed in terms of a particular frequency. According to *EINSTEIN'S RELATION,* this frequency is equal to the difference energy experienced between the two divided by Planck's Constant.

When a number of magnetons exposed to a common magnetic field align themselves with that field, two alignments are possible: parallel (with the imposed field) and antiparallel (opposite to the imposed field). A slight difference in the energy content of these two directions can be detected when a magneton is made to flip from one orientation to the other. This energy difference derives from the distortion of the aligning field that occurs in the presence of each magneton field. In this case, parallel orientations represent a little less energy than antiparallel orientations do.

When immersed in weak magnetic fields, some magneton axes remain thermally excited in random directions, some aligning with the field and some against the field. There are always more magnetons aligned with than against, however. In stronger fields, all susceptible magnetons are aligned either with or against the common field. As field intensities increase, the number aligned with increases, whereas the number aligned against decreases.

All of the magnetons in a material are exposed simultaneously to both thermal forces and aligning magnetic-field forces. These two forces interrelate. Flipping back and forth between parallelism and antiparallelism is part of a material's normal thermal activities. Although the time at which a particular magneton reverses direction is of a random derivation, the relative density of parallel and antiparallel populations is a constant that depends both on the strength of the imposed field and on temperature.

As a result of the flipping back and forth between parallelism and antiparallelism, energy is continually radiated from a body of magnetized material at the frequency represented by the energy difference of these two alignments. In other words, the oscillatory activities of a magneton suspended in electric/magnetic fields can be envisioned in either one of two different ways. One is that of a precession and the other of a flipping back and forth between two energy states. Since both of these phenomena occur at exactly the same frequency, they are really two different ways of looking at the same phenomenon!

## 8.3 LARMOR SPECTRUM

A magneton's precessional frequency (and its flipping back and forth, energy-difference frequency) is called the *LARMOR FREQUENCY*. This is a resonant condition derived from the combined effects of magnetic and mechanical moments and is calculated as follows:

$$F_L = kg\ (e/m)\ B \qquad \text{(Eq. 8.1)}$$

where $F_L$ is the Larmor frequency; $g$, the ratio of magnetic moment to mechanical moment; $e/m$, the ratio of electric charge to mass; $B$, the strength of the magnetic field seen by the magneton; and $k$, a constant.

In this equation, $g$ is called the *MAGNETOGYRIC RATIO* (or the *GYROMAGNETIC RATIO*). It is an immutable characteristic of a particular type of magneton. The ratio $e/m$ is also a particular magneton characteristic. Because both $g$ and $e/m$ are constant for a particular type of magneton, each type precesses at one, and only one, specific frequency in each magnetic field.

In terms of fundamental particles, the Larmor frequency of a proton (or *PROTON RESONANCE*) is calculated by

$$F_p = 42.58\ B \qquad \text{(Eq. 8.2)}$$

where the frequency $F_p$ is given in megahertz when the magnetic field strength $B$ is given in tesla.

Other nuclei resonate at other frequencies, for instance: sodium at 11.3 megahertz and chlorine at 4.2 megahertz in a 1-tesla field.

The Larmor frequency of an electron is calculated by

$$F_e = 28.0\ B \qquad \text{(Eq. 8.3)}$$

where the frequency $F_e$ is given in gigahertz when the magnetic field strength $B$ is given in telsa. Both $F_p$ and $F_e$ can be used to measure field strengths with great precision.

Following the dictates of Eqs. 8.2 and 8.3, protons always resonate at exactly the same frequency when they are exposed to the same field, and electrons always resonate at some other exact frequency when they are exposed to the same field. In a molecular environment, however, the magnetic field strength $B$, as seen by an individual magneton, is the vector sum of an imposed field and the microenvironmental chemical fields created by other moving charged particles that are in the general vicinity of the one whose precessions are of interest.

Returning to Eq. 8.1, if the strength of an imposed field is substituted for the strength of the field $B$ actually seen by a magneton, the preceding equations have a number of possible solutions for each particular type of magneton. In fact, there is an individual solution for each microenvironment experienced by each particular type of magneton.

The number and character of the microenvironments that exist in a molecule are determined by its chemical nature. The number of atoms and the number of electrons in a molecule are both finite quantities. In addition, the bonds between the various charged particles represent finite increments of energy. As a result, each molecule contains a finite number of microenvironments for each type of magneton present. Figures 8.3 and 8.4 purport to show the two basic sources of these microenvironments. One source is the Lenz fields that are derived from orbiting electrons. The other is the attraction, or repulsion, between magneton pairs.

Based on the finite nature of microenvironment/magneton possibilities, if an imposed field is held at constant strength and the frequency of directional change is varied continuously over a range, precessions for the same type of magnetons can be detected at a number of discrete frequencies. There is, then, a frequency spectrum over which resonant conditions can occur. Conversely, if the frequency of field change is held constant and the field strength is swept over a range, precessions can be detected at a number of different field strengths.

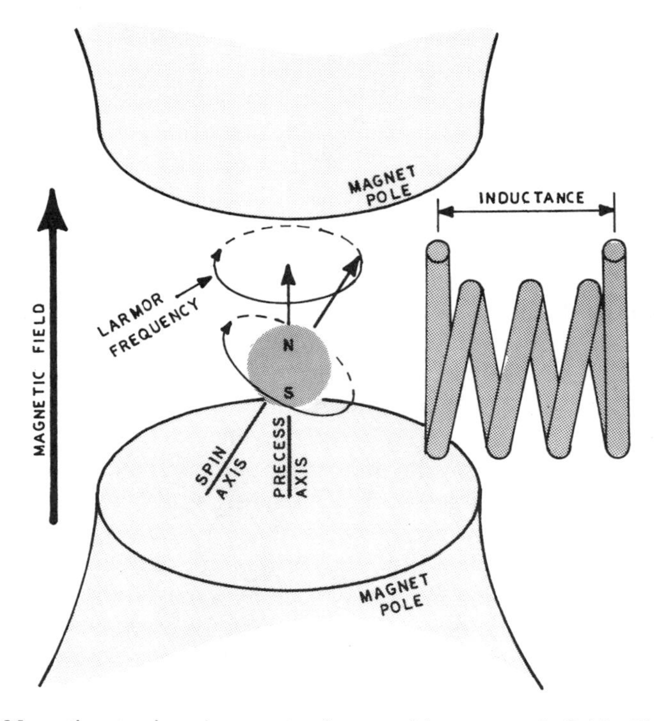


Fig. 8.1. Magnetic pumping: A magneton immersed in a magnetic field will precess at a frequency that is characteristic of that field. This precession frequency may be detected by noting the change in the inductance of a precession-stimulating coil when that coil is exposed to a swept frequency.

## 8.4 MAGNETIC PUMPING

A specimen of material can be exposed simultaneously to both an alternating field and a constant field, where one of these magnetic fields is directed at right angles to the other. The constant field is used to establish directional coherence through magneton alignments, whereas the alternating field is used to stimulate magneton oscillatory activities and, in the process, establish coherence of both frequency and phase.

The actual aligning field seen by all specimen magnetons is the vector sum of all the fields—constant-imposed, alternating-imposed, and microenvironmental. Since the direction of the aligning field is continually undergoing change at the alternating-field frequency, there is an alternating force on each directionally coherent magneton that seeks to change that direction. If the frequency of this alternating force is the same as a Larmor frequency, the magnitude of Larmor activities is maximized. This process is call *MAGNETIC PUMPING.*

Figure 8.1 makes use of one magneton to illustrate the principles involved. As the activity of one magneton is too small to be detected, the coherence of a great many similar magnetons in identical environments is required if the consequences of this phenomenon are to be observed. An examination of Fig. 8.1 should take into account the many coherent magnetons symbolically represented by the one shown.

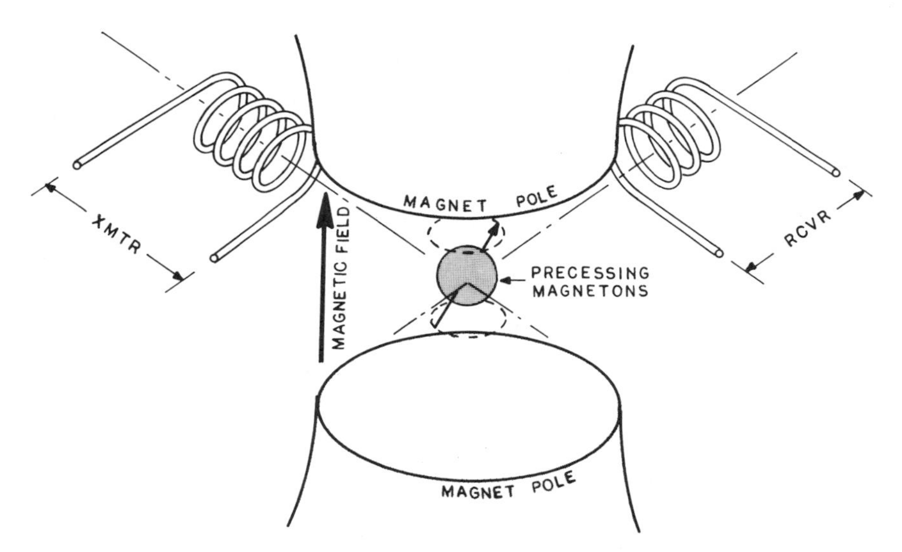


Fig. 8.2. Crossed-fields detection: Precessing magnetons radiate energy that may be detected as a voltage generated in a nearby "receiver" coil.

When a specimen is subjected to this magnetically variable-force configuration, energy is consumed in the interaction between a material's magnetons and the alternating field. This energy consumption is maximized by a Larmor frequency. As is usually the case, it is manifested by a rise in sample temperature. This temperature rise is maximized when a material is driven at a Larmor frequency during field exposure.

If an alternating field is supplied by a coil driven with an alternating current, the absorbed energy can be detected as a measure of coil impedance. Since the driven frequency is continually varied over a range, the apparent coil impedance decreases as each Larmor frequency is encountered.

In Fig. 8.2, crossed transmitting/receiving coils are used for detection in contrast to the one-coil arrangement of Fig. 8.1. In a manner equivalent to that of the configuration in Fig. 14.5, a transmitting coil is used to illuminate a specimen. A receiving coil is then used to detect energy reflected from that specimen. Since the reflected energy is maximized by a Larmor resonance, the signal increase detected in the receiving coil is more or less equivalent to the impedance decrease in a one-coil system.

## 8.5 MAGNITUDE OF LARMOR SIGNAL

For purposes of measurement by external means, the precessions of parallel magnetons are cancelled by those of an antiparallel nature. As a result, only the surplus population in the parallel direction contributes to mensuration processes. The magnitude of a Larmor signal is proportional to this surplus population.

As the strength of an imposed field increases, more magnetons are aligned in the field direction and fewer against it. The magnitude of a Larmor Signal then increases with increasing field strength. As a first approximation the signal-to-noise ratio of a Larmor resonant response is determined as follows:

$$S/N_L = knd\,(f/T)^{3/2}(Q/b)^{1/2} \qquad \text{(Eq. 8.4)}$$

where $S/N$ is the Larmor signal-to-noise ratio; $n$, the number of magnetons present in each cubic centimeter of the sample; $d$, the coil diameter (assuming the sample fills the coil); $f$, the operating frequency at Larmor resonance; $T$, the absolute temperature; $Q$, the coil figure of merit; $b$, the bandwidth of the detector; and $k$, a constant.

Magnetic field strength $B$ enters into this equation by establishing the Larmor frequency, $f$. Both the temperature and the size of sample are also significant considerations. Larger samples provide larger "surplus populations,'' whereas higher temperatures reduce that surplus through added incoherence and greater antiparallel populations.

The *Q* of the signal, which is obtained from the Larmor resonance of an individual magneton, is very sharp. If the *Q*'s of many such resonances are to be summed without losing this sharpness, the magnetic field seen by all participating magnetons in a specimen must be exactly the same. If the imposed field is not homogeneous, the *Q* of the summed signal is blunted even to the point of losing the signal altogether. In addition, the various molecular relaxation times must be considered when determining the speed with which the frequency of an alternating field may be changed. Again, the detected signal is blunted and may be lost if the magnetic environments are changed at too fast a rate.

## 8.6 CHEMICAL SHIFTING

*NUCLEAR MAGNETIC RESONANCE,* commonly abreviated as NMR, is concerned with the precessional activities of paramagnetic nuclear magnetons in their microenvironments. Using NMR measuring techniques, the frequency resonance experienced in an imposed field is different from that which would be computed with Eq. 8.1 that uses the strength of the imposed field in the calculations. The difference exists because each nuclear magneton is exposed to the vector sum of both imposed and local fields, not just the former.

Although the nuclei from dozens of elements exhibit resonance to some degree, protons (the nuclei of the odd, and by far the most abundant, isotope of hydrogen) are the most prevalent. Since protons are present in all organic substances and in water, the term ''nuclear magnetic resonance'', as this term is most commonly used, might be more precisely expressed as *PROTON PARAMAGNETIC RESONANCE.* In any event, although this discussion stresses protonic activities, the same principles also apply to other nuclear particles to varying degrees.

The establishment of a nuclear resonance for a nucleus in a molecular bond which occurs at a different frequency than would be expected in the same imposed field for an unbonded nucleus is called a *CHEMICAL SHIFT*. This shift is caused by the effects of orbiting electrons that shield the nuclei from external fields. When an electron is in spherical orbit, the diamagnetic changes to that orbit reduce the field seen by the nucleus. This effect is dependent on the Lenz field, which as shown in Figs. 8.3 and 8.4 is opposed to the imposed field. Since chemical shifts are caused by Lenz fields, they are also called *DIAMAGNETIC SHIFTS*.

Delocalized electrons, which circulate widely through complex molecular structures of which a nucleus is a part, have similar diamgnetic shielding effects. Nonspherical orbits create opposite *PARAMAGNETIC SHIELDING* effects.

Chemical fields cause frequency shifts away from those frequencies that would otherwise be computed with Eq. 8.1. The degree of such shifts is an indication of the strength of the force that holds electrons in nuclear orbit.

Because of the cancellation of parallel effects by those magnetons in an antiparallel mode in an imposed field of fifteen kilogauss at room temperature, only about five protons out of every million present can be usefully employed for mensuration purposes. To complicate matters, chemical shifting for protons is of an order of magnitude of ten parts per million of the imposed field strength. Since only five protons out of every million present are available for measurements that must be detected over a magnetic field range spanning only ten parts in one million, it is obvious that NMR instruments must be very sensitive, very stable, and very precise. Material samples are often rotated in order to average out the effects of small variations in the flux density of imposed fields.

The resonant frequency of an isolated proton is described by Eq. 8.2. On the other hand, it must be taken into account that the resonant frequency of a proton is shifted by orbiting electrons whenever the proton forms chemical alliances either in an atom or in a molecule. The amount of this shift is a function of the strength of the bonds on the orbit electron that causes the shift. When an atom gives up, or gains, an orbit electron in a chemical relationship, diamagnetic shielding is either decreased or increased, as the case may be. In order for chemical-shift values to have the same meaning in different laboratories, the position of the one strong resonant line from tetramethylsilane (TMS) is designated as the internationally accepted zero reference.

When a proton is incorporated in a solid, energy is absorbed over such a wide range of frequencies that sharp resonant effects are not usually encountered. It is then necessary to dissolve sampled materials in a solvent whose own resonant responses can easily be separated from those of the sample. Deuterochloroform is a useful solvent since it provides a minimum of background signals that can be easily distinguished from those of many other materials.

Because an NMR signal is proportional to the size of a sample, or to the total number of responding magnetons present in a sample, NMR techniques can be used in moisture metering as well as other processes in which there is a proportional mixture of two different materials. In addition, under certain circumstances, it is possible to obtain signals from some nucleonic magnetons in solids. In such cases, *KNIGHT SHIFT* is the microdifference in the field seen by a nucleus in a metal compared to what it would see in a diamagnetic insulator. The Knight shift is a result of the diamagnetism of conduction electrons rather than of orbit electrons.

## 8.7 SPIN-SPIN SPLITTING

Diamagnetism from electron orbits is not the only magnetic microfield seen by structural protons. Whenever several protons are bound together in one molecule, the interactions of their toroidal magnetic fields caused by their spins must be added to those of the diamagnetic chemical shifts. Such multiproton interaction is called *SPIN-SPIN COUPLING.*

In spin-spin coupling, three relationships between two protonic magnetons are possible. In the first, if the directional orientation between two magnetons is random and constantly changing, their fields do not interact, and the energy content of one is unaffected by its neighbors. However, if two magnetons are aligned by an imposed field, the random nature of

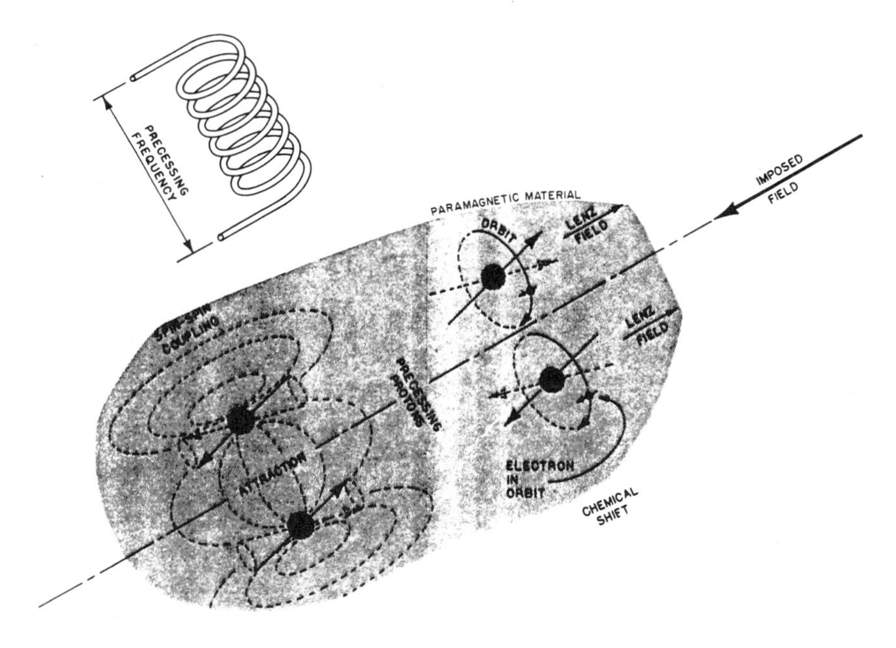


Fig. 8.3. Spin-spin attraction: A magneton precesses at a rate characteristic of the magnetic field actually experienced. This field is the sum of an imposed field and the fields seen at the local level. The strength of a local field may be deduced by subtracting the effect of the known imposed field and depends on whether adjacent magnetons are aligned in a pattern of attraction (as shown here) or repulsion. Since both conditions commonly exist, the changes in the inductance of a coil used to induce precessions have a response at two different frequencies.

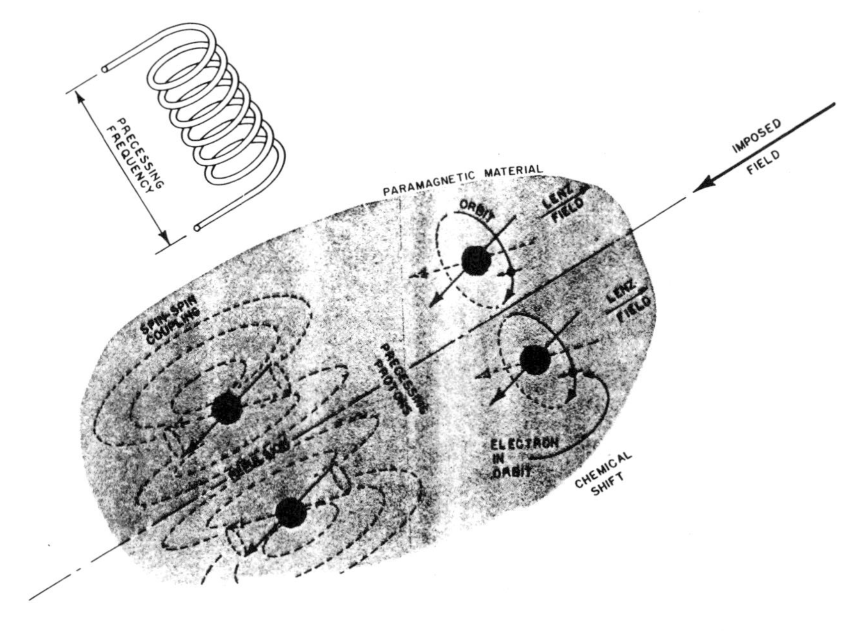


Fig. 8.4. Spin-spin repulsion. (See Fig. 8.3 for the attraction mode.)

the directional orientation is eliminated. Under the conditions of coherence, toroidal microfields have a chance to interact. In such interactions, two protonic magnetons are oriented either in the same direction or opposite directions. These two possible configurations are illustrated by Fig. 8.3 and 8.4. In one illustrated circumstance, the field seen by an individual proton is increased slightly, and in the other, decreased slightly. As a result, the absorption peak, otherwise caused by resonance after chemical-shifting, is split into a doublet. Here a "doublet" refers to two resonance responses centered on a chemical-shift frequency but separated by the difference in microfield caused by the two possible orientations. If more than two interactive protons are present within one molecule, each absorption peak is split into a number of peaks, or *MULTIPLETS,* with predictable numbers of lines, spacings, and relative intensities.

The "widths" of chemically shifted resonant lines are determined by the multiplet structure. As this structure is more complex in solids, than it is in liquids, solids have much broader line widths. In fact, line widths in solids can be several orders of magnitude greater than they are in liquids.

In solids, the interactions between protonic magnetons in a number of different molecules are involved in forming multiplets. In liquids, on the other hand, one molecule is subject to thermal tumbling in relation to all other molecules so that the total intermolecular activity is incoherent. As a result, any formation of multiplets *between* molecules is also incoherent and cannot be detected. Only those multiplets formed *within* individual molecules are coherent and hence detectable. This phenomenon is called *MOTIONAL NARROWING* since it is derived from relative movements between molecules.

It is possible to add an auxiliary coil to the magnetic structure of Fig. 8.2 so that a particular peak can be resonated continually while the other peaks are swept by the main drive coil. Under these circumstances, the particular peak is destroyed by motional narrowing and removed from the spectrum that would be otherwise obtained. This process is called *DECOUPLING*.

## 8.8 SECOND ORDER SPIN-SPIN SPLITTING

Additional peak splitting into a *HYPERFINE STRUCTURE* is a result of electron-proton spin-spin interactions. Here a protonic magneton sees a slightly different microfield because of nearby similarly aligned electronic magnetons. Since spin-spin couplings are transmitted through chemical bonds, it is possible to deduce bond angles and bond spacings from an analysis of multiple hyperfine spacings and relative intensities. Line widths and other chemical shifts are indications of various other processes. For instance, exchange rates, equilibrium constants, and barriers to rotation can be evaluated with the same technique.

In *NUCLEAR MAGNETIC RESONANCE SPECTROSCOPY*, absorption traces are the envelopes of the intensities of each unit that make up a multiple hyperfine structure. As such, the area under a trace reflects the relative numbers of nuclei present in each chemical association. Since molecules contain only integral numbers of nuclei, relative intensities vary according to these integers.

## 8.9 NUCLEAR MAGNETIC RESONANCE FLOW METER

The velocity of a moving particle can be determined from a measurement of the time it takes for that particle to traverse a given distance. In those circumstances in which a quantity of material moves from one location to another, it is possible to determine the velocity of its motion from a measurement of the time it takes an individual particle (or rather a small group of particles) imbedded in that material to traverse a given distance.

If the material is normally homogenous, there must be some way of imposing a difference between one particle and another in order to select

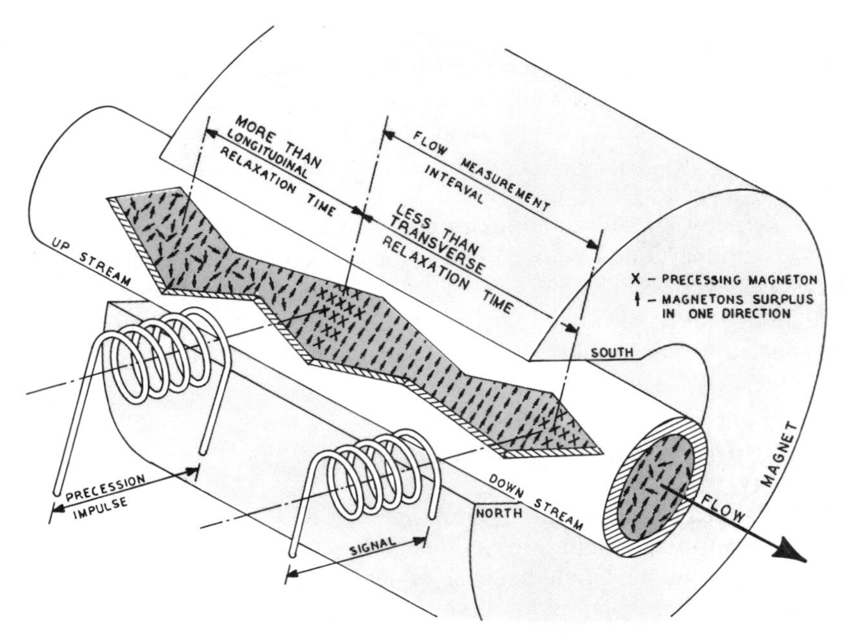


Fig. 8.5. Nuclear magnetic resonance (NMR) flow meter: The velocity of a fluid may be deduced by the time it takes a particle of that fluid to traverse a standard distance. In an NMR flow meter, individual particles are "tagged" for a short period of time by stimulating them into precession.

the particular particles that are to be used in the process of determining traverse time. Some means of "tagging" a small group of particles is required.

Localized molecular precession in a small volume of material is a measurable anomaly that will last for the period of the transverse relaxation time. As the transverse relaxation time for nuclear magnetons can be measured in seconds, or at least fractions of seconds, it is possible to tag the required group of molecules for a long enough period to make them usable in some flow measurements.

Figure 8.5 shows a configuration in which a precession-detection station is established downstream from a precession-stimulation station. The separation between these two stations must be represented by a distance that is less than can be traversed in the transverse relaxation time. The actual traverse time of the material, between the stimulation and the detection points, can then be measured and the velocity determined from this measurement.

Two exposures to relaxation times are involved in this process. The first is the time taken to establish directional coherence so that precessional coherence is possible, known as longitudinal relaxation time. The second is the time taken for precessional coherence to decay to a nondetectable minimum, which is the transverse relaxation time.

If water is placed in a 1-kilogauss field, its induced magnetism reaches an equilibrium of 4.11 microgauss. This magnetization has a transverse relaxation time of approximately 2 seconds. In these same circumstances, the hydrogen nuclei (the protons) in the water precess at a frequency of 4.258 megahertz.

## 8.10 ELECTRON PARAMAGNETIC RESONANCE

*ELECTRON PARAMAGNETIC RESONANCE* (EPR) is a property related to the magnetic microfields experienced by unpaired orbit electrons functioning as magnetons. It represents an expansion of nuclear magnetic resonance, which explores the microfields seen by nuclear magnetons. As indicated by the difference between Eqs. 8.2 and 8.3, *EPR SPECTROSCOPY* is an investigation of absorption patterns over a range of higher frequency than that examined in NMR Spectroscopy. However, the same factors of line position, intensity, and width (the hyperfine structure) are available for analysis.

Although most electrons in molecular associations occur as pairs, in a few circumstances single electrons in nuclear orbit are sufficiently remote from other magnetons to give absorption responses. In particular, electronic magneton precessions can be detected in free radicals, biradicals, complex systems functioning in the triplet state, transition metals, rare-earth ions, and crystals with certain types of point defects.

Since electronic reaction times are much shorter with EPR than they are with NMR, EPR can be used to monitor various chemical mechanisms such as equilibria, reaction kinetics, potential barriers, and the like. Although EPR reaction times are much too short for tagging molecular groups using NMR flow meter techniques, it is possible to use paramagnetic resonance patterns to "label" certain chemical combinations. For instance, a free radical with a distinctive paramagnetic spectrum can be attached to certain chemical compounds that would otherwise not have one. Those compounds with the radical attached can then be identified after they have been passed through processes of various kinds. A nitroxide radical is commonly used with biological molecules for this purpose.

## 8.11 FERROMAGNETIC RESONANCE

In ferromagnetic insulating materials (such as ferrites), all of the electronic magnetons present in a specimen contribute to resonant processes because all these magnetons are prealigned by exchange forces. The resonant phenomena that accompany the exchange-force alignments are termed *FERROMAGNETIC RESONANCES* (commonly called FMR). An FMR response to all aligned magnetons contrasts with the NMR response in that the difference between parallel and antiparallel magneton populations is the only thing that contributes to measurable effects.

Since ferromagnetic absorption is much greater than paramagnetic absorption (because of exchange-force alignments), energy/material interactions can be detected at lower energy levels. In addition, ferromagnetic resonances are very sharp because exchange coupling between ferromagnetic magnetons is strong enough to overcome the effects of those microfields that would otherwise broaden paramagnetic responses.

Since all magnetons precess about the direction of an imposed field, energy is absorbed strongly from transverse radio-frequency fields when these are driven at precessional frequencies. A thin film of ferromagnetic material placed in an aligning magnetic field is capable of strongly absorbing an oscillating field whose magnetic vector is perpendicular to the static field.

Because magnetization in ferromagnetic materials is large, demagnetization fields are also large. As a result, a common magnetic field seen by all magnetons in all directions is found only in spherical specimens. Spheres of yttrium-indium-garnet (YIG) are commonly used in microwave circuits as ferromagnetic resonators. The bias magnetic field required to make a YIG sphere of 375-micrometer diameter oscillate is related to the frequency of that sphere by the ratio of 2.8 megahertz per oersted. A 3.6-gigahertz resonance requires a bias field of about 1300 oersted.

## 8.12 WINSLOW/AULD/ELSTON CONFIGURATION

In the *WINSLOW/AULD/ELSTON CONFIGURATION*, the frequency of a YIG sphere is changed by eddy currents induced in nearby surfaces. Generally speaking, exchanges of energy between microwave oscillators and conducting surfaces introduce shifts in the resonant frequencies of these oscillators. In the resonant condition of a YIG sphere (when exposed to an aligning magnetic field), the precessional activities of all electronic magnetons induce eddy currents in the surface of nearby conductors. The

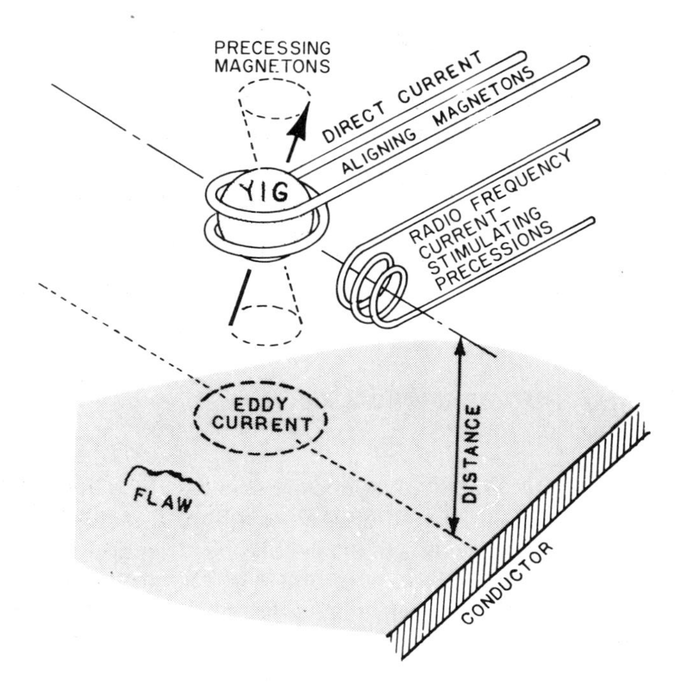


Fig. 8.6. Winslow/Auld/Elston configuration: The precessing magnetons of a YIG sphere may be used to induce eddy currents into nearby conducting bodies. Some of the characteristics of these bodies may be deduced through impedance measurements of the precession-stimulating coil.

energy reflected from these eddy currents changes the magnetic field seen by the YIG magnetons and shifts the resonant frequency of the YIG sphere. The actual cycles of this resonant frequency can be counted and presented in the form of a numeric display.

The amount of energy reflected under these circumstances depends on the electric/magnetic characteristics of the conducting surface and on the distance of the YIG sphere from this surface. If all variables except one are kept constant, the frequency of the YIG sphere can be calibrated in terms of that one variable.

A YIG probe can be used to discriminate between various types of conducting materials, to detect variations in the surfaces of these materials (flaws), and to measure the distance of a YIG sphere from such surfaces. The aligning magnetic field required by Eq. 8.1 can be oriented either perpendicularly (as shown in Fig. 8.6), or tangentially to test surfaces. The perpendicular configuration is required for magnetic materials.

## 8.13 SPIN-WAVE RESONANCE

All magnetons in thin ferromagnetic films are aligned by magnetic fields imposed as film-plane perpendiculars. If one magneton in such a sensitized film is displaced from its perpendicular orientation and then allowed to return to its preferred direction, this angular displacement is propagated as a spin wave both over the film surface and through the film thickness.

The resonance phenomenon experienced by spin waves propagated from one magneton to another is called *SPIN-WAVE RESONANCE* (commonly abreviated as SWR).

## 8.14 RADIANT-ENERGY/SPIN-WAVE RESONANCE

In the configuration of Fig. 8.7, an imposed magnetic field aligns magnetons in a direction perpendicular to the plane of a film. It is possible for these magnetons to oscillate about this perpendicular direction in a spin wave that has a resonant frequency determined by the strengths of both the imposed and local fields. Electromagnetic energy can then be directed along a path tangential to the film's surface. Under the following very special conditions this energy will stimulate the spin wave's resonant condition: First, the electromagnetic energy must have the same frequency

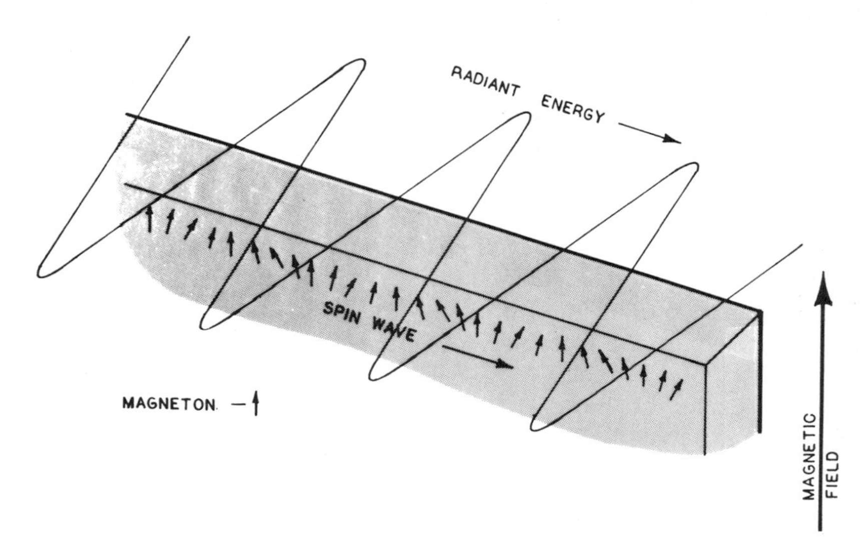


Fig. 8.7. Radiant-energy/spin-wave resonance: Radiant energy of a particular frequency is significantly absorbed when it is exposed to spin waves of the same frequency.

as the frequency of spin-wave resonance. Then, if the film's thickness, measured in terms of the electromagnetic energy's wavelength, is exactly some odd number of half wavelengths, the magnetons will be excited into spin-wave resonance and electromagnetic energy will be absorbed as a result.

## 8.15 ANTIFERROMAGNETIC RESONANCE

In *ANTIFERROMAGNETIC RESONANCE* (abreviated AFMR), the interactions of the spins of antiferromagnetic magnetons, which are aligned in their two characteristic directions by exchange forces, shift the frequency of these magnetons from the frequency they would experience in either paramagnetic or ferromagnetic materials.

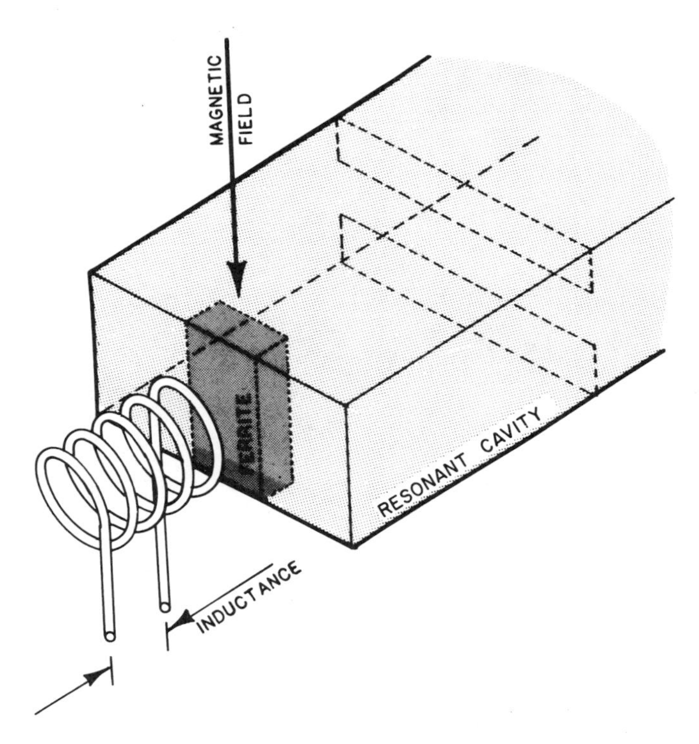


Fig. 8.8. Magnetization detection effect: The presence of electromagnetic energy of a particular frequency may be deduced by noting the inductance of a coil exposed to a sample of ferrite material whose magnetons have been stimulated into precession by that very same energy.

## 8.16 NUCLEAR QUADRUPOLE RESONANCE

In some molecules the electrostatic field is not symmetrical about all axes (see Fig. 3.14). Under these circumstances, the energy content of a particular magneton alignment depends both on magnetic directions and on the directions of those electric fields that cause electric-dipole alignments. The resonant frequency between any two alignment possibilities that results from these two aligning criteria is called *NUCLEAR QUADRUPOLE RESONANCE* (abreviated as NQR).

## 8.17 MAGNETIZATION DETECTION EFFECT

The magnetization of a ferrite material, which would otherwise be achieved in response to a steadily imposed magnetic field, is reduced whenever the Larmor resonant frequency of the constituent magnetons is stimulated by an alternating field. The decrease in steady magnetization is proportional to the square of the alternating field amplitude regardless of the relative directions of the two fields.

Since magnetization is a consequence of magneton alignments in re-

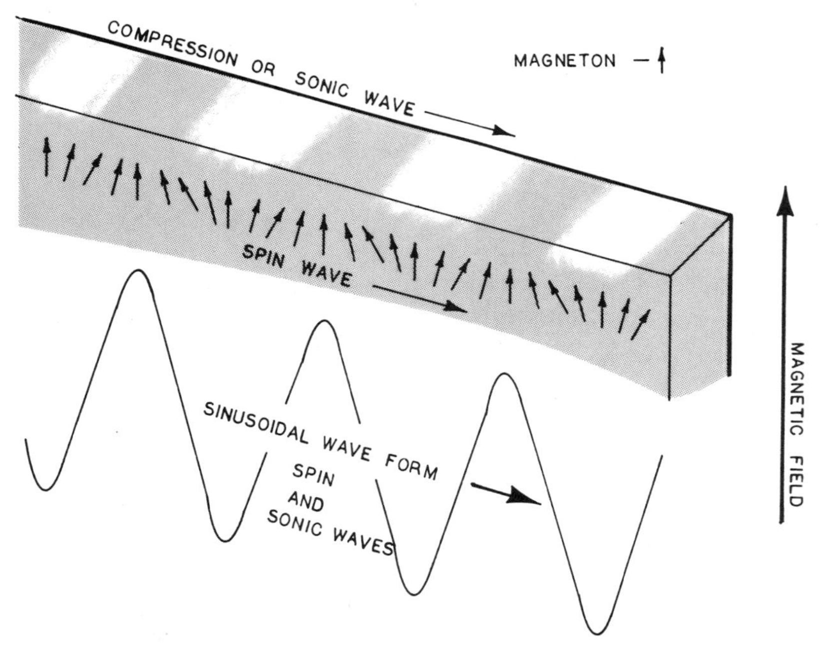


Fig. 8.9. Sonic-wave/spin-wave resonance: Sonic energy of a particular frequency is significantly absorbed when it is exposed to spin waves of the same frequency.

sponse to imposed magnetic fields, any force that limits magneton alignments reduces magnetization. If the magnetons experience Larmor oscillations under these circumstances, their ability to respond to a steady field is limited by the degree of the oscillations.

In Fig. 8.8, a sample of ferrite material is shown exposed simultaneously to an aligning magnetic field and a precession stimulation field. In this configuration, the inductance of a sensing coil is a function of the strength of the resonating electromagnetic energy.

## 8.18 SONIC-WAVE/SPIN-WAVE RESONANCE

As shown by Fig. 8.9, both sonic compression waves and magneton spin waves can exist simultaneously in a given volume of magnetized material. If the compression wave occurs at the resonant frequency of the spin wave, interaction will occur. This interaction even makes it possible to extract energy from the spin wave to amplify the sonic wave.

# 9
# RADIANT ENERGY

The emission of electromagnetic radiation from a quantity of material is one consequence of particle movement between molecular potential-energy states. Einstein's Relation determines the frequency of the radiant energy produced when particles move from higher quantized energy levels to lower ones.

Electromagnetic radiation has both electric and magnetic components, and these sustain each other in the propagation processes through interactions between oscillating electric and magnetic fields.

Frequency, phase, direction of propagation, and direction of polarization are basic characteristics of electromagnetic radiation. *POLARIZATION* is defined by the orientation of the plane in which the oscillations of the electric component take place.

By pictorializing the electric component alone, Fig. 9.1 illustrates radiant energy in its purest form. Here, since monochromatic (single-frequency) radiation propagates in-phase, the plane of polarization for each photon lies parallel to the plane of every other. Taking all factors into consideration, this is called *COHERENT RADIATION*. Coherent radiation includes coherence of frequency, coherence of phase, coherence of propagating direction, and coherence of polarization.

In tracing coherent radiation back to its points of origin, particle movements are allowed only between two energy states (or levels), and these movements must be synchronized in terms of both time and space; that is, all particles must move the same "distance" (from an energy standpoint), in the same direction, and at the same time.

A source of radiant energy producing complete coherence is called a *LASER*. Lasing can occur only under very special circumstances. As a result of the many energy levels present in all material, the random nature of thermal activities and of particle movements in both time and space, and the more or less spherical configuration of atomic orbits, radiation is usually polychromic with both incoherent phase and incoherent polarization.

Crystalline periodicity, electric-dipole alignments in electric fields, and

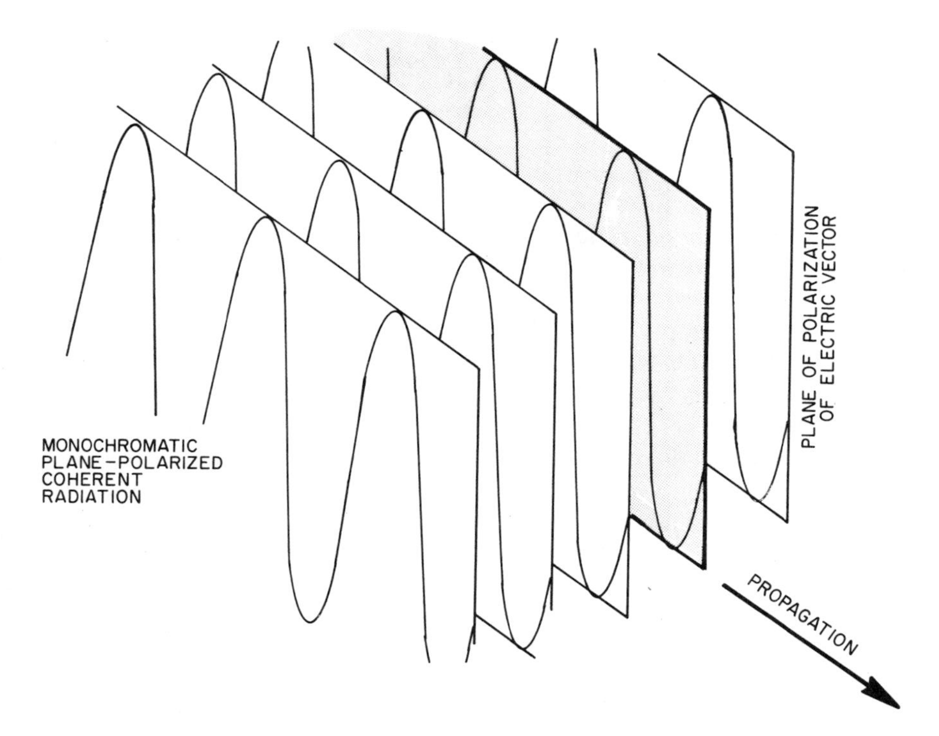


Fig. 9.1. Coherent radiation includes coherence of frequency, phase, direction, and polarization.

magneton alignments in magnetic fields can influence coherence of polarization in some circumstances. Absolute frequency and phase coordination can be achieved only by laser action, however, and it cannot be derived from thermally excited sources of any kind.

## 9.1 POSSIBLE VARIABLES

Magnetic fields have no direct effect on electromagnetic radiation, but they do affect material structures. Since material structures establish the character of electromagnetic radiation, various effects are observed in radiation patterns that are derived from magnetic influence.

The material/radiant-energy relationships potentially affected by magnetic fields include the following:

1. Radiant energy is emitted from all materials at temperatures above absolute zero

2. Emitted radiation can be either polarized or not polarized
3. Radiant energy is reflected from the surfaces of, absorbed by the molecular structure of, and transmitted through all materials to some degree
4. Radiant energy absorbed by materials can be reradiated either at the same or at lower frequencies
5. Radiant energy reflected from material surfaces can be polarized; if the radiant energy is already polarized, its polar angle can be changed
6. Radiant energy can be diffracted by using the molecular periodicity of thin films as a grating
7. Radiant energy transmitted through materials can be refracted
8. Radiant energy transmitted through materials can be double refracted (even triple refracted)
9. Radiant energy transmitted through materials can be polarized; if the radiation is already polarized, its polar angle can be changed
10. The velocity of radiant energy in certain materials depends on the polar angle
11. Radiant energy has a wavelength that can interact both with the dimensions of thin films and with crystalline periodicity
12. When two identical waves interact with one another, the result can be either a reinforcement or diminution. If two identical waves are out of phase by 180 degrees, their effects are completely cancelled.
13. A *RAY* is the rectilinear path in a homogeneous medium along which electromagnetic energy is propagated. Although ray directions are nearly always random, they can exist in a configuration where all are parallel. This configuration is called *COLUMNATION*.
14. A *BEAM* consists of many rays traveling more or less in the same direction. In a columnated beam, all rays are parallel. When beams interact with nonhomogeneous media, they can be completely destroyed as a result of their constituent rays being scattered.

Each of the above listed relationships is a result of radiant energy interacting either with molecular positions in, or molecular activities of, various materials. Since magnetic fields have potential effects on both activities and positions, they may also affect these relationships.

## 9.2 LUMINESCENCE

Various relationships that exist between atomic particles represent quantized potential energy states. A particle in each such state can be described as representing a unique level, or quantity, of energy. When a relationship is changed by moving a particle from one of the possible potential energy levels to another, energy is either emitted or absorbed in the form of an electromagnetic photon of a specific frequency. Materials that emit photons from causes other than high temperature are said to be *LUMINESCENT*.

The frequency of a photon is determined by the difference in energy levels encountered when a particle changes from one state to another. As level differences are quantized, so are frequencies specified. The so-called "corpuscular" nature of a photon, as it was conceived a few years back, was derived from the fact that each change in energy level represents a separate event that takes place over a very short period of time. This type of "event" is also characteristic of the *GEIGER EFFECT*, in which a pulse of light is emitted when the movement of one high-speed subatomic particle is stopped by its collision with a molecular structure. Each such pulse is related to a single collision as a unique event.

Continuous electromagnetic radiation of one frequency is constructed from a very large number of identical events that transpire over an extended time interval. The energy content of radiant energy is defined in terms both of the number of photons that occur in this time interval and of the frequencies of these photons.

In an individual atom, electrons circle a nucleus in one or more of a finite number of specific orbits. The number and energy content of these orbits are characteristic of each type of atom. The only possible change in relation between a nucleus and an electron occurs in the movement of the electron from one orbit to another. The photons derived from such movements are limited to those frequencies represented by the difference in energy that exists when an electron moves between two orbits. As a result, an individual atom can emit electromagnetic energy only at a few specific frequencies. These frequencies are observable as finite lines in either refractive or diffractive spectra.

When a number of atoms are assembled together into a monatomic gas at low pressure, the emission frequencies of individual atoms are unchanged. A heated, or otherwise stimulated, monatomic gas emits electromagnetic energy at those frequencies which are characteristic of its individual atoms. As the pressure of the gas is increased, interactions between its adjacent atoms take place, and the spectral lines broaden. This broadening of spectral lines indicates that the energy content of each orbit in a close association of two or more atoms is modified slightly.

When atoms are assembled into molecules, the relationships between orbit electrons and their various nuclei become much more complex. The number of possible energy levels increases significantly over that possible with individual atoms. As a result, molecular gases emit electromagnetic energy at many more frequencies than do monatomic gases. Molecular frequencies are also finite in number, however, and their locations in refractive/diffractive spectra are still characteristic of the types of molecules contained within a particular gas.

So many energy levels are created when molecules associate as crystalline solids that emitted frequencies extend more or less continuously over distinct bands. If these bands overlap, as they do in metals, radiation frequencies are continuous over a broad spectral range. In insulating materials, bands do not overlap, and radiation frequencies then occur in bands separated by other bands in which there is no radiation.

When several nuclei are associated as molecules or in more complex crystals, one nucleus can move slightly in relation to the others. A nucleus can vibrate within its chemical bonds, and asymmetrical molecules can rotate within their bonds. These movements represent additional quantized-energy forms of lower frequencies than those that result from electron orbit changes.

The following may be considered as an imprecise, but still generally valid, statement: Tumbling asymmetrical molecules create microwaves; vibrating nuclei produce infrared energy; electrons moving between outer orbits generate visible light; electrons moving from inner orbits relate to X rays; and splitting nuclei form gamma rays.

Considering the wide variety of molecular possibilities, radiant energy is present in the environment in a more or less continuous spectrum even though this spectrum is constructed from discontinuous components. In this case, the discontinuities are expressed in terms both of photon frequencies and of incidences of photon occurrence.

## 9.3 STIMULATION

All materials generate electromagnetic energy at temperatures above absolute zero. The creation of this energy is a result of the quantized movements of atomic particles participating in various thermal activities. At low temperatures, changes from one quantized energy state to another represent small energy increments. As a result, the accompanying electromagnetic energy is primarily of low frequency. At higher temperatures, thermal agitations become more frantic, higher energy levels are more often involved, and the generated energy exhibits higher frequency components in greater quantity.

A curve can be plotted showing the quantity of energy present at each frequency. The total energy present in a system is then represented by the area under this curve from zero to infinite frequency. The *STEPHAN-BOLTZMAN LAW* describes the integral involved.

With increasing temperature, the quantity of radiation increases at all frequencies, whereas the maximum frequency of the Stephan-Boltzman distribution curve moves to a higher frequency. This movement toward a higher frequency peak is described by *WEIN'S LAW* and explains why materials at 700°C are called "red" hot. At this temperature, the maximum of the wave-length distribution curve lies in the 700-nanometer region. Wave lengths of about 700 nanometers are perceived as "red" by the human eye.

When temperatures rise to 1500°C, energies whose wave lengths are as short as 400 nanometers become significant, and materials at this temperature appear to be "white" hot. The combination of all radiation between 400 and 700 nanometers loosely describes the visible spectrum. If all frequencies from red to violet are present, the emitted energy looks "white" to the human eye. Materials producing energy in visible-light frequencies because of an elevated temperature are said to be *INCANDESCENT*.

Emission frequencies can be stimulated by material interactions with photon, electron, or ion beams. Since beams tend to have a limited energy composition in contrast to the all-encompassing constitution of thermal energy, only portions of the possible electromagnetic spectrum are beam-stimulated. As a result, luminescent energy has fewer frequency components than does incandescent energy.

In *BREMSTRAHLUNG RADIATION*, X rays are emitted when beams of electrons or other fast-moving charged particles are slowed by absorbing media.

An *ENERGY GAP* is the energy difference between a particle occupying one or another of two possible quantized states. If impinging energy from a beam is greater than the gap energy, a gap transition can occur and electromagnetic radiation be created. On the other hand, if beam energy is less than the gap energy, electromagnetic energy cannot exist. This leads to the conclusions of *STOKES LAW*, which states that the wave-length of beam-stimulated energy can only be equal to, or longer than, that of the absorbed electromagnetic energy.

Electromagnetic energy emitted from materials at frequencies that are less than the excitation frequencies, or as a result of activation by other than electromagnetic means, is called *PHOSPHORESCENCE* if it lasts for a noticeable time after stimulation is terminated and *FLUORESCENCE* if it does not last. These phenomena can be found at any frequency, visible or invisible. For instance, X rays of lower frequency fluoresce when some materials interact with X rays of higher frequency, whereas visible light is emitted from some materials undergoing ultraviolet stimulation.

The possibilities of electromagnetic emission as a result of material/

energy interactions are extensive and can appear a bit exotic. In the *GUDDEN-POHL EFFECT*, a momentary flash of light is produced when an electric field is either imposed on, or removed from, certain phosphors. In the *TSCHERENKOV EFFECT*, on the other hand, water luminesces with a "weird" bluish glow when activated by either gamma rays (as from an atomic reactor) or from electron beams.

Luminescence can also be encountered in some materials undergoing mechanical strain, electric strain, or certain chemical processes. *TRIBOLUMINESCENCE* accompanies mechanical strain; *ELECTROLUMINESCENCE* is produced by electric strain; *CHEMILUMINESCENCE* is produced in chemical reactions; *BIOLUMINESCENCE* is produced by biological processes; and *GALVANOLUMINESCENCE* occurs in certain electrolytic rectifiers. *CRYSTALLUMINESCENCE* is even observed in some solutions during crystal formations.

## 9.4 SURFACE EMISSION

In *TRANSPARENT* materials, no significant energy interactions take place between adjacent molecules at the frequencies of transparency. Any energy that enters a material at a transparent frequency passes on through the material without significant modification.

Radiant energy emerging from the surface of a body constructed from a transparent material must originate at some source in back of that body. Although the polar angle of energy can be affected by passage through a transparent material, the characteristics of the emerging radiation are otherwise determined by the source and not by the material.

In *OPAQUE* materials, on the other hand, frequencies of molecular-energy interchange exist in which a photon emitted by one molecule is immediately absorbed by a neighbor. A photon of an opaque frequency never moves through a material very far. Energy is transmitted through such materials only by raising the energy level of one molecule above that of its neighbors. That molecule then interchanges energy back and forth with its neighbors until all achieve a common level.

As long as a particular molecule in an opaque material sees the same environment in all directions, energy interchanges will take place in a constant back-and-forth process. Those molecules near a surface, however, do not see the same environment in all directions. A photon emitted by a surface molecule in a direction toward the surface is not absorbed and radiated back by a neighbor but propagates through space as an element of electromagnetic radiation.

The passage of radiant energy through a transparent medium does not effect that medium. In contrast, the loss of a photon by radiation from a surface molecule lowers the energy content of the medium, and, of course, a photon entering it raises the energy level.

All energy emitted from the surface of opaque materials is a result of the activities of a very few molecules and electrons immediately adjacent to the surface. In turn, the activities of the surface molecules are determined by the energy they receive from interchanges with internal molecules. The higher the frequency of radiation and the greater the opaqueness of a material, the thinner the layer of surface molecules that participates in emitting radiant energy.

The characteristics of *TRANSLUCENT* materials lie in between those of transparency and opaqueness. In translucent materials, the radiant energy also comes from the surface molecular activities of the material, but the thin layer found in opaque materials now has depth. The amount of depth determines the degree of translucency.

In thin films, where the depth of translucency is greater than the film thickness, the amount of radiation emanating from the surface of the film is a function of the film thickness. In fact, this phenomenon can be used to measure film thickness.

## 9.5 FREQUENCY SHIFTING

Radiation is emitted from luminescent materials as stimulated particles move back and forth between the various energy levels available to them. Each movement of a particle between higher and lower discrete levels, generates one photon of a discrete frequency.

In thermally stimulated environments, magnetons are directed in a random fashion at any particular time and are continually changing their directions over a period of time. These random arrangements represent minimum energy configurations.

If a magnetic field is impressed on these chaotic movements, some magnetons will be aligned long enough for their toroidal fields to interact. Two such interactions are possible. One represents a little more energy than a random arrangement would, whereas the other represents a little less. Figures 8.3 and 8.4 illustrate this principle (in addition to the doublets of nuclear magnetic resonance).

In the *ZEEMAN EFFECT*, as in nuclear magnetic resonance, the alignment of magnetons by magnetic fields splits a single frequency derived from random activities into two frequencies centered on the random frequency. The deviations of these two frequencies from the center frequency is proportional to the strength of the applied magnetic field.

The *PASCHEN-BACK EFFECT* is a shifting of energy levels by strong magnetic fields that exceeds the split of the Zeeman Effect.

In the *STARK EFFECT,* spectral lines are similarly split by electric-dipole alignments that take place at the molecular level in electric fields.

The *DUFOUR EFFECT* is an abnormal Zeeman Effect that is observed in a direction parallel to the field direction if circularly polarized waves are changed to plane-polarized waves by means of quarter-wave plates.

Actually, there are two Zeeman Effects. One is sighted in the direction of an imposed field, whereas the other is sighted at right angles to it. As it is relatively difficult to sight in the field direction with most laboratory instruments, the second effect is the one most often referred to as ''the'' Zeeman Effect.

When sighted longitudinally in the field direction, the frequency derived from random activities is not polarized; the two shifted frequencies are polarized, however. One is polarized in the longitudinal direction and the other in the transverse direction.

A second means of frequency shifting is available when the distance between a source of radiation and an observation station is undergoing change. Following the *DOPPLER-FIZEAU PRINCIPLE,* a spectral line is shifted in frequency by an amount equal to the velocity of movement between generation and observation points.

## 9.6 POLARIZATION

Although ''polarization'' is really a unitary concept, it can be considered from either of two points of view. One of these is envisioned in terms of plane polarization, as shown in Fig. 9.1, whereas the other is envisioned in terms of elliptical polarization, as shown in Fig. 9.2.

Elliptical polarization can be envisioned as if it were constructed from two plane-polarized waves of the same frequency, where the polarization plane of one is at right angles to the polarization plane of the other. Plane polarization, on the other hand, can be envisioned as if it were constructed from two circularly polarized waves of the same frequency, where one wave rotates in the opposite direction from the other.

Circular polarization is a special form of elliptical polarization. If two circularly polarized waves of the same frequency rotate in opposite directions, their vector sum is a plane-polarized wave whose angle of polarization depends on the relative phase angle between the two circularly polarized waves.

If two plane-polarized waves of the same frequency are added vectorially, the result will be an elliptically polarized wave that can be represented as a vector that both rotates and changes in magnitude as it propagates through space. If one of the plane-polarized waves leads the other by a certain phase angle, the elliptical vector will rotate in one direction. If

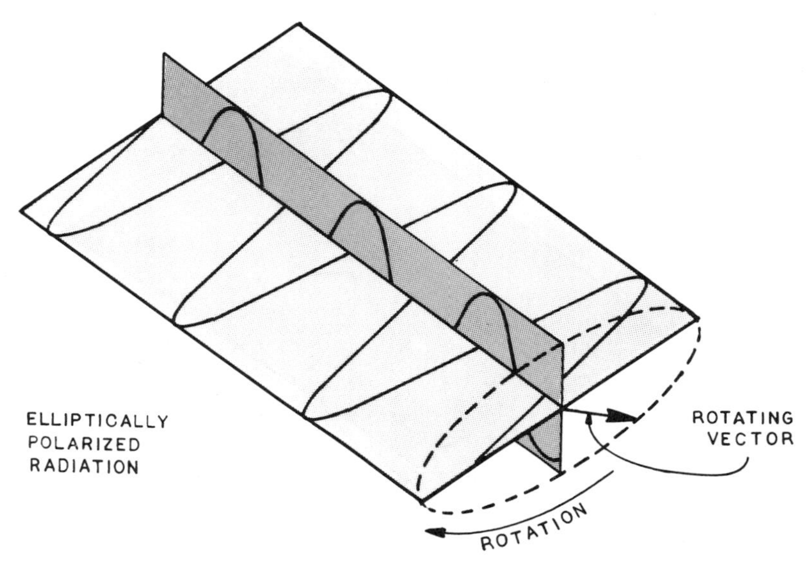


Fig. 9.2. Two plane-polarized waves of the same frequency but of different planes of polarization combine vectorially into an elliptically polarized pattern.

that same first wave lags the second by another phase angle, the elliptical vector will rotate in the opposite direction. If the two plane-polarized waves are either in-phase or out-of-phase, the ellipse will be reduced to a third plane-polarized wave.

A photon of electromagnetic energy is created each time a particle moves from a higher energy state to a lower energy state. The frequency of the photon is determined by the energy difference between these two states; the phase of the photon is determined by the time of particle movement; and the plane of the photon is determined by the direction the particle happens to be moving at the time of photon creation. In the most general circumstance, photon frequencies are fixed by the available energy states, but the time and direction are established by completely random processes.

Electromagnetic energy derived from a directionally random process is not polarized; that is, the polarization of an individual photon is incoherent in relation to all other photons. Polarized electromagnetic radiation can be fabricated in three ways: by regularizing the creative forces, by eliminating the random radiation that occurs in all phases other than the one desired, or by imposing regularization on the radiation after it has been emitted by refractive, diffractive, reflective, or scatter forces. Since polarization results from movement alignments as particles change from one energy state to another, anything that affects these motional directions will affect polarization.

In the *STARK-LUNELUND EFFECT*, the light emitted by a beam of moving atoms is polarized by the unidirectional movements of the atoms. This unidirectional activity tends to regularize the directions of particles moving back and forth between energy states.

Molecular alignments in crystal lattices are also capable of polarizing emitted radiation.

## 9.7 MAGNETIC QUENCHING

Whenever magneton alignments force all particle movements into one common direction, the electromagnetic radiation emitted as a result of these movements is polarized. As discussed previously in regard to the Zeeman Effect, an imposed field splits incoherent activities into two sets of directionally coherent activities on the one hand and polarizes the radiation that results from the two changes in energy level on the other.

When a magnetic field establishes coherent magneton directions, the energy content of the molecules that support these magnetons is changed. Sometimes this change is enough to eliminate one energy level. As a result, one or more spectral lines in the radiation that emanates from the affected material is also eliminated. This is called *MAGNETIC QUENCHING*.

In the *WOOD-ELLETT EFFECT*, a spectral line of polarized radiant energy is quenched by the imposition of a magnetic field because of the elimination of the energy state represented by this spectral line as a result of the magnetically changed order. In an extension of the Wood-Ellett Effect, a magnetic field quenches fluorescence in some vapors by altering those energy levels that create fluorescence.

## 9.8 ABSORPTION

To the degree that magnetic fields affect reflectivity, transmissibility, and scatter, they will also affect *ABSORPTION*. Absorption is a process that is the reverse of emission. If a photon of electromagnetic energy is emitted when a particle moves from a high energy state to a second state of lower energy, an identical photon will be absorbed when the same particle makes the reverse move.

Absorption, however, is not the *exact* reverse of emission. According to the *FRANCK-CONDON PRINCIPLE*, the energy spectrum *emitted* from an excited body is slightly different from that *absorbed* because of the changes in lattice spacings caused by thermal activities during emission that are not necessarily present in the same degree during absorption. In addition, when energy and materials interact, some energy is reflected from surfaces, some is transmitted through materials, and some is scattered by inhomogeneities in the transmitting path. To the degree that these functions are subject to spectral variations, the characteristics of absorption will show spectral variations from those of emission. As a result, absorption is a spectrally variant process that has a somewhat different spectrum from that of emission.

Absorption is proportional both to the time and the intensity of exposure. Reflections tend to reduce intensity, whereas scatter tends to increase exposure time. The path of a scattered ray through a material is much longer than the path of a rectilinear ray. With a longer path, more opportunities for molecular interactions arise.

It may be generally stated that if the scatter centers of a material take on an incoherent pattern, more energy will be absorbed than would be if the pattern had a regularity of some kind, at least for all frequencies except those whose wavelengths interact with the dimensional periodicity of a regular state. This principle is illustrated when a normally clouded colloidal solution is clarified in the presence of a magnetic field. The colloidal particles act as scatter centers. If these particles happen to have a magnetic moment, they are aligned by the magnetic field into a regular pattern, with accompanying minimal scatter.

Absorption can be selective in terms of frequency, direction of propagation, and plane of polarization. In *PLEOCHROIC MATERIALS*, the frequency selectivities of absorption differ along different crystal axes.

Just as the radiation emitted from solid materials covers broad spectral bands of frequency, the absorption of most materials also follows these same broad bands. In *DIDYMIUM MATERIALS*, however, there are very narrow absorption/emission bands in those parts of the spectrum that are otherwise transparent.

Just as various energy/material interactions can affect the absorption of electromagnetic radiation, moreover, the absorption of electromagnetic radiation can affect other expressions of a material's energy state. Phosphorescence is one example of a phenomenon that persists after absorption of radiation has ceased.

## 9.9 PHOTOMAGNETIC EFFECT

Paramagnetism is the process whereby thermally dispersed magnetons tend to align themselves either with, or against, an imposed magnetic field. The magnitude of this phenomenon is measured as the difference in population density between the magnetons aligned in each of the two directions.

In some materials, primarily certain salts, paramagnetism is enhanced by exposure to electromagnetic radiation. This is called the *PHOTO-*

*MAGNETIC EFFECT*. The absorption of electromagnetic radiation by a crystal lattice changes the energy content of the individual particles in such a lattice and also changes the distribution of energy between these particles. Just as the addition of energy in the form of heat alters the response of a paramagnetic material, so will the addition of energy in the form of electromagnetic radiation.

## 9.10 REFRACTION

A ray of electromagnetic radiation that passes obliquely from one medium to another changes its direction at the interface. This change in direction is called *REFRACTION*. Refraction is derived from the fact that electromagnetic radiation propagates at different velocities in different materials.

The direction of a refracted ray is referenced to a line that is normal to the surface separating two media at the point at which an incident ray strikes that surface. In proceeding from a medium of high velocity to one of lower velocity, a refracted ray is always bent toward the normal, and away from the normal when proceeding in the opposite direction.

The *REFRACTIVE INDEX* of a material is defined as the velocity of electromagnetic radiation in a vacuum divided by the velocity in the material. According to *SNELL'S LAW*, the ratio of the sine of the angle between incident ray and normal divided by the sine of the angle between refracted ray and normal is a constant. Furthermore, the incident ray, the normal, the reflected ray, and the refracted ray all lie in the same plane.

If radiant energy travels from a material of high index to one of low index, it is totally reflected at the interface if the angle of incidence exceeds some *CRITICAL ANGLE*. This critical angle is a characteristic of the velocity difference between two media.

The velocity at which radiant energy travels in a dense medium depends on frequency as well as the density of the transporting medium. As a result, high frequencies are refracted more than low frequencies (e.g., blue light is bent more than red light). The separation of electromagnetic radiation into its constituent colors, or frequencies, is called *DISPERSION*. The measure of the dispersive angular spread is called *BIFRINGENCE*. The *ABBE NUMBER*, which is used to rate dispersive phenomena, is the reciprocal of bifringence.

In most materials, the relation between frequency and refraction is a continuous function. In materials with an *ANOMALOUS DISPERSION*, however, there is a break in this relationship.

As shown by Fig. 17.4, the packing density of a crystal can be significantly different along different crystalline axes. When this condition exists, ray propagation can occur at different velocities in different directions since propagating velocity has a tendency to be proportional to density.

In the *BARTHOLINUS EFFECT*, a single ray is split into two rays when it is exposed to a propagating medium with different densities in different directions. The ray least affected by splitting is called the "ordinary" ray, whereas the ray most affected is called the "extraordinary" ray. The process of splitting rays by this means is also called *DOUBLE REFRACTION*. (As implied by the three axes of different densities shown in Fig. 17.4, triple refraction is also encountered.)

When a ray experiences double refraction, the two resulting rays are polarized at right angles to each other. This polarization is here called the *HUYGENS EFFECT*.

*DICHROIC ABSORPTION* is a term used to describe that condition in which absorption in a medium depends on the direction of propagation through the medium. Dichroic absorption is commonly associated with double refraction whenever one refracted ray is absorbed more than the other. In this case, absorption differences can be traced back to the crystalline packing density differences encountered in different crystallographic directions.

According to the *GLADSTONE-DALE EFFECT*, the refractive index of a material changes with changes in the density of that material. If a material expands, its refractive index decreases and vice versa. Since density can be affected by mechanical stress, temperature, various chemical phenomena (crystal packing densities), electric fields and magnetic fields, a number of effects stem directly from the Gladstone-Dale Effect that depend on the cause of the density change, among them the Kerr Effect and the Mirage Effect.

In the *KERR EFFECT*, mechanical strain is induced by electric fields in a manner analogous to the magnetic field compression of the Cotton-Mouton Effect. In the *MIRAGE EFFECT*, a distortion of images results from atmospheric refraction caused by air density changes over hot surfaces.

The axes of molecules contained in a fluid are randomly oriented by thermal forces. If these molecules happen to be anisotropic, the directions of their axes can be changed into some pattern of regularity when they are exposed to shear forces. Light passing through such a pattern is subject to double refraction, the degree of which refraction is a function of shear intensity. As shear, in turn, is a function of fluid velocity, the amount of double refraction can be used as an indication of fluid velocity. A condition exhibiting this phenomenon is called *BIREFRINGENT FLOW*.

In a *SCHLIEREN PROJECTION*, a lens is used to project a columnated light beam onto a screen. All of the columnated rays are brought to a single focal point midway between the projection lens and the screen. If the medium through which the columnated rays pass is exposed to any refractive anomalies, the refracted rays will not be a part of the columnation and will not pass through the common focal point. If a knife edge is placed near that focal point in such a way that it interrupts the refracted rays but does not interfere with the columnated rays, the refracted rays will be removed from the image of columnation projected onto the screen. As a result, the refracting anomalies are shown as shadow on the screen.

## 9.11 MAGNETOREFRACTION

*MAGNETOREFRACTION* describes those circumstances in which the refractive index of a material is changed upon exposure to a magnetic field.

In the *COTTON-MOUTON EFFECT*, certain dielectric materials become double-refractive when they are placed in magnetic fields so that the ordinary ray is retarded in relation to the extraordinary ray. The Cotton-Mouton Effect results from the compression of a dielectric material by a magnetic field, as shown in Fig. 6.1.

In fact, double refraction results from any pattern of regularity that exhibits different densities in different directions. In the *VOIGHT EFFECT*, strong magnetic fields cause certain vapors to become double-refractive when light rays are directed at right angles to the magnetic field direction.

## 9.12 REFLECTION

A portion of a radiant-energy ray incident on a smooth surface is reflected from that surface by either of two mechanisms. In the first, radiation traveling from a material of high index of refraction to one of lower index is totally reflected if the angle of incidence exceeds the critical angle. In the second, energy is absorbed by surface molecules and by conduction electrons that happen to be near the surface and simply reradiated to a degree that may well approach the total energy absorbed. Although surface particles are stimulated by their absorption of energy, this increase in energy is of little consequence in most circumstances since it is dissipated in the form of energy reradiated at the same frequency. On the other hand, if conduction electrons near the surface are manipulated by some means, as shown in Figs. 9.10, 9.11, 9.12, and 20.1, both the amount and direction of radiation can be modified.

Only the particles in a surface layer no more than a few atomic spacings deep take part in this process. Such particles are exposed to two completely different environments. Looking inward, they see a maze of fellow particles all busily radiating energy to, and absorbing energy from, each other. Any energy added to this morass is immediately returned in kind. Looking outward, they see an environment into which they can dump their excessive photons without having them returned.

If the material back of a surface is metallic, all of the incident energy is absorbed by the surface particles. Since these particles include a great many conduction electrons, which are easily excited to almost any frequency, most of the energy is reradiated, and little is passed on to the inner particles as heat. On the other hand, if the material back of a surface is not metal, some of the energy will be reflected and some refracted.

Because of their involvement in the absorption/reradiation of energy, the surface molecules of either a metal or a nonmetal can affect the character of the reflected ray. If the surface molecules are organized into some kind of orderly array in which the order is different in one direction than in some other direction, the character of the reflection may depend on the direction of ray incidence. In addition, the reflected ray is likely to be polarized by its interaction with the array.

According to *BREWSTER'S LAW*, the reflected and refracted rays from nonmetallic materials contain the maximum amount of polarization when the tangent of the incident angle equals the index of refraction.

Energy interactions with surface particles depend on the amount of energy impacting a surface as well as on other factors. According to *LAMBERT'S COSINE LAW*, the illumination falling on a surface per unit area varies with the cosine of the angle of incidence.

Relative motion between an incident ray and a reflecting surface also has an effect on reflected energy. In the *ESCLANGON EFFECT*, oblique movements of a reflecting surface causes a deviation in the reflective angle of an impacting ray.

## 9.13 KERR MAGNETO-OPTIC EFFECT

In the *KERR MAGNETO-OPTIC EFFECT* of Fig. 9.3, the light reflected from the surface of a magnetized material is elliptically polarized. This polarization is a result of interactions between photons and the molecular order imposed by the magnetizing field. The shape and the orientation of this ellipse depends on the density of the magnetic flux emerging from the reflecting surface; that is, the stronger the magnetic field, the greater the molecular order and the more this order influences the photons.

When the incident ray is already polarized and its plane of polarization and magnetizing vector both lie in the same plane, the reflected ray is also polarized, but now its major axis is rotated slightly. If the reflecting surface is covered by a film whose thickness is one-quarter of the wavelength of the incident radiation, multiple reflections increase the rotation of the major axis. In all circumstances, the amount of rotation is a function of both frequency and angle of incidence.

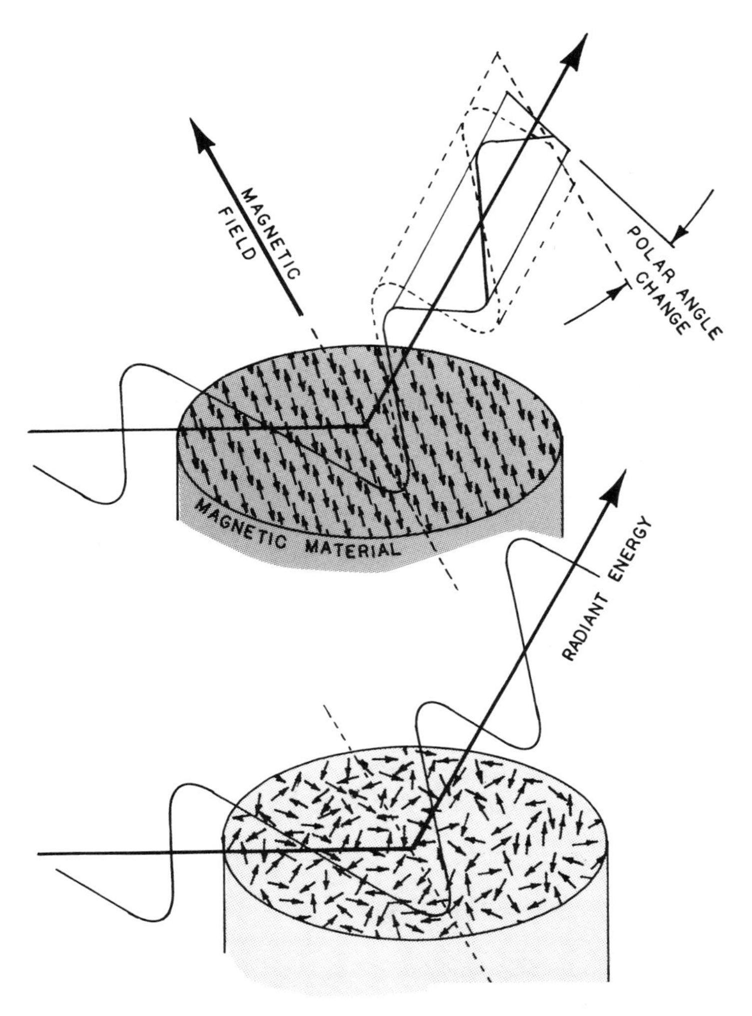


Fig. 9.3. Kerr Magneto-optic Effect: The polar angle of electromagnetic radiation reflected from the surface of a magnetized material is a function of magnetization.

## 9.14 INTERFERENCE

Electromagnetic radiation propagates through space in the form of waves of an oscillatory nature. An interaction of two or more of these waves at any point in space is embodied as a vector sum that takes into consideration frequency, phase, direction, and plane of polarization.

According to the *FRESNEL-ARAGO LAW,* two beams cannot interfere if their polar angles are at right angles to each other. Two waves on the other hand, can oscillate at practically the same frequency and in the same plane. If these oscillations are in-phase, the vector sum represents an energy increase; if out-of-phase, an energy decrease.

*INTERFERENCE* is an interaction of two waves to produce a vector sum. *INTERFERENCE FRINGES* are spatial variations in this sum whereas BEATS are temporal variations. As an example, *NEWTON'S RINGS* are interference patterns achieved when monochromatic radiation is projected onto semitransparent thin films. The wave reflected from the back surface of such film interacts with the wave reflected from the front surface. Depending on the film thickness, the resulting reflected energy can be either a reinforcement or diminution of the incident energy. If the film thickness is equal to one-quarter of the wavelength of the incident energy, the two reflections will be 180 degrees out-of-phase because the back-surface reflection passes through the film twice.

Newton's Rings thus reveal variations in film thickness by the light and dark bands observed in the pattern of reflected energy. If magnetic fields change film thicknesses as a result of magneton alignments, these changes can also be measured by noting the variations in the Newton Ring pattern.

## 9.15 DIFFRACTION

*DIFFRACTION* is a process that bends energy rays by interfering with one or more of their edges.

When a portion of a propagating wave front is blocked by an obstacle of some kind, the wave so affected continues to propagate but is now bent in the direction of the obstacle; that is, it is bent around and behind the interrupting obstacle. That the nature of the continuing wave is changed by the interruption is demonstrated by the fact that energy is carried into regions it would not otherwise reach in the absence of diffraction.

According to a *HUYGENS CONSTRUCTION*, every point on a propagating wave front can be considered to be the center of a new "wavelet." A continuing wave is then constructed from the interference patterns obtained from many preceding wavelets. This interference results in a continuing wave that propagates in a direction at right angles to the wave front. If some of these wavelets are removed from the wave front, their interference with their neighbors is also removed. As a result, energy propagates in all directions from the center of an edge wavelet. Since the energy propagating towards a diffracting obstacle is not removed by interference with neighboring wavelets, it propagates behind obstacles.

If a sphere or other round object is used to interrupt a columnated wave front, the diffraction caused by the edges of the sphere forms a white spot in the center of the sphere's shadow. This is called the *ARAGO WHITE SPOT*.

A ray of energy passing through a small slit is diffracted (bent) in two directions since both sides of the ray are interrupted by the slit's two edges. In other words, a ray passing through such a slit is split into two rays, each moving along a linear path that differs from the path of the original ray. The angle of diffraction, that is, the angle between the direction of the original ray and the direction of the diffracted ray, is a function of the width of the slit and of the wave length of the energy. Consider, for instance, a line extending from one edge of a slit to a point one wavelength away from the backside of the other edge. The diffracted ray will propagate along a path that is at right angles to this line. Following this same logic, a small circular aperture generates a conical diffraction beam because all of the edges of a circular beam are subject to interruption.

Both transmissive (as just described) and reflective diffractions are possible; in the latter, light reflected from a reflective surface that consists of a narrow strip (analogous to a narrow slit) is diffracted by the same mechanism as is transmissive diffraction.

Since the angle of diffraction is determined by wavelength, each "color" (or frequency of energy) is diffracted by a different amount. Longer wavelengths are bent more than shorter ones. (Note how this differs from refraction, in which shorter wavelengths are bent more than longer ones.)

Figure 9.4 shows a *DIFFRACTION GRATING* formed by the periodicity of molecular positions in a crystalline film. Since this periodicity has the effect of a number of slits, a single ray is combined with the diffractions of the many other rays that originally constituted an incident beam to form a diffracted beam.

Any periodic pattern of density difference can be used to accomplish diffraction under appropriate conditions. For instance, sonic vibrations in the form of a standing wave can be established in fluids. Here the density of the fluid is a function of its position within such a standing wave. The *DEBBY-SEARS EFFECT* makes use of the anomalic periodicity caused by these standing waves as a diffraction grating. In this case, changes in the standing wave frequency cause changes in the angle of diffraction for a particular radiant energy frequency.

There are two general classes of diffraction. In *FRESNEL DIFFRACTION,* both the light source and the projected diffraction pattern are observed very close to the diffracting obstacle. *FRAUNHOFFER DIFFRACTION,* on the other hand, takes place in columnated beams where the diffraction pattern is observed at the focal plane of a lens.

*HEILIGENSCHEIN* is a phenomenon of light emanating from vegetation. It is attributed to both the refractions and diffractions of light reflected from dew drops condensed on leaves.

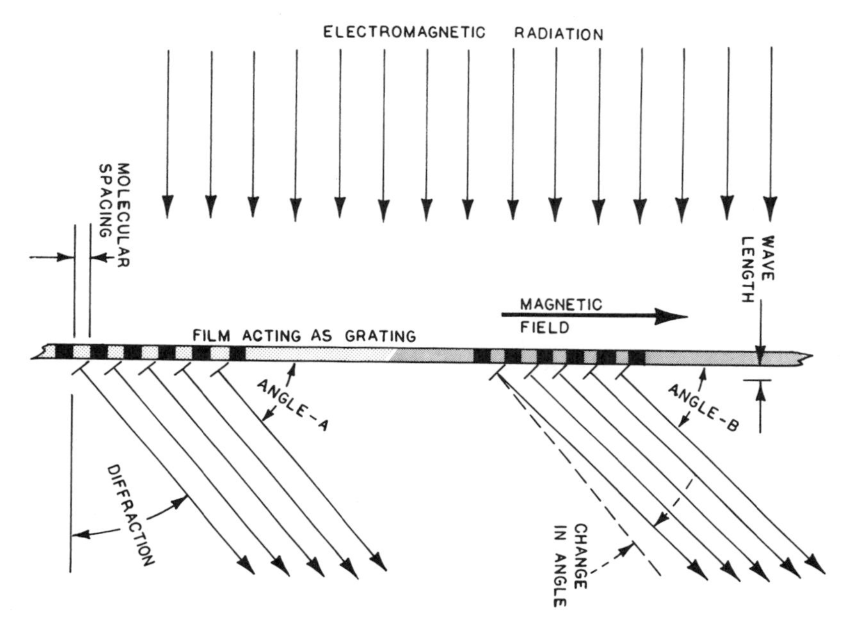


Fig. 9.4. Magnetodiffraction: Where molecular spacing functions as a diffraction grating, the diffracting angle is a function of magnetization.

## 9.16 MAGNETODIFFRACTION

The regularity of molecular spacings in crystal lattices is commonly used as a diffractive grating for X rays. Just as magnetic fields can change crystal periodicity, so can they affect diffractive angles by changes in spacings between diffracting molecules (see Fig. 9.4). This spacing change can be accomplished in magnetic fields if molecules are magnetically polarized. In the *MAGNETODIFFRACTIVE EFFECT,* spacings in a diffractive grating are changed by an imposed field.

## 9.17 SCATTER

*SCATTER,* in general, is defined as a process that brings about a condition of less order from a previous condition of more order. Order is here defined in terms of either position or direction.

*PHOTON SCATTER,* in particular, is defined as directional changes of the columnated parts of relatively large-diameter beams into many smaller beams or rays traveling in different directions. It is caused by the interaction of electromagnetic radiation with scatter centers, the latter being defined as particles, rods, or surfaces of small radii and small volumes of varying density in otherwise homogeneous media.

The amount of scattered energy depends on the size of the scatter centers in relation to the wavelengths of the scattered energy. For wavelengths that are shorter than the radii of scatter centers, the scattered energy is nearly independent of wavelength. For wavelengths that are much longer than the radii of scatter centers, the amount of scattered energy falls off in inverse proportion to the fourth power of the wavelength. For particles where the radii and the wavelength are equal, the scattered energy is at its maximum.

The shorter wave lengths of blue light are scattered more by atmospheric particles than the longer wavelengths of red light. As a result, the sky looks blue when viewed by scattered light and red when viewed by directly transmitted light. Sunsets and sunrises tend to be red because they are viewed by the latter. The sky directly overhead is always blue unless the sun is also overhead or there is a cloud cover. (The *GREEN FLASH* is a refractive phenomenon that occurs under some atmospheric conditions at sunrise and sunset.)

This principle can be used to monitor certain processes. For instance, precipitating particles in solution progressively increase in diameter. Their size, at any time, can be judged by the color of the scattered light. In another example, the density of scatter centers in transparent media can be monitored by observing the intensity of light scattered at right angles to a stimulating energy beam. A *NEPHELOMETER* is an instrument responding to this principle.

Scatter is a process that has different characteristics in directions viewed as either longitudinal or transverse to the stimulating energy beam. This is called the *PDOTNIKOW EFFECT*.

Scattering is commonly classified by the type of scatter center. In *THOMSON SCATTER*, electromagnetic radiation is scattered by electron clouds, whereas in *DELBRUCK SCATTER* light beams are scattered by electrostatic strains established in dielectric materials.

In some processes, scattered energy is of a different frequency than incident energy. In *COMPTON SCATTER*, elastic collisions of photons with electrons change the frequency of the photons. The change in frequency depends on the scatter angle. In fact, *DOUBLE COMPTON SCATTER* is possible, in which the change is consequential enough for two new frequencies to result from a single collision.

In the *RAMAN EFFECT*, scattered radiation adds new frequencies to incident light passing through transparent gases, liquids, or solids. The spectrum of these new frequencies can be used to judge the molecular constituents of the medium.

Scatter and refraction can interrelate. According to the *CHRISTIANSEN EFFECT,* when finely powdered transparent substances are immersed in transparent liquids, the mixture achieves transparency only for that frequency at which the index of refraction for both substances is exactly the same. For all other frequencies, the powder granules act as scatter centers. A Christiansen filter functions as a very sharp band-pass filter.

Both coherent and incoherent scatters are encountered. In the latter, there are no phase relationships between different parts of a scattered beam. In *RAYLEIGH SCATTER,* on the other hand, there are definite phase relationships between incoming and scattered waves. As a result, interference between scattered waves from two centers can be detected as occurring at specific angles. If the scatter centers are organized into some kind of regularity of pattern, the scattered energy is polarized by the scatter process.

In the *TYNDALL EFFECT,* a portion of a powerful beam of light sent through a colloidal solution is polarized, the amount of polarization depending on the size of the colloidal particles. Polarization is complete if the particles are much smaller than the wavelength of the scattered radiation.

In the *BRILLOUIN EFFECT,* a doublet of scatter lines appears on each side of a monochromatic beam passing through certain liquids. These lines are polarized in planes at right angles to each other.

In each of the above effects, scatter is a function of the relative order/disorder of the scatter centers. To the degree that magnetic fields affect this order, they will also affect scatter and the phenomena associated with scatter.

Light constructed from scattered rays is said to be diffuse.

## 9.18 NEMATIC CRYSTALS

*NEMATIC CRYSTALS* are mobile, threadlike structures capable of alignment in magnetic fields. A fluid containing such crystals has one optical axis that coincides with the magnetic field direction and others that do not. Along the field-induced axis, light scattering is minimal and light transmission, maximum. In any other direction, or in the absence of magnetic fields, light scatter is maximum.

## 9.19 POLAR ANGLE ROTATION

Electromagnetic radiation is modified by passage through dielectric materials. Its propagating velocity is changed, and some energy is absorbed as heat. This absorption results from an increase in the thermal activities of the material's molecular structure. The dispersive angular spread of velocity encountered as a function of frequency is the bifringence discussed in relation to refraction.

In addition to refraction and heat loss, passage of a plane-polarized wave through certain materials rotates the orientation of its plane. The amount of rotation is also a function of frequency, *ALLOGYRIC BIREFRINGENCE* being a measure of the spread of the polar angle change.

As observed in the *ARAGO EFFECT,* the polar angle is rotated by an amount, characteristic of a material, that is proportional to the distance the wave travels through that material. In the *BIOT EFFECT,* rotation is proportional to the amount of active material dissolved in an inactive solvent. *BEER'S LAW* stipulates that a long column of vapor has the same rotary power as the short column into which that vapor condenses.

Polar angle rotation in crystalline structures is not a simple process. In double-refracting materials exposed to a plane-polarized, monochromatic energy beam, energy is transported as two plane-polarized beams. The two planes of polarization have angles that are determined by the molecular structure of the material and are at right angles to each other. When the two transporting beams emerge from the double-refracting material, they recombine into a single beam, but the recombined beam may be quite different from the original.

Double-refracting materials are characterized both by different absorptivities and different propagating velocities for different directions and for different polar angles.

Transport characteristics vary when a double-refracting material is rotated both around its double-refracting axis and in a plane parallel to that axis. The incident polarized energy can be made to enter the double-refracting material so that its plane of polarization is oriented at 45 degrees to the double-refracting propagating planes. The two plane-polarized transport beams then emerge from the double-refracting material with phase angles determined both by the propagating velocities of the two planes and by the thickness of the material. Nevertheless, their planes of polarization remain the same on emergence as they were on entry.

If the thickness of a double-refracting material is such that the emerging plane-polarized waves have changed their phase relationships by 180 degrees, they combine into a plane-polarized wave that has been rotated 90 degrees from the polar angle of the incident wave. Such a crystalline thickness is called a *HALF-WAVE PLATE.* At any other crystal thickness, the emerging radiation is reconstituted as an elliptically polarized wave that approaches circular polarization whenever the thicknesses are multiples of one quarter of a wavelength.

Since blue light is about half the wavelength of red light, a blue half-wave plate is approximately the equivalent of a red quarter-wave plate. A blue wave thus emerges from such a plate as a plane-polarized wave whereas the red wave emerges in circular polarization. Since plane-polarized waves are easy to suppress with cross-polarized plates, the energy combination of red and blue in plane polarization can emerge from a "system" as a red wave in circular polarization. This combination of plane-polarized and circularly polarized waves emerging from double-refracting materials is one factor responsible for color fringing in polarized radiation. Another color fringing mechanism is the allogyric birefringence of polar angle variation with frequency.

## 9.20 FARADAY EFFECT

The polar angle rotation of a plane-polarized wave caused by its passage through a solid dielectric that has been magnetically polarized in the direction of propagation is called the *FARADAY EFFECT*, a phenomenon shown in Fig. 9.5. The KUNDT EFFECT is the same phenomenon observed in liquid and gaseous dielectrics. The KERR ELECTROOPTIC

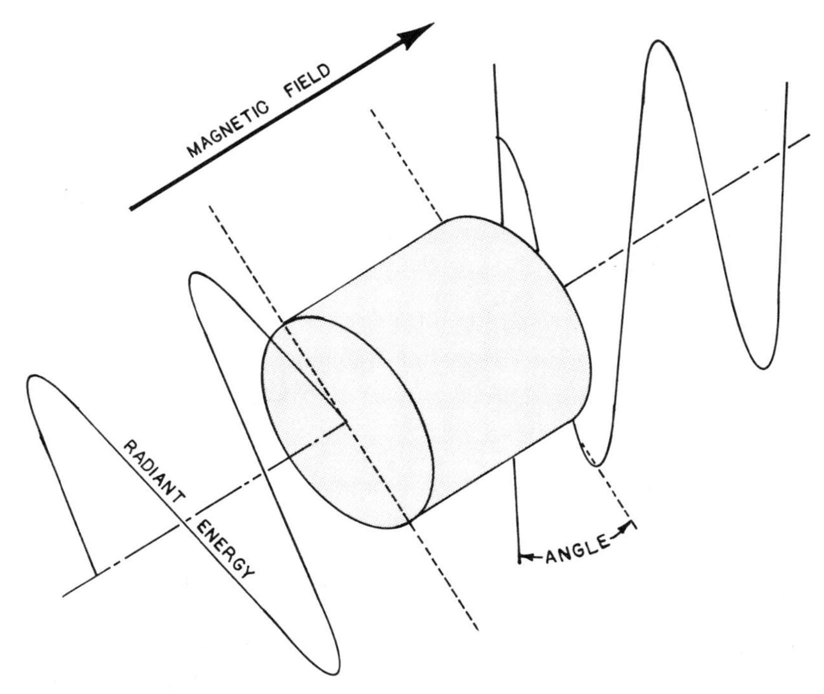


Fig. 9.5. Faraday Effect: The polar angle of a plane-polarized wave is changed when it passes through a magnetized dielectric.

EFFECT is an analogous phenomenon in which the polarization of the dielectric is established by electric instead of magnetic fields. All these effects are merely expansions of the basic Arago Effect. In the Arago Effect, a material's polar-angle rotating characteristics are intrinsic. In these effects, a material's characteristics are modified by imposed magnetic or electrostatic fields.

A plane-polarized wave can be represented as the addition of two circularly polarized waves rotating in opposite directions. As a plane-polarized wave passes through a dielectric, its two oppositely rotating components can be exposed to different environments. If one of these components is affected by its environment more than the other is by its environment, the reconstructed plane-polarized wave is modified in terms of both magnitude and polar angle.

The Faraday Effect stems from an interaction between electromagnetic energy and the magnetic nature of a material's structure. As dielectric materials can be diamagnetic, paramagnetic, ferrimagnetic, or antiferrimagnetic, the mechanisms that cause the Faraday Effect are somewhat different in each type of medium. In fact, the Faraday Effect can be subclassified in terms of the dielectric type.

Magnetons in some dielectrics can be aligned by imposed magnetic fields and stimulated into precession. The relation between precession and magnetic field direction in this case is a tensor property. If precession is clockwise when looking with the field, it is counterclockwise against the field, and not possible when viewed from any direction at right angles to the field.

Magnetons can be stimulated into precession by interacting with electromagnetic fields. This interaction has both frequency and directional characteristics. The absorption of electromagnetic energy by a polarized dielectric is a function of both the direction of a plane-polarized wave in terms of the imposed magnetic field direction and the frequency of that wave in terms of magneton precessing frequency. When energy and materials interact to stimulate precession, a great deal more of that energy is converted into heat and less transmitted through the material than would be the case if precession were not stimulated.

If one of the circularly polarized components of a plane-polarized wave stimulates precession and the other does not, their propagation through a material will be at different rates. The angle of the plane-polarized wave is then shifted by this rate difference. The amount of shift depends on the degree of interaction between one component and the precessing magnetons, in relation to the other component's interaction.

The angular rotation of a plane-polarized beam is proportional to the length of path through the dielectric, the strength of the aligning field, the cosine of the angle between the direction of the magnetic field and the direction of the polar angle of the beam, and Verdet's Constant.

*VERDET'S CONSTANT* is a function of the type of material, the temperature, and the frequency of the radiant energy. It is approximately proportional to the square of the frequency.

As magnetons take time to align themselves with an imposed field, a time delay is experienced between the time of field application and the establishment of the Faraday Effect.

The Faraday Effect can be observed in transparent materials at visible frequencies when precession is stimulated in paramagnetic magnetons. In ferrites, ferrimagnetic magnetons can be made to precess at microwave frequencies.

## 9.21 FARADAY ISOLATION

A microwave isolator eliminates reflections in microwave circuits by allowing the passage of electromagnetic energy in one direction while rejecting it in the other. The principle of *FARADAY ISOLATION* is shown in Fig. 9.6.

In a Faraday Isolator, the polar angle of transmitted energy is rotated by 45 degrees in a body constructed from a magnetized ferrite. Any reflection of this energy will be rotated by an additional 45 degrees when it returns through the ferrite from the opposite direction. After passing through the ferrite body twice, such reflections have a 90-degree difference in polar angle from the transmitted energy and can then be selectively absorbed on the basis of polar angle difference without affecting the transmitted wave.

Since the rotation of the polar angle is a function of magneton precessing frequency, an isolator is a frequency-sensitive device, and since the precessing frequency is a function of the aligning-field strength, such a device can be tuned to operate at a specific frequency by adjusting this strength.

## 9.22 ABSORPTION BY MAGNETON RESONANCE

All magnetons aligned by an imposed magnetic field precess in one direction, which is either clockwise or counterclockwise, depending on the imposed field direction. Such precession always takes place in a plane at right angles to the imposed field.

The unidirectional nature of precessing magnetons creates a condition in which electromagnetic energy propagates through polarized dielectrics

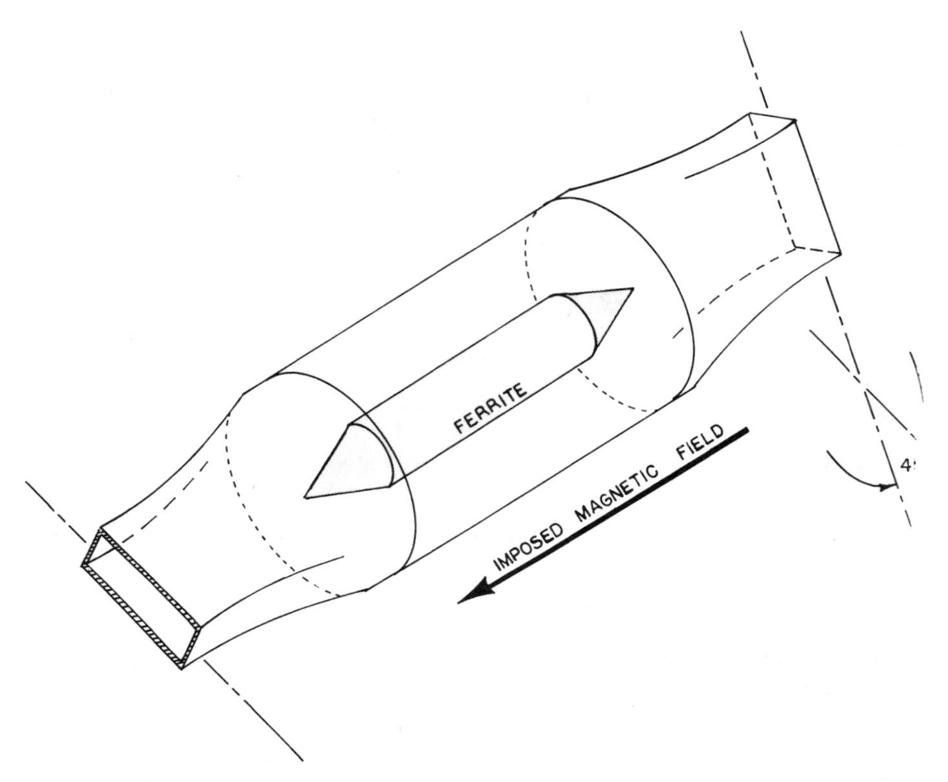


Fig. 9.6. Faraday isolator: The polar angle of a plane-polarized wave can be changed by 45 degrees when it passes through an appropriately configured magnetized dielectric.

with different characteristics in different directions since it is absorbed by the dielectric if it stimulates precession and passes relatively unaffected if it doesn't.

In Faraday propagation, in which the imposed field and the direction of propagation are the same, one circular component of a plane-polarized ray stimulates precession whereas the other does not. Therefore, one circular component is absorbed by a dielectric more than the other is. As a result, the polar angle of the reconstructed plane-polarized ray is rotated by an amount that is a function of this absorption.

If the imposed field is at right angles to the direction of propagation on the other hand (see Fig. 9.7), a plane-polarized ray stimulates precession in one direction and passes unaffected in the other. As a result, radiation is passed by a polarized dielectric in one direction but is blocked in the other direction.

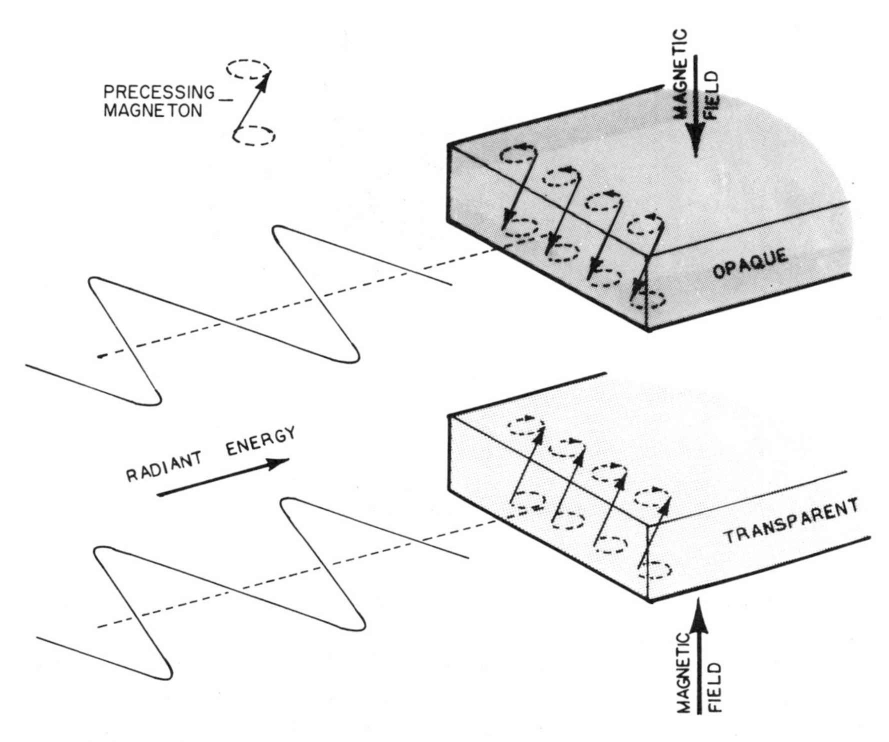


Fig. 9.7. Absorption by magneton resonance: Polarized radiant energy is absorbed by a dielectric if it stimulates the precession of magnetons that are a part of the dielectric's structure.

## 9.23 CIRCULATOR

A device making use of Faraday rotation in two directions is called a *CIRCULATOR*.

Consider those circumstances in which energy is delivered by a transmission line from a transmitter to an antenna. If the transmission line includes Faraday propagation with 45 degrees of rotation (as shown in Fig. 9.6), any reflected radiation is rotated by an additional 45 degrees, thus creating a 90-degree relationship between transmitted and reflected energy.

If the reflected energy is selectively absorbed by a load of some kind, the transmitter sees only the impedance of the antenna, without regard for the reflections it returns, whereas the antenna is influenced only by the absorbing load. If a receiver is substituted for the latter, the outgoing energy is delivered to the antenna and the incoming energy to the receiver.

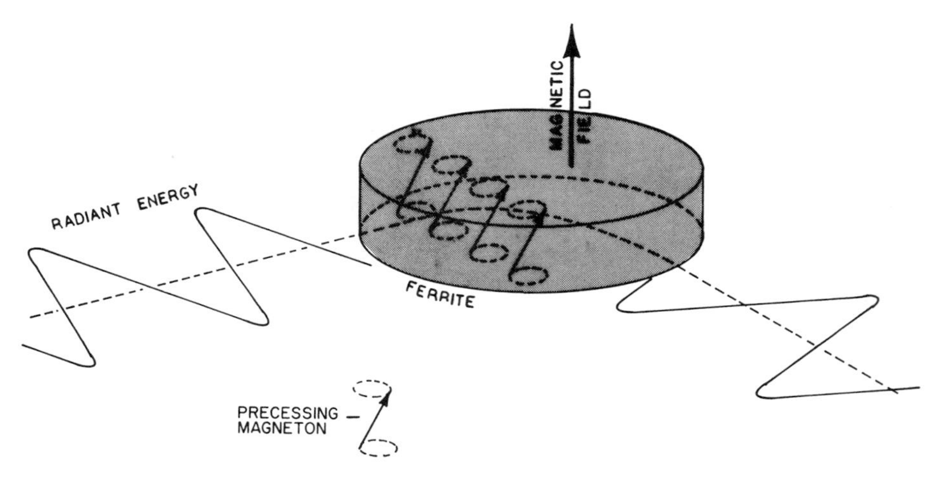


Fig. 9.8. Absorption circulator: Energy is refracted when it encounters an appropriately configured dielectric body containing resonant magnetons.

Circulation can also be accomplished using energy absorption at right angles to wave propagation (see Fig. 9.7) so that energy passes in one direction while being blocked in the opposite direction. As shown by Fig. 9.8, energy is refracted, or bent from a linear path, when it encounters an appropriately shaped dielectric body containing resonant magnetons. In this case, the energy is bent away from the direction of no transmission toward the direction of transmission.

## 9.24 OPTICAL PUMPING

In *OPTICAL PUMPING*, monochromatic electromagnetic radiation is absorbed by low-pressure gas magnetons that are experiencing coherent precession (see Fig. 9.9).

Both the emission and absorption of electromagnetic energy by a particular material involve the movements of particles from one energy state to another. Radiation is produced when the particles move from higher to lower energy states, whereas absorption returns these same particles to higher energy states. Because the energy gap between two states is the same regardless of the direction of particle movement, both emission and absorption spectra of low-pressure gasses are essentially the same.

In exploring radiation/absorption relationships, the activities of one sample of a gas can be used to stimulate, or irradiate, a second sample of the same gas. One sample acts as a source of radiation whereas the

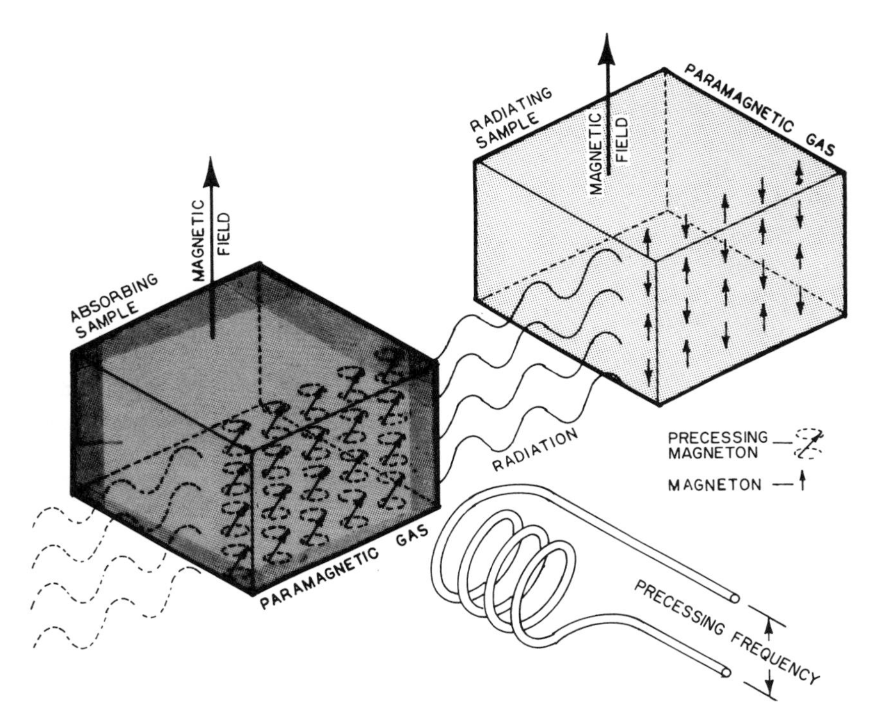


Fig. 9.9. Optical pumping: Two identical samples of paramagnetic gas may be configured so that the energy emitted by one is transmitted through the other. The transmission will be blocked if the magnetons of the transmitting sample precess at the emitted frequency. A measure of the precessing frequency of the absorbing sample is a very sensitive indicator of the magnetic aligning field applied to both samples.

other absorbs that radiation. In such a system relationship, if either the polarization of the impinging radiant energy is incoherent or the orientation of the absorbing gas magnetons random, the interaction between radiation and material is minimal and the radiant energy will pass through the absorbing sample with maximum intensity.

On the other hand, if two samples of the same gas are exposed to a common magnetic field, the magnetons in both samples will be unidirectionalized by that field. Under these circumstances, the radiation created by one sample will closely match the absorption characteristics of the other, and the direction of radiant polarization will exactly match the absorbing magneton directions. Although these conditions alone do not materially affect transmissibility, they do "sensitize" a gas for a precessional interaction.

If a time-direction-varying magnetic field is now added to a sensitized gas to stimulate precession in the absorbing sample, the absorption will be maximized, the transmissibility minimized, and the absorbing gas sample becomes opaque for the one precessing frequency. The Larmor frequency that causes the absorbing gas to become opaque is, as always, a function of the strength of the aligning magnetic field.

If the output of a photocell, detecting the through-radiation of the absorbing sample, is amplified, phase shifted, and fed back to the driving coil in a manner that produces opaqueness, the system so constructed will oscillate at the opaque, or Larmor, frequency.

It should be stressed here that, in the Faraday Effect (or other NMR phenomena), precession is stimulated by the frequency of electromagnetic radiation. In optical pumping, on the other hand, precession is stimulated by an auxiliary varying magnetic field, and the frequency of the electromagnetic radiation is not of prime importance.

## 9.25 MAGNETIC PHOSPHENE EFFECT

If a human puts his head into a magnetic field that is varying at a rate between 10 and 100 hertz and whose strength is between 200 and 1000 gauss, he will see flashes of light. This phenomenon is called the *MAGNETIC PHOSPHENE EFFECT*.

## 9.26 GANTMAHKER EFFECT

In the *GANTMAHKER EFFECT*, a very thin film of thickness equal to a magnetron diameter and carrying an electric current passes electromagnetic energy of cyclotron frequency whenever it is exposed to a magnetic field directed parallel to the film surface. This occurs only when the frequency of the radiant energy and the cyclotron frequency are the same.

The Gantmahker Effect proves that radio-frequency radiant energy and carrier electrons interact under controlled conditions.

If a magnetic field is oriented in the plane of a conducting sheet whose magnetic and electric fields are codirected, conducting electrons flow along flux lines. When energy is added to the conducting electrons at right angles to the current flow, flux-line-directed electrons follow the spiral paths of Fig. 2.4. In this figure, the magnetron trajectories of Fig. 2.3 have been added to linear paths.

In a reflection process, electrons just beneath a metal surface receive energy from electromagnetic radiation in addition to their normal thermal

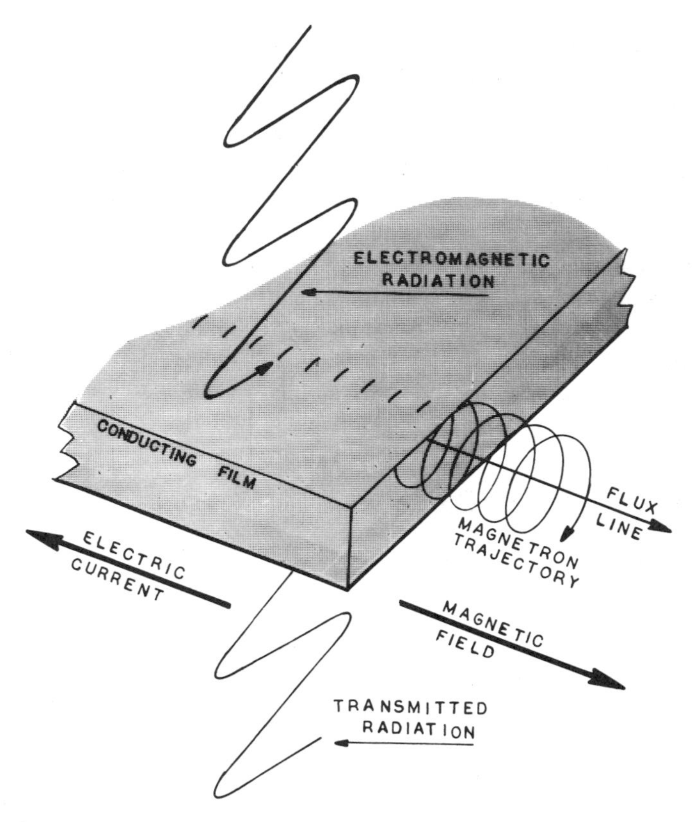


Fig. 9.10. Gantmahker Effect: The electric carriers in a conducting film may be stimulated into helical paths of given magnetron radius and cyclotron frequency. Electromagnetic energy of the same frequency will be passed through such a film when its thickness is equal to twice the magnetron radius. The film is opaque to energy of other frequencies.

energies. They promptly relieve themselves of this added energy through electromagnetic reradiation. There is a time delay, or phase shift, however, between absorption and reradiation.

Electromagnetic radiation impinging on conducting sheets adds energy to the conducting electrons flowing in those sheets, and in a direction at right angles to the conducting paths. If the conducting paths are flux-line directed, the added energy stimulates magnetron activities. As shown in Fig. 9.10, if the frequency of radiation is equivalent to the time of magnetron circulation and if the thickness of the sheet is equal to the magnetron diameter, the radiation adds energy to a magnetron traverse each time a conducting electron reaches the surface of the conducting sheet.

Under these circumstances, each conduction electron periodically functions as a surface electron on both sides of a conducting sheet. As a surface electron stimulated by electromagnetic radiation, it can relieve itself of the added energy by reradiation from both front and back surfaces! The conducting sheet is then transparent to radiant energy of that particular frequency capable of maintaining spiral trajectories that touch both surfaces.

Whenever magnetron diameters are integral fractions of sheet thickness, radiation is also passed at transit-time frequencies. The higher the transmission frequency for a given sheet thickness, however, the less the energy transmitted. At all other frequencies, a Gantmahker sheet is resistant to the passage of radiant energy. As a result, it functions as a very sharp band-pass filter.

## 9.27 TUNED CONDUCTING WIRE

The resistance of a conducting wire changes when it is exposed to electromagnetic radiation under circumstances in which most of the carrier electrons circulate in magnetron trajectories near the wire's surface and the cyclotron frequency of these electrons is the same as the frequency of the electromagnetic radiation.

An electron exposed to an accelerating voltage in a crystal lattice reaches a terminal velocity. Consistent with distribution theory, the terminal velocity of the majority of carrier electrons is about the same.

Electrons with both transverse and axial components of velocity circle around flux lines in spiral paths. Such paths include a "magnetron radius" and a "cyclotron frequency" that are fixed both by terminal velocity and the imposed magnetic field strength. (See Fig. 9.11.)

As indicated in the discussion of fine wire longitudinal magnetoresistance in Sec. 20.2, thermal forces contribute enough transverse energy so that if a wire is exposed to an axial magnetic field whose radius is about the same as the magnetron radius, most of the carrier electrons will tend to travel very near the surface of the wire.

In the process of reflection from metallic surfaces, most of the impinging energy is absorbed by carrier electrons near the surface and then reradiated. Since surface electrons in metallic bodies can function simultaneously as carriers and reradiators, there can be an interaction between impinging energy and carrier efficiency. This interaction modifies the electric resistance that would otherwise be seen by the carrier electrons.

The frequency at which carrier electrons are also surface electrons is determined by voltage, magnetic field strength, and wire diameter.

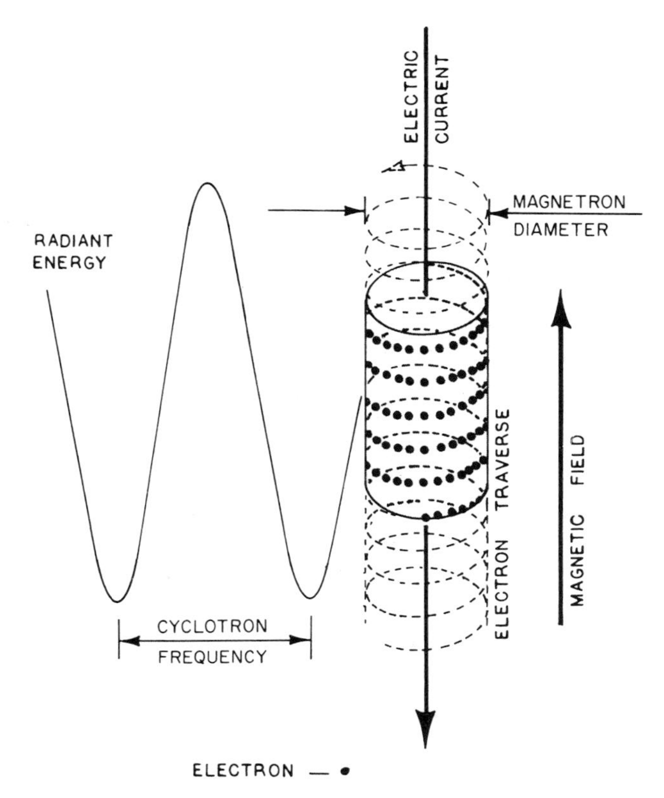


Fig. 9.11. Tuned conducting wire: If the carriers in a circular conductor are forced into a helical magnetron path whose diameter is the same as that of the conductor, the resistance of the conductor is increased when it is exposed to radiation of cyclotron frequency.

## 9.28 TUNABLE MICROWAVE BAND-PASS FILTER

A wire grid constructed with a spacing less than the wavelength of the energy to which it will be exposed, which is normally enough to block the passage of electromagnetic radiation when its current-carrying components are exposed to an axial magnetic field, is transparent to radiant energy of that particular frequency capable of maintaining spiral trajectories of wire-diameter size. Furthermore, the resistance of this grid changes significantly at the same frequency.

The device shown in Fig. 9.12 combines the Gantmahker Effect and the principles used in tuning conducting wires.

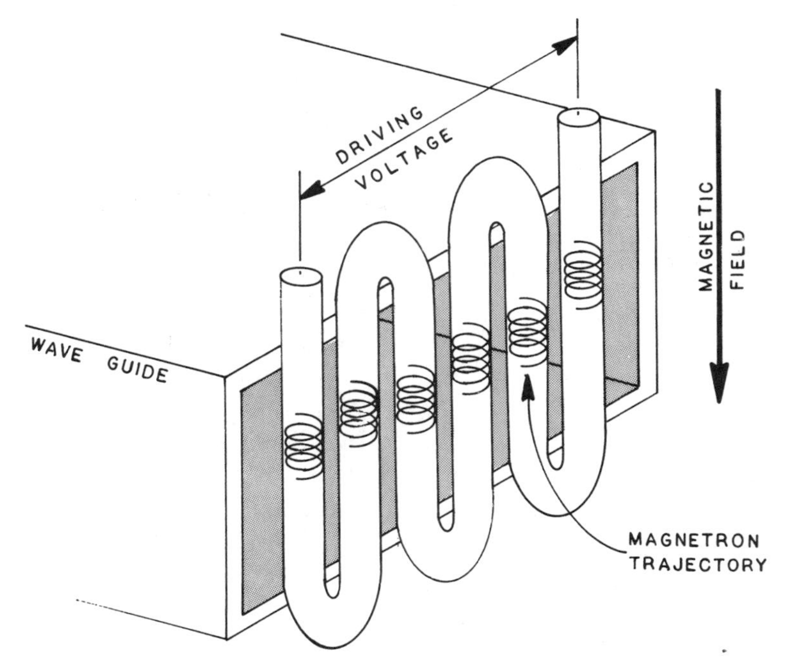


Fig. 9.12. Tunable microwave band-pass filter: A wire grid can be used to block the flow of electromagnetic radiation from the end of a waveguide if the spacing between wires is significantly less than the radiation's wave length. This blockage is relieved when the carriers in the grid are stimulated into helical paths whose magnetron radius is equal to the radius of the wires and whose cyclotron frequency is the same as the frequency of radiation.

## 9.29 TUNED MAGNETIC SEMICONDUCTOR

Resistivities of magnetic semiconducting materials are increased by precessions of the constituent magnetons. In such a material, particles functioning as magnetons act as scatter centers for carrier electrons, and thus the greater the scatter action, the greater the material's resistivity.

As discussed in Sec. 5.5, magneton-imposed scatter is minimized when magnetons are in an ordered array. Above the Curie temperature, where thermal forces impose disorder, carrier electrons experience an increase in magneton-imposed scatter. Similarly, precessing magnetons in an ordered array scatter carrier electrons with more vigor than nonprecessing magnetons would in the same array.

Since the frequency of Larmor precessions stimulated by electromagnetic radiation is determined by the strength of the imposed magnetic field, a magnetic semiconductor can be tuned to have a maximum resistivity when it experiences electromagnetic radiation of a particular frequency.

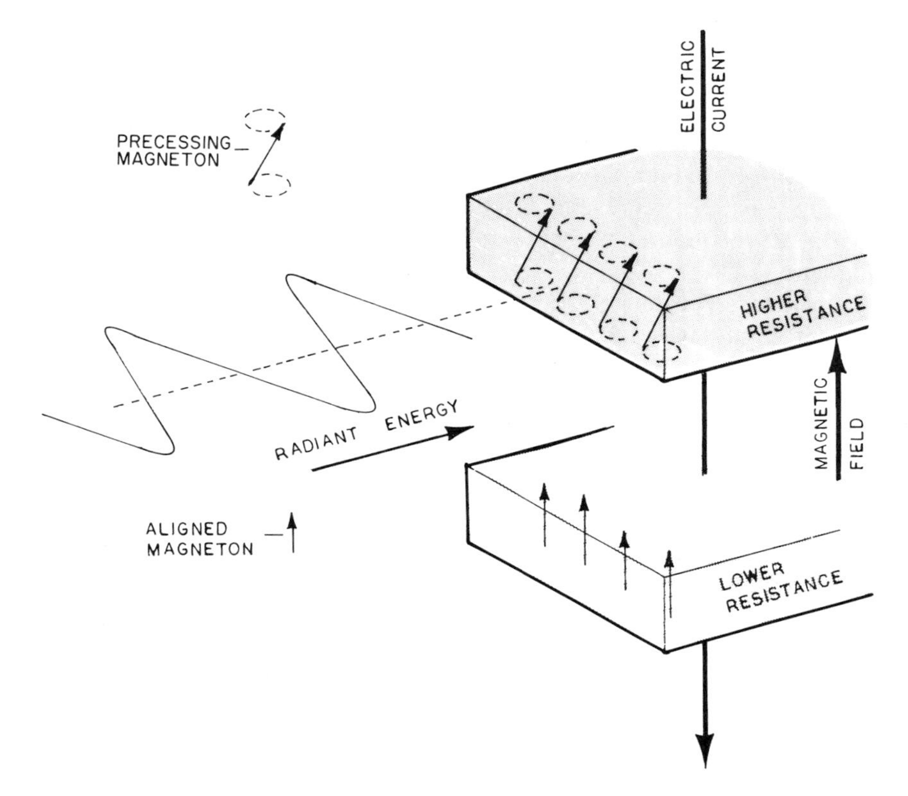


Fig. 9.13. Tuned magnetic semiconductor: The resistivities of materials are increased when constituent magnetons are stimulated into precessions.

The configuration of Fig. 9.13 is somewhat analogous to the tuned conducting wire of Sec. 9.27, although in this case energy is extracted from carrier electrons by the increase in scatter whereas in Sec. 9.27 it is added to them.

As an alternative, precession can be stimulated by an auxiliary changing field in a system arrangement somewhat equivalent to optical pumping.

## 9.30 HOLOGRAPHY

A space-filling, three-dimensional interference pattern results when two coherent beams interact. If the beams are of the same frequency, the interference pattern does not change with time providing that the coherence of direction, phase, and polarity remains unchanged for both beams. Such an interference pattern is completely stable in three-dimensional space.

There is a different interference pattern, however, for each difference of direction, phase, or polarity of either beam.

If a photographic film is immersed in such a three-dimensional interference pattern, a two-dimensional record is possible for one particular plane in this three-dimensional space. This photographic record of an interference pattern represents a unique description of the conditions that determined the differences between the beams in terms of direction, phase, or polarity.

Consider, for instance, a coherent beam divided by a beam splitter into two parts. One part can be used to illuminate an object. The other part can then be used as a reference to interact with the light reflected from, transmitted through, or otherwise modified by that object. If a resulting photographic record of the interference pattern is illuminated by a coherent beam that reproduces the reference beam, the anomaly that modified that part of the original beam used to illuminate the object is revealed as a virtual image by the beam/photograph interaction.

A virtual image recovered by this beam/photograph interactive process is called a *HOLOGRAPH*.

# PART II

# THE EFFECTS OF MAGNETIC FIELD CHANGES ON MOVING CHARGED PARTICLES

Part I discussed basic magnetic theory, showing how individual magnetic fields achieve balances, each with every other and also with the chemical, thermal, mechanical, electrostatic, and electromagnetic energy components of given environments. If any one of these components—mechanical stress, an increase in temperature, exposure to electromagnetic radiation, for instance—is changed, the total environment arrives at a new balance.

In addition, if a given environment is exposed to a magnetic field, the component balance will be different from what it would be in the absence of such a field. Although a magnetic field is not an energy form in its own right but rather a derivation of the energy of moving electrically charged particles, a *changing* magnetic field does transmit energy. In these circumstances, such energy transmission sets up an interaction between the environment so exposed and the moving electrostatic charge that creates the field, and both are affected in the process. Part II will explore this interaction.

# 10
# MOVING CONDUCTOR

When quantities of material move through magnetic fields, the constituent charged particles (electrons, ions, etc.) are individually exposed to the Lorentz forces of Eq. 2.4.

Exposing a charged particle to a Lorentz force is equivalent to exposing that same particle to an accelerating voltage. As is true of voltages derived from any source, such a force is capable of driving a current so long as carriers are present and a complete circuit is available in which current can flow. If a particular configuration of material movement makes current flow impossible, a redistribution of carrier population densities creates voltage gradients within the body of such materials that oppose, and exactly balance, the Lorentz forces.

When quantities of material move through magnetic fields, either the kinetic energy represented by current flow, or the potential energy of a voltage gradient is extracted from the material-moving force.

## 10.1 WIRE IN A MAGNETIC FIELD

A voltage is generated between the two ends of a conducting wire when that wire is moved through a magnetic field. Now a wire can be moved in any direction through space, but the voltage generated by this motion is only a function of the actual vector of the motion that occurs at right angles to the field direction; that is, the wire must actually cut across flux lines if it is to produce a voltage. Mathematically stated, the expression for a voltage generated in a conductor moving in a magnetic field—the *LORENTZ VOLTAGE*—is as follows:

LORENTZ VOLTAGE $$E = LB\, dx/dt \sin \phi \qquad \text{(Eq. 10.1)}$$

where $dx/dt$ is the velocity of the moving conductor; $B$, the strength of the magnetic field; $L$, the length of the wire (its effective length at right angles to the field direction), and $\phi$, the angle the direction of motion makes with respect to the magnetic field direction.

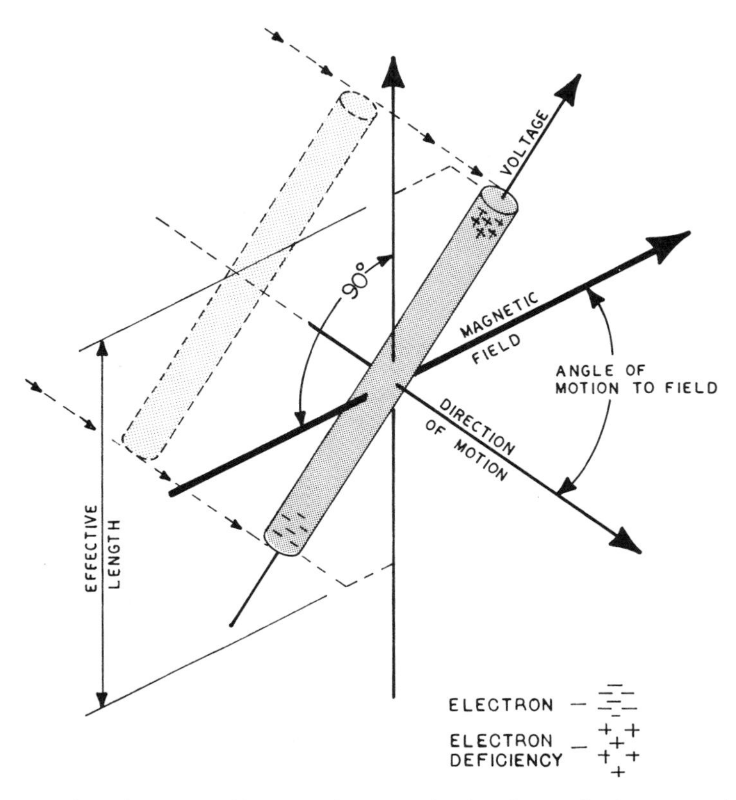


Fig. 10.1. A voltage is generated between the two ends of a wire moving in a magnetic field.

Although a voltage is generated by a wire moving in a magnetic field, a metering current cannot be extracted from this wire unless there is a return path. If the return path is another wire moving in a field, it generates a voltage of its own. If the two wires move in the same field at the same rate and in the same direction, the two voltages exactly oppose each other, and there is no usable transductive output signal. On the other hand, if one of the wires moves at a different velocity, through a different field, or in a different direction, a "difference" voltage is generated between the two wires that can be put to a useful task.

If a return path is provided for the wire generator shown in Fig. 10.1, electrons flow from the bottom of the wire through the return path and back into the top of the wire. This is equivalent to a battery with a positive terminal at the top and a negative terminal at the bottom. Under these circumstances, the generated voltage derived from the forces on the electrons (the Lorentz voltage) is considered to extend from the negative to the positive end of the wire.

If current does flow through such a wire, the flow of electrons downward is equivalent to the flow of current upward. As this current flows through the resistance of the wire, it generates an opposing voltage that extends in the same direction as the current flow.

If no current flows through such a wire, the charges build up at each end until there is a voltage between the two ends that opposes the Lorentz voltage. In this case, the Lorentz voltage tries to separate the charges, whereas the voltage caused by the charge separation tries to bring them back together again.

## 10.2 ACCELERATED CONDUCTOR

Consider the circumstance of a conductor moving through field-free space at a constant velocity. This movement does not affect any particular particle within the conductor more than any other particle since all particles are moving at the same speed.

When a conductor is accelerated, however, a force of acceleration is applied to each particle that differs in accordance with the mass of that particle. (Force equals mass times acceleration.) In addition, each particle responds to this force within the constraints of the chemical bonds that help create the material from which the conductor is fabricated. As the bonds on conduction electrons tend to be less restrictive than those on other particles, the force of electron inertia displaces electrons slightly within the material, and this displacement establishes a voltage gradient within the conductor as a result of the force of acceleration.

As conduction electrons see a different environment in each type of conductor, the effect of acceleration sets up a different voltage in each type of conductor. If a complete electric circuit is constructed from two different types of conducting material, a circulating current will result when this circuit is accelerated because of the difference voltage generated within the two types of conducting material.

In the *TOLMAN EFFECT*, a voltage appears in an accelerated conductor as a result of electron inertia.

## 10.3 COIL OF WIRE

As an extension of the concepts of Sec. 10.1, a voltage is generated in a coil of wire if the flux density through that coil is changed in any way. The coil of Fig. 10.2 is first of all a *CHARGE GENERATOR*. The number of coulombs of electric charge generated in such a coil is proportional to the change in flux density through the coil.

If such a coil is connected to a ballistic galvanometer, moreover, and moved from one area of field intensity to another area of different intensity,

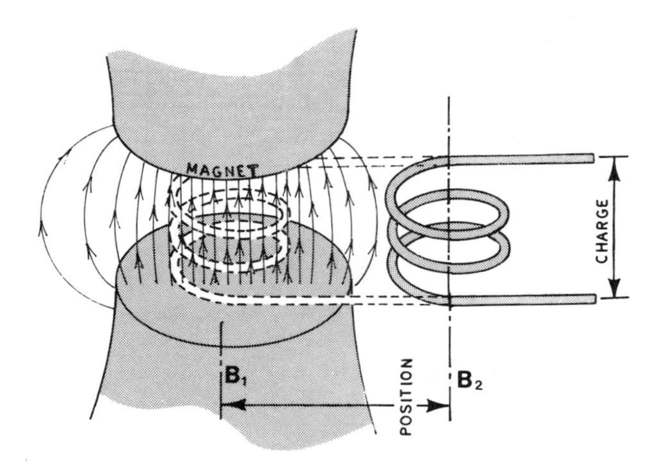


Fig. 10.2. Charge generator: An electric charge is generated in a coil that is proportional to the change in flux density that is passing through the coil.

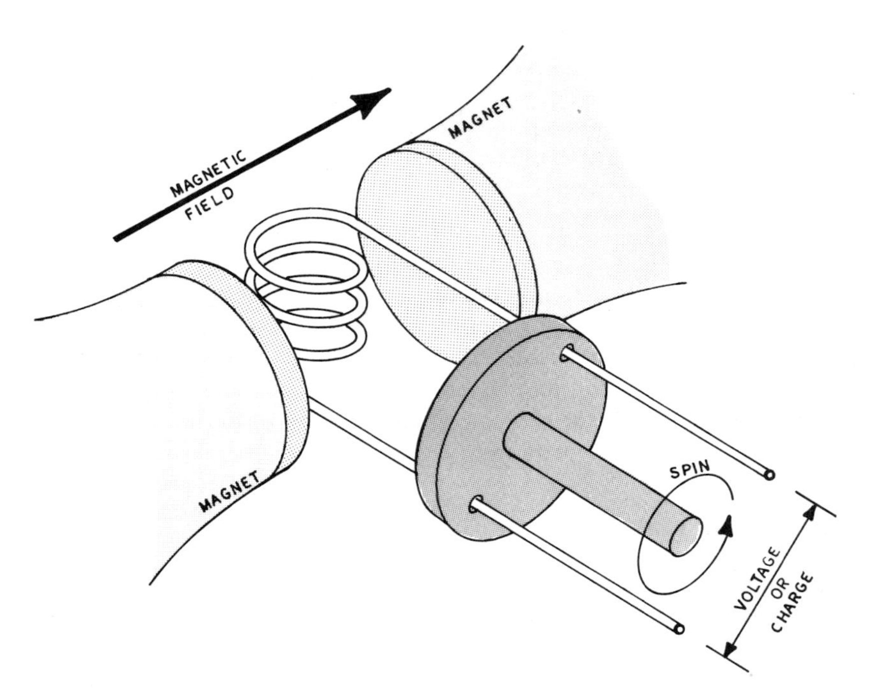


Fig. 10.3. A sinusoidal voltage is generated in a coil that is continually rotating in a magnetic field.

the ballistic throw is proportional to the difference between the two field intensities. By standardizing one of the two fields, it is possible to use such an arrangement for field-intensity measurements. If the coil is moved rapidly back and forth between two field intensities, however, the coil output is an alternating voltage proportional both to the difference in field intensities and to the frequency of change.

In Fig. 10.3, the coil of wire shown in a magnetic field can be rotated by 180 degrees. Doing so is equivalent to moving the same coil from a field of one intensity to another of equal intensity but opposite direction. It is also equivalent to moving the coil from a field equal to twice the field intensity to one of zero intensity. If this coil is rotated continually at a constant velocity, a sinsusoidal voltage will be generated. Figure 12.3 illustrates the relative voltages generated at each instant of such rotation.

## 10.4 HOMOPOLAR "DC" GENERATOR

A somewhat different version of the moving conductor principle is found in the *HOMOPOLAR "DC" GENERATOR* of Fig. 10.4. This device is based on a rotating disk rather than a moving wire. Such a disk functions, however, as if it were constructed from a number of radial strips or wires, each of which is moved in a manner to cut flux lines. When such a disk rotates in a magnetic field, a voltage is generated between its center and its outer edge. This voltage is proportional to the angular rate of change caused by rotation, as well as to the total number of interrupted flux lines.

This device also acts as a charge generator; that is, it produces a given number of electric charge units for each degree of angle rotated. This

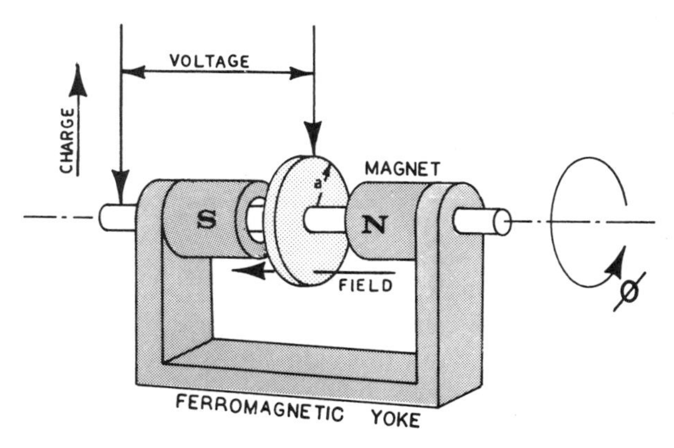


Fig. 10.4. Homopolar dc generator: A voltage is generated between the center of a conducting disk and its outer rim when the disk rotates in a magnetic field.

number is proportional to the number of interrupted flux lines and inversely proportional to circuit impedance. It does not depend on the velocity of rotation.

A homopolar "dc" generator provides an expeditious means of bringing a ballistic galvanometer back to zero once it has been deflected.

## 10.5 FOUCAULT CURRENTS

As discussed in Sec. 10.1, a voltage is generated whenever a wire is moved through a magnetic field. We shall now consider the electric current induced in a conducting sheet when it moves in a nonuniform magnetic field.

It is possible to think of a sheet of metal or other conducting material as having been constructed from a number of strips or wires lying side by side. Voltages will then be induced in each one of these "strips" whenever the sheet is drawn through a magnetic field, just as they would be induced in individual wires similarly moved. If the magnetic field in question is uniform, the voltage generated in each strip is equal to, and opposed by, that of every other strip. No current flows in such a sheet moving in a uniform magnetic field. On the other hand, if the field is nonuniform, the voltages induced in each strip are not equal, and a lower-voltage strip can act as a return path for the current driven by a higher-voltage strip. Currents then flow in conducting sheets moving in a nonuniform field.

Currents so induced in conducting sheets are here called *FOUCAULT CURRENTS*. Sometimes these currents are called "eddy currents," but it is felt that this latter term is best restricted to currents induced by changing fields rather than by moving conductors.

As postulated by Lenz, forces are created in opposition to any motion that results in the generation of an electric current. *FOUCAULT DAMPING* is a force opposed to the velocity of conducting sheets passing through nonuniform magnetic fields. Figure 10.5 is a sketch illustrating a Foucault pendulum capable of critical, or any other, damping.

## 10.6 METALLIC LIQUID FLOW

The principle of moving conductors in magnetic fields can be used to monitor the flow of metallic liquids. In this case, it is the liquid itself, rather than a solid disk or wire, that interrupts flux lines.

Metallic liquids contain a great many electrons that are free to act as carriers; that is, electrons move freely throughout a body of liquid and are subject only to the forces of electric fields that are created should they happen to achieve a nonhomogeneous spatial distribution.

Carriers in a fluid are subject to Lorentz forces as they are carried

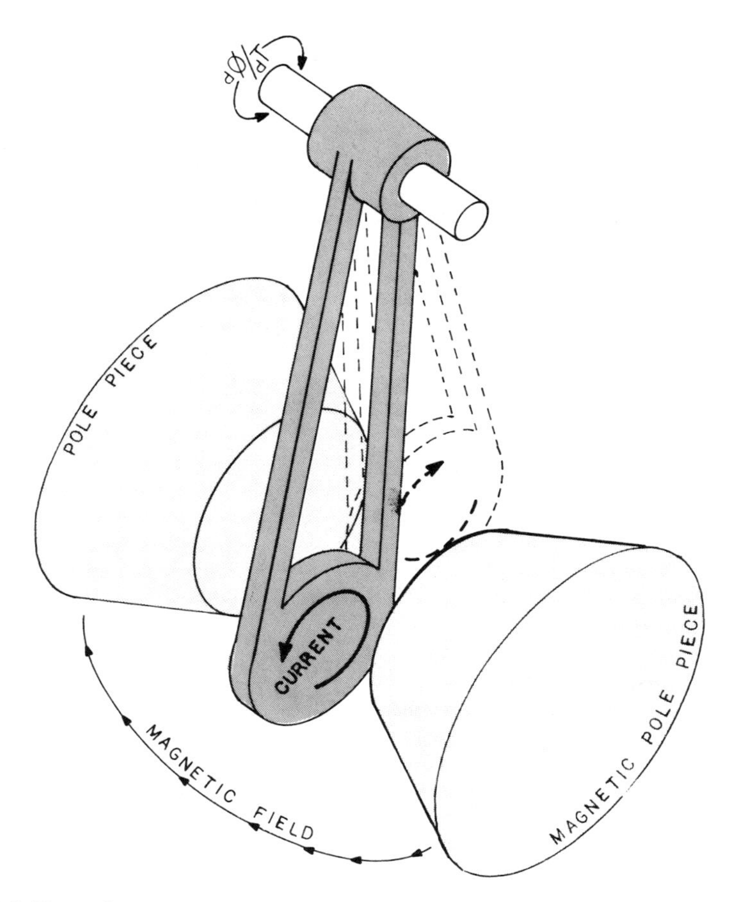


Fig. 10.5. Foucault currents: Electric currents are induced in a conducting body when it is moved from an area of one field strength to that of some other.

through a magnetic field by the motion of that fluid. These Lorentz forces push electrons in a direction at right angles to the magnetic field direction. As shown in Fig. 10.6, this push causes a concentration of electrons on one side of a fluid stream greater than the concentration on the other side. The voltage that results from this concentration gradient balances the Lorentz forces that resulted from the liquid flow. This voltage gradient is proportional to the velocity of liquid flow.

As shown in Fig. 10.7, a practical flow meter allows a liquid to flow through an insulated nonmagnetic tube placed in the gap of a permanent magnetic circuit. Electrodes inserted in the liquid at right angles both to the direction of the flow and to the direction of the magnetic field extract a voltage that is proportional both to the average velocity of the liquid and to the length of the gap between the electrodes.

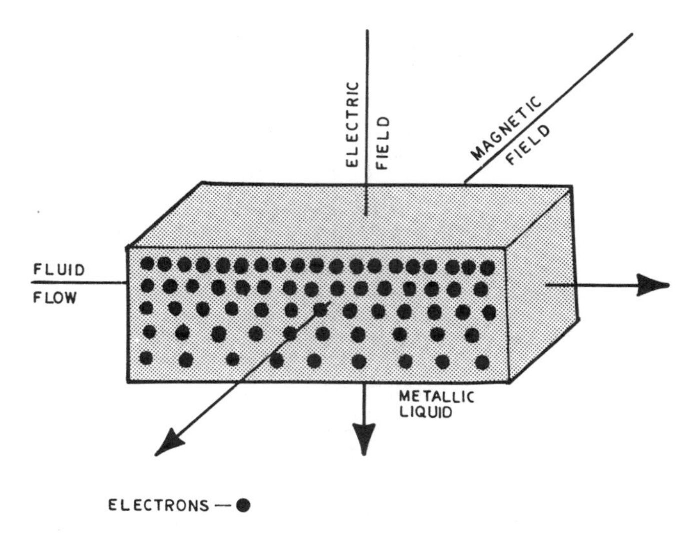


Fig. 10.6. Metallic liquid flow: An electric field is generated in an orthogonal relation with a magnetic field and with electron movement.

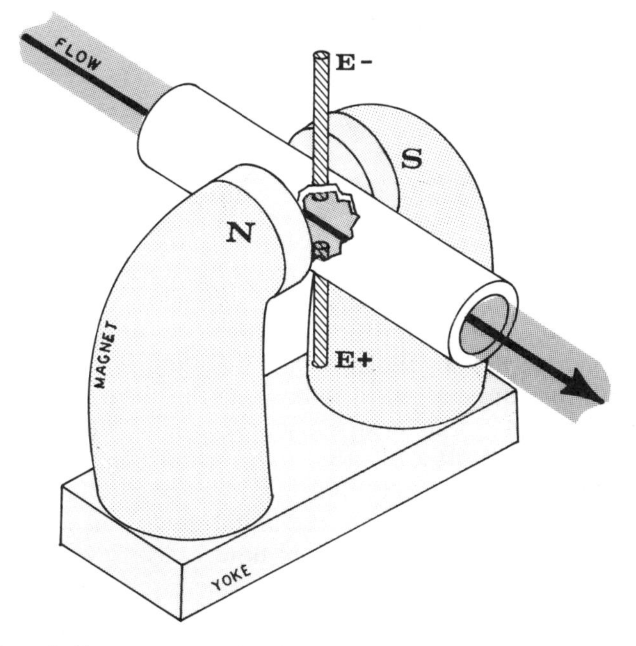


Fig. 10.7. Magnetic flow meter: An electric gradient is established within a magnetic fluid when it flows through a magnetic field. Electrodes placed at right angles to both the magnetic field and the direction of flow are used to extract a voltage that is proportional to the velocity of flow.

## 10.7 ELECTROLYTIC FLOW

Lorentz forces create voltage gradients in all conducting liquids moving through magnetic fields as a result of nonhomogeneous distributions of charge carriers.

In metallic liquids, the charge carriers are electrons that move through the liquids, pass over the liquid–solid electrode interfaces, and continue on through the external metering circuits. In nonmetallic liquids, the charge carriers are ions. From an electric charge standpoint, the ionic distribution of Fig. 10.8 is the same as the electron distribution of Fig. 10.6.

Ions, however, cannot pass over liquid–solid interfaces. As a result, the passage of even the minute currents needed to service metering circuits quickly accumulate a concentration of ions at each electrode. Positive ions accumulate at one electrode and negative ions at the other. This ionic accumulation is called *POLARIZATION*. Polarization creates voltage gradients at the interfaces that block the flow of measuring currents. It also offsets Lorentz forces thereby effectively eliminating signal voltages.

The blocking effect of polarization is reduced if alternating magnetic fields are used instead of nonvarying fields. Such oscillating fields must have half cycles that change direction fast enough to limit the accumulation of ions. Alternating fluxes of high frequency, however, are likely to induce voltages of considerable magnitude into various parts of metering circuits. These tend to mask the flow signals and cause the output of a measuring

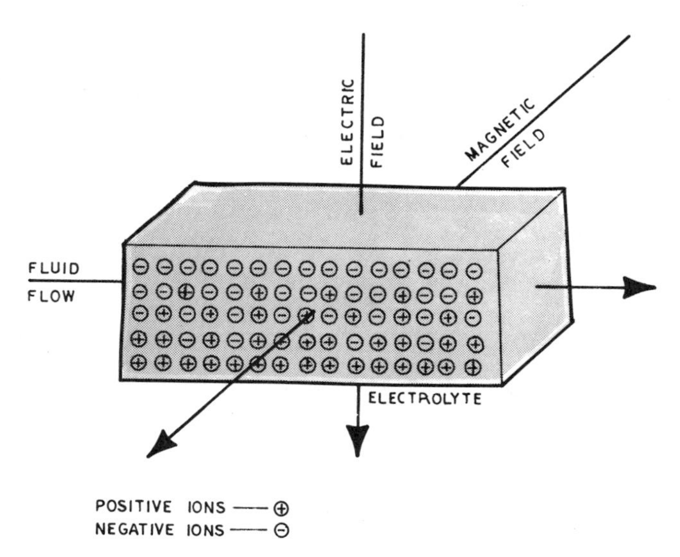


Fig. 10.8. Electrolytic flow: An electric field is generated in an orthogonal relationship with a magnetic field and with ionic carrier movements.

circuit to register other than zero voltage for zero flow. If a zero bucking voltage is added to balance the zero error, the amount of voltage needed is likely to be large in respect to the flow signal, and bucking voltage stability problems can be a major source of metering error.

In addition, the output voltage with an alternating field is dependent on the fluid's dielectric characteristics as well as its velocity. A single zero-offset voltage of given phase and amplitude will not take care of changes to both the dielectric constant and induced voltage.

## 10.8 MAGNETOHYDRODYNAMIC FLOW

Any type of conducting fluid can be used to generate the Lorentz voltages described in the previous sections. Gases, on the other hand, are relatively nonconducting media with high source impedances, making it difficult to extract useful signals. If a gas is heated until it becomes a plasma, however, the situation becomes quite different. In a plasma, electrons are stripped from the molecules, leaving a seething mass of free electrons and positive ions. As shown in Fig. 10.9, these carriers can be separated by Lorentz forces in the conventional manner.

The conductivity of a plasma is much higher than that of any metal. As a result, the transfer function of velocity to voltage is very efficient, and

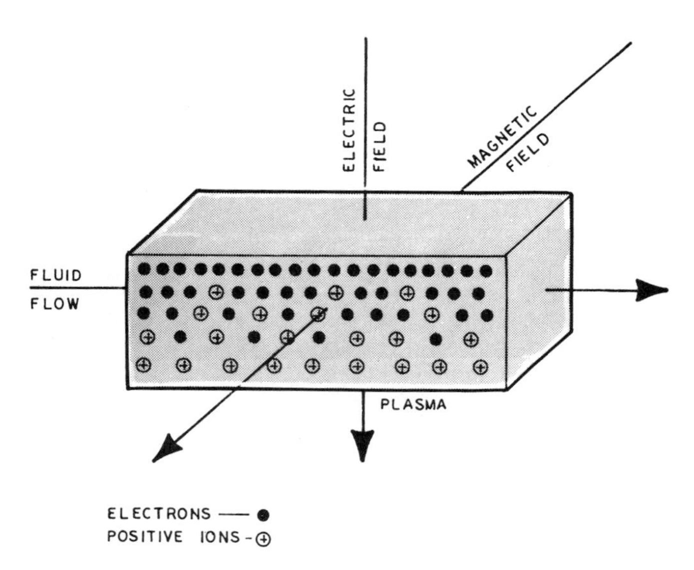


Fig. 10.9. Magnetohydrodynamic flow: An electric field is generated in an orthogonal relationship with a magnetic field and with carrier movements that result from both ion density and electron density gradients.

this technique offers a promising approach to power generation. A major problem is that of containing the very-high-temperature plasmas.

The extraction of heavy currents from such a system affects the placement of voltage-collecting electrodes, resulting in two different types of magnetohydrodynamic generators.

In a *FARADAY MAGNETOHYDRODYNAMIC GENERATOR*, current flows in response to the Lorentz voltages generated at right angles to plasma flow. Several pairs of electrodes are placed at right angles to plasma flow, and each pair is used to drive an individual load.

In a *HALL MAGNETOHYDRODYNAMIC GENERATOR*, current flows in response to the Hall voltages generated at right angles to the Lorentz currents. In such generators, the generated voltage is the vector sum of the Hall and Lorentz voltages. Voltage-collecting electrodes are placed at oblique angles to plasma flow, and the collected voltages are operated in series to drive one load.

## 10.9 ION SLIP

When a plasma moves at high speed, the electrons tend to move faster than the ions. The velocity difference is called *ION SLIP*. Ion slip creates a voltage that is axial to the direction of plasma flow.

## 10.10 SOUND WAVE STIFFENING

Sound travels through propagating media as a series of pressure-stress alternations. These alternations assume the form of waves in which each constituent particle of the propagating medium is subject to displacements with cyclically varying velocities. Such displacements normally follow linear paths.

If the propagating medium is electrically conductive, some of the oscillating particles carry electric charges. When these oscillating charged particles are placed in a magnetic field, they are subject to Lorentz forces, and individual paths are distorted to enclose areas, each such enclosure generating a Lenz field. A charged particle continuously traversing a path that encloses an area represents a circulating electric current.

In other words, sound waves traveling through electrically conductive media in magnetic fields generate electric currents. Following Lenz's conclusions, these circulating currents create new forces that oppose the forces that created the currents in the first place. The result is a reduction in the amplitude of a sound wave in the presence of a magnetic field in comparison to what the amplitude of that wave would have been without the magnetic field. This is called the *MAGNETIC ACOUSTIC EFFECT*.

The magnetic acoustic effect has several significant aspects. In one, the phase velocity of the wave is increased in the presence of a magnetic field. In another, the mechanical stiffness of the material is increased by a sound wave's interaction with a magnetic field.

This effect is masked in ferromagnetic materials by the Villari Effect of Sec. 16.2.

## 10.11 CONCENTRIC FLOW METER

In the flow meter of Fig. 10.7, a unidirectional magnetic field is imposed at right angles to the movement of a conducting fluid. Lorentz forces create voltages by pushing carriers orthogonally to both field and movement. The cylindrical Biot-Savart field of Fig. 2.1 can also be used in flow measurement.

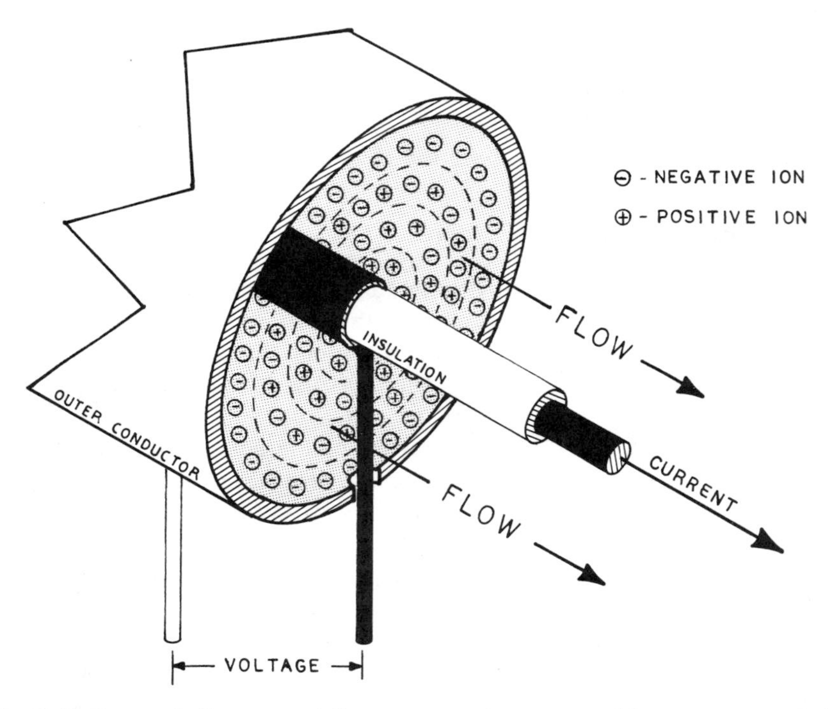


Fig. 10.10. Concentric flow meter: A flow meter may be constructed from two concentrically mounted conductors. A Biot-Savart field will surround the inner conductor when it carries an electric current. A voltage is then generated between the inner and outer conductors that is proportional to fluid velocity when the outer conductor is used to channel the flow of a fluid.

In the configuration of Fig. 10.10, a conductor is concentrically located in the conduit of an electrically conducting fluid. A circular magnetic field surrounds this conductor when it is carrying an electric current. The movement of a conducting fluid through this circular Biot-Savart field forces carriers either outward, toward the outer conductor, or inward, toward the inner conductor. This carrier distribution is detected as a voltage between the two conductors.

## 10.12 CHAPIN TRANSDUCER

A *CHAPIN TRANSDUCER*, as shown in Fig. 10.11, consists of a coil of wire mounted on the face of a piezoelectric crystal. If such a configuration

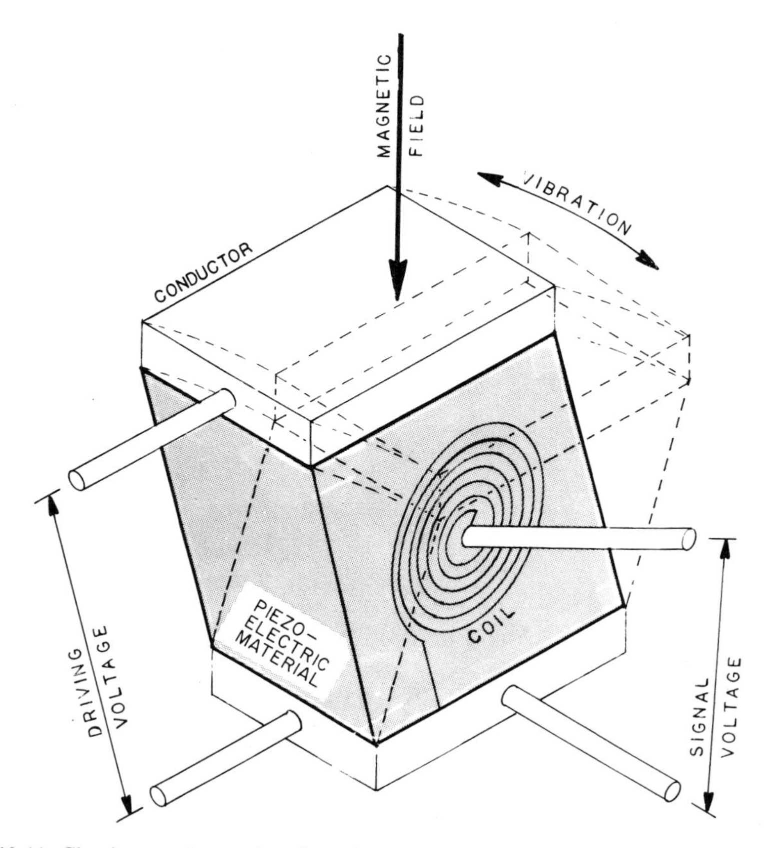


Fig. 10.11. Chapin transducer: A voltage is generated in a coil when the flux through the coil is changed. This voltage is proportional to the time rate of change. A significant voltage can be extracted from a very small coil if that coil is oscillated at a very high frequency.

is driven with the coil oscillating in a direction that maximizes flux-line interceptions, an alternating voltage is generated within the coil whose magnitude is proportional to magnetic field strength. In such an arrangement, the relatively high speed of flux-line interceptions compensates for the relatively few flux lines intercepted.

## 10.13 HYDROMAGNETIC EFFECTS

Constituent electric carriers move as a part of the movements of conducting fluids. When conducting fluids move through magnetic fields, these carriers are exposed to Lorentz forces. The electric currents that flow in response to the latter interact with, and modify, the originating magnetic fields.

The dynamic effects of fluids under hydromagnetic conditions are predictably different in the presence of magnetic fields than in the absence of such fields. *HYDROMAGNETIC WAVES* move with different magnitudes and with different frequencies than would waves in the same fluids in the absence of magnetic fields. In addition, the perturbations of turbulent motion can be either amplified or damped by hydromagnetic forces. Vortices can be self-perpetuating in the presence of magnetic fields of particular strengths that would be quickly dissipated under other circumstances.

# 11
# ELECTROMAGNETIC INDUCTION

The changing magnetic fields associated with changing currents in one conductor induce voltages in other nearby conductors. This phenomenon, known as *ELECROMAGNETIC INDUCTION*, combines the effects of Ampere's and Faraday's Laws.

The experience of a conductor immersed in a magnetic field can be considered in a more general light than it was in Sec. 10.1. As far as a generated voltage is concerned, the motion of a conductor through a magnetic field is only a special case of a more general situation. It doesn't really matter whether a conductor is moving in a magnetic field or a field is otherwise changing around a conductor. Flux lines are geing cut in both circumstances, and, if flux lines are cut by a conductor, a voltage is generated in that conductor.

If a conductor moves through a constant magnetic field, or if a source of magnetic field is moved in relation to a stationary conductor, the induced voltage is proportional to the time rate of field change seen by that conductor as this change is expressed in terms of the product of field strength and velocity of motion.

In electromagnetic induction, a secondary conductor is immersed in the magnetic field associated with an electric current flowing in a primary conductor. A voltage is induced in this secondary conductor whenever the primary current changes because, whenever the current changes, the surrounding magnetic field also changes. In this case as well, the voltage induced in a secondary conductor is proportional to the time rate of field-change.

## 11.1 INDUCED VOLTAGE

Figure 11.1 illustrates the circumstance in which a changing current in a primary conductor induces a voltage in a secondary conductor. The voltage so induced is proportional to the length of the secondary conductor. As shown in the figure, an induced voltage causes a nonhomogeneous distribution of charge along the length of a secondary conductor. This charge

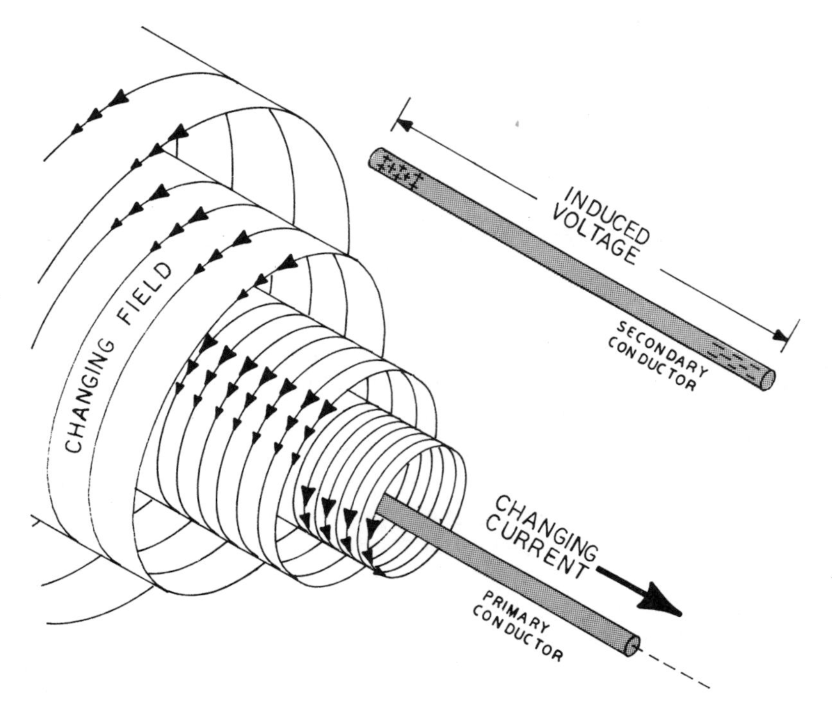


Fig. 11.1. Induced voltage: An induced voltage is generated between the two ends of a wire exposed to a changing magnetic field.

redistribution, in turn, creates a voltage that exactly balances the inducing force.

To look at this phenomenon in another way, the kinetic energy of a primary current as transmitted through space by a changing Biot-Savart field induces the potential energies of charge distribution. The magnitude of the voltage so induced is proportional to the magnitude of the inducing current's time rate of change. If that current is following a time variant sinusoidal pattern, the rate of change is maximum whenever the absolute value of the current is zero. As a result, the induced sinusoidal voltage is maximum whenever the inducing current is zero.

In vector terminology, an induced sinusoidal voltage always lags the inducing sinusoidal current by 90 degrees.

## 11.2 CONSEQUENTIAL CURRENTS

Voltages are induced in response to Faraday's Law. Currents flow in response to these induced voltages. Although they are not induced directly,

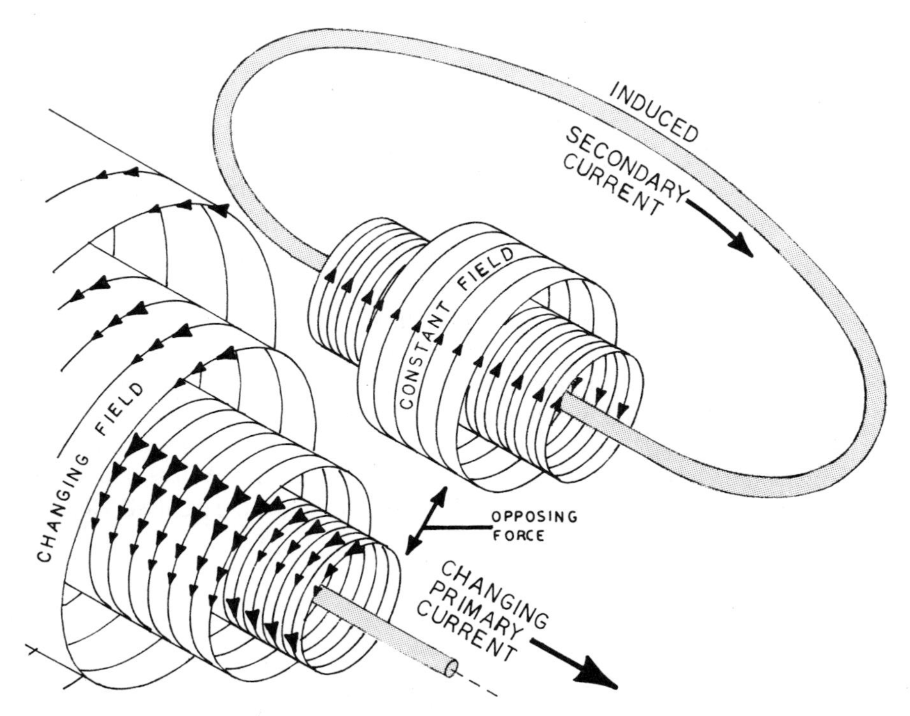


Fig. 11.2. Induced current: When a voltage is induced in a coil exposed to a changing magnetic field, an induced current flows in response to it.

currents that flow in response to induced voltages are commonly called "induced currents."

Figure 11.2 shows the two ends of the secondary conductor of Fig. 11.1 brought together in the form of a loop. This configuration provides a complete circuit in which a secondary current flows in response to an induced secondary voltage and both the induced secondary voltage and the resulting secondary current are proportional to the primary current's time rate of change.

Since an induced voltage always lags the inducing current by 90 degrees and the induced current lags the induced voltage by some number of degrees between 0 and 90 (in the absence of capacitive reactance), a current induced under such circumstances flows in a direction opposite to that of an inducing current. When inducing and induced currents flow in opposite directions, their individual fields add to strengthen the composite field between primary and secondary conductors. As one consequence when current induction is in process, there is an opposing force experienced between a primary and secondary conductor. These conductors will move apart if there is no mechanical force present to keep them together.

## 11.3 SINGLE LOOP

In Fig. 11.3, the primary and secondary conductors of Figs. 11.1 and 11.2 are combined into one loop driven by one primary current. This current always flows in the same direction and with the same magnitude in every part of the loop. If it changes in any one part, it also changes by exactly the same amount and at exactly the same time in every other part.

In addition, the Biot-Savart-shaped fields of each small unit length of loop interact with every other unit length, and changes in one unit length influence the conditions in every other unit length. In this interaction, a change in current through one unit length generates a voltage in every other unit length. This voltage is directed to drive a current through these other unit lengths in that direction in which current is already flowing. In other words, a voltage created by changes to an electric current is directed to prevent the changes that caused it to exist in the first place. This tendency of a current flowing in a coil to maintain itself is called *SELF-INDUCTANCE*.

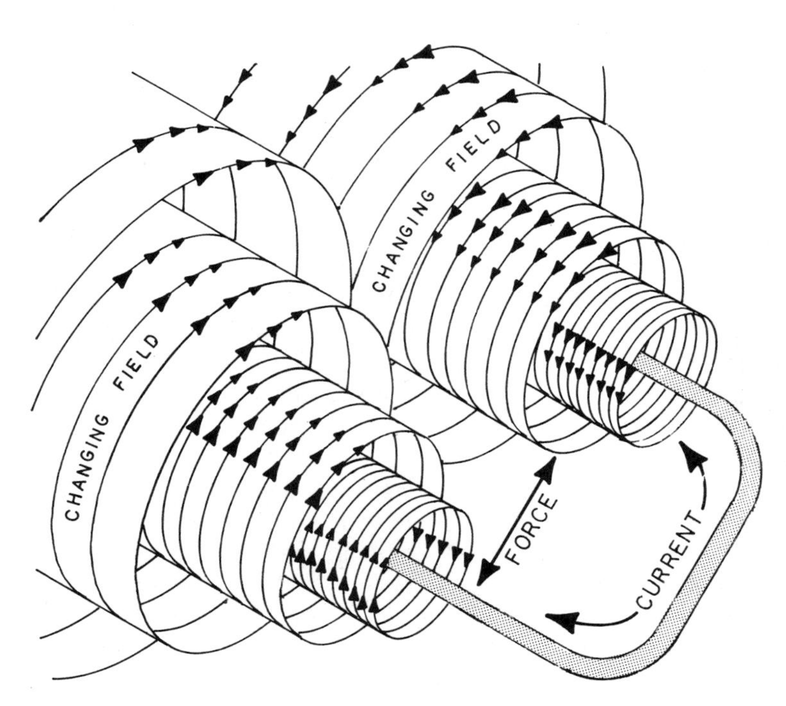


Fig. 11.3. Single loop: A coil of wire induces forces on its own moving carriers that oppose any changes in the current flowing through it.

As a current flows through opposite sides of a loop in opposite directions, an opposing force exists between these opposite sides. In the absence of restraining forces, one loop expands to a perfect circle, whereas several loops of small diameters expand into one loop of larger diameter.

## 11.4 FARADAY'S LAW

The two ends of a secondary conductor can be brought together after completing more than one loop. If each such loop is arranged coaxially with every other loop, the assembly is called a *coil*, and each loop is considered to be one turn of that coil. Under these circumstances, a voltage is generated if the magnetic flux changes through the coil. Faraday's Law describes this voltage as follows:

$$E_f = -\mathrm{k}N\, d\phi/dt \qquad \text{(Eq. 11.1)}$$

where $E_f$ is an electromotive force (emf) called the *FARADAY VOLTAGE*; $\phi$, the number of flux lines linking the coil; $d\phi/dt$, the time rate of change of magnetic field strength; $N$, the number of turns in the coil; and k, a constant. If $d\phi/dt$ is expressed in terms of a sinusoidal frequency, the Faraday voltage is proportional to that frequency.

The principle that current changes in a coil generate voltages that oppose these changes is called *LENZ'S LAW*. Lenz's Law is stated mathematically as the minus sign in Faraday's Law.

## 11.5 SELF-INDUCTANCE

*SELF-INDUCTANCE* reflects an interaction between Faraday's and Ampere's Laws in which changes in current are opposed by the voltages that are induced by these changes.

A coil is constructed from a number of coaxial turns sharing the same current and occupying the same general space. A part of the magnetic flux produced by the current in one turn of the coil links with, or cuts through, an adjacent turn. If the current in the first turn changes, the flux it produces also changes and a Faraday voltage is generated in the adjacent turn. According to Lenz, voltages so induced cause currents to flow in the second turn and these, when linked with the loop of the first turn, oppose the force creating the original change. This induced voltage attempts to maintain the current in the first turn at whatever level it was flowing before the change was initiated.

Since this induced voltage is proportional to the time rate of current change, it would become infinite under instantaneous change conditions.

As infinity cannot be achieved under any circumstances, it is impossible to change the current flowing in a coil instantaneously.

Generally speaking, a coil consists of many neighboring single turns connected in series. The magnetic flux produced by any one turn passes through all other turns. If there is a change in the current flowing through a coil, there will be a change in the flux linking all turns and an electromotive force generated in each turn. All of these individual electromotive forces add up as one single voltage across the entire coil. Since these electromotive forces oppose the changes that cause them, the summed voltage generated by a changing current is called a *BACK EMF*.

Mathematically stated, the relationship between current flowing through a coil and voltage across that coil is as follows:

$$E = -L\, dI/dt \qquad \text{(Eq. 11.2)}$$

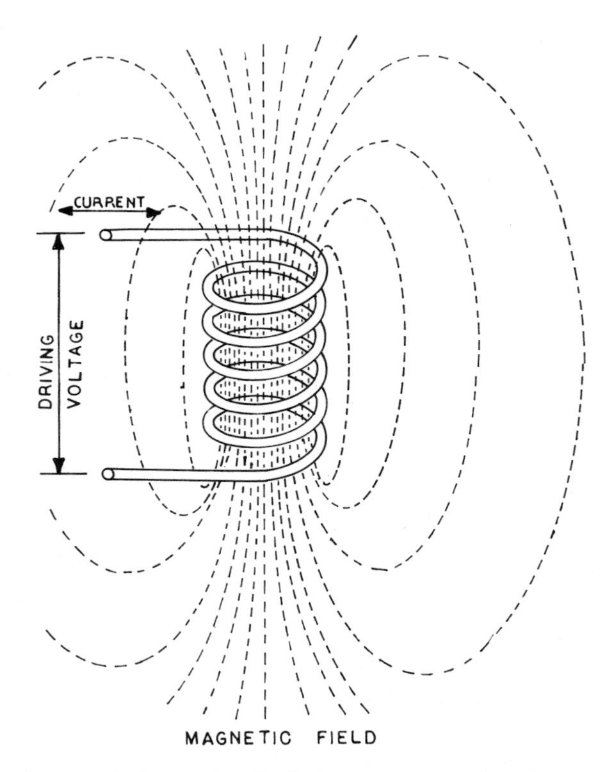


Fig. 11.4. Coil acting as an inductor: A coil of wire represents an impedance to the flow of an alternating electric current.

where $E$ is the voltage across the coil; $I$, the current flowing through the coil; $dI/dt$, the time rate of change of this current; and $L$, a constant called the *COEFFICIENT OF SELF-INDUCTANCE.*

Figure 11.4 illustrates the essential components of a coil acting as an inductor. If a current flows through this coil and is suddenly interrupted, the energy contained in the collapsing magnetic field will generate a back emf. This voltage can easily be large enough to perforate the coil insulation and to otherwise destroy the coil's electric integrity.

## 11.6 INDUCTIVE REACTANCE

Inductance is a characteristic of a coil that opposes any change in current flowing through that coil. The actual current-change limiting value of a coil depends on the current's rate of change as well as its inductance. The ability of a coil to impede, or limit, the flow of an alternating current depends, then, on the frequency of the alternating current. This ability is called *INDUCTIVE REACTANCE* and is mathematically stated as follows:

$$X = 2\pi f L \qquad \text{(Eq. 11.3)}$$

where $X$ is the inductive reactance; $f$, the frequency of alternating current; $L$, the coil's inductance; and $\pi$, the universal constant "3.14."

## 11.7 TIME CONSTANT

The current in a coil cannot be changed instantaneously; when a voltage is applied to the coil, therefore, the current increases from zero to some maximum over a period of time.

Consider a voltage generator driving an inductance connected in series with a resistance as shown in Fig. 11.5. If a constant voltage is suddenly intitiated by this generator, all the voltage first appears across the inductance and none across the resistance because the back emf generated by any sudden change in current makes it impossible to change the current instantaneously. In these circumstances, no current flows when the voltage transient is first applied.

Current immediately starts to flow in this circuit, however, at some later time when equilibrium has been achieved and all the voltage appears across the resistance and none across the inductance. In other words, although the voltage applied to the circuit is changed instantaneously, the current flowing through the circuit takes some time to respond to that change. For convenience sake, this current transient response is identified

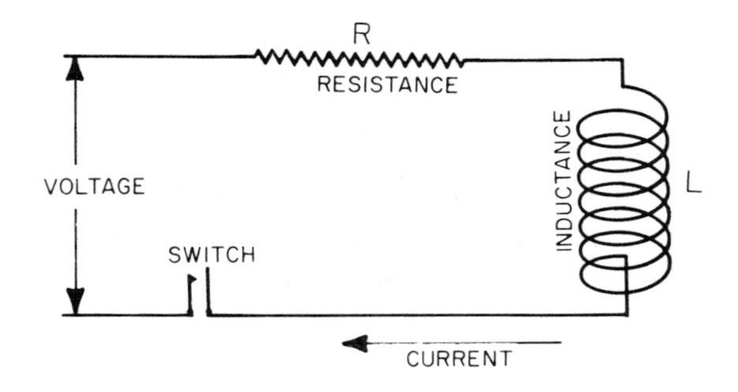


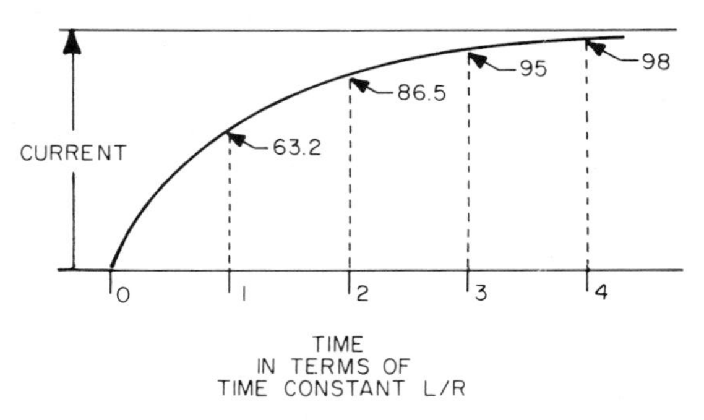


Fig. 11.5. *L/R* time constant: The build-up of an electric current flowing in a circuit is impeded by the presence of an inductor. The rate of build-up is defined by the time constant *L/R*.

by a time constant. The *L/R TIME CONSTANT* is the time taken for the current in Fig. 11.5 to reach 63.2 percent of its final value after the switch is closed. It is mathematically stated as follows:

$$T = L/R \qquad \text{(Eq. 11.4)}$$

where $T$ is the time, in seconds; $L$, the inductance, in henrys; and $R$, the resistance, in ohms. $L/R$ also describes the time taken for the current to pass from 63.2 to 86.5 percent (63.2 percent of the remaining 36.8 percent), from 86.5 to 95 percent (63.2 percent of the remaining 13.5 percent), and so on.

Just as the current through an inductance cannot be changed instantaneously, so the voltage across a capacitance cannot be changed instantaneously. There is, then, an *RC TIME CONSTANT* that is analogous to the L/R time constant. In this case, 63.2 percent of the final voltage is achieved after a period of time $T$ equal to the product of $R$ and $C$, where $T$ is the time in seconds; $C$, the capacitance, in farads; and $R$, the resistance, in ohms.

## 11.8 IMPEDANCE

The term *IMPEDANCE* is used generally to describe the response of a system to a sinusoidally varying stimulus. In an electric context, a voltage is used to drive a current through a coil that provides an impedance to the flow of current. Following Ohm's Law, the stimulating voltage $E$ is equal to the product $ZI$, where the current $I$ is the response to the circumstance of the impedance $Z$.

All coils have resistive as well as inductive characteristics. Both of these factors are commonly distributed more or less uniformly throughout the coil turn by turn and unit length by unit length. For the purposes of circuit analyses as shown in Fig. 11.5, however, a coil can be represented by a pure resistor in series with a pure inductor. The *INDUCTIVE IMPEDANCE* of this "coil" to the flow of a sinusoidally varying electric current may then be stated as follows:

$$Z = R + i(2\pi fL) \qquad \text{(Eq. 11.5)}$$

where $Z$ is the impedance; $f$, the frequency of alternating voltage; $L$, the coil's inductance; $R$, the coil's resistance; and $\pi$, the universal constant "3.14." The symbol $i$ (implying that the term that follows is imaginary) indicates that the two components of $Z$ are vector additions in which $(2\pi fL)$ is always at right angles to $R$.

When an alternating voltage is applied to a coil, a current flows, and the product of the current and the impedance is equal to the *INDUCTIVE VOLTAGE* as follows:

$$E = ZI = RI + i(2\pi fL)I \qquad \text{(Eq. 11.6)}$$

where $E$ is the voltage applied to a coil expressed by the product $ZI$; $RI$, the voltage drop across the resistive element; and $(2\pi fL)I$, the voltage drop across the inductive element. Here, again, the symbol $i$ indicates that the two component voltages are added in a vector sense where $RI$ is always at right angles to $(2\pi fL)I$.

Figure 11.6 depicts the components of Eq. 11.6. Here the resistive vector $RI$ and the inductive vector $(2\pi fL)I$ are at right angles to each other and add up to the driving voltage $E$ for all values of $E$. As shown, the current $I$ always flows in the direction of the vector voltage $RI$.

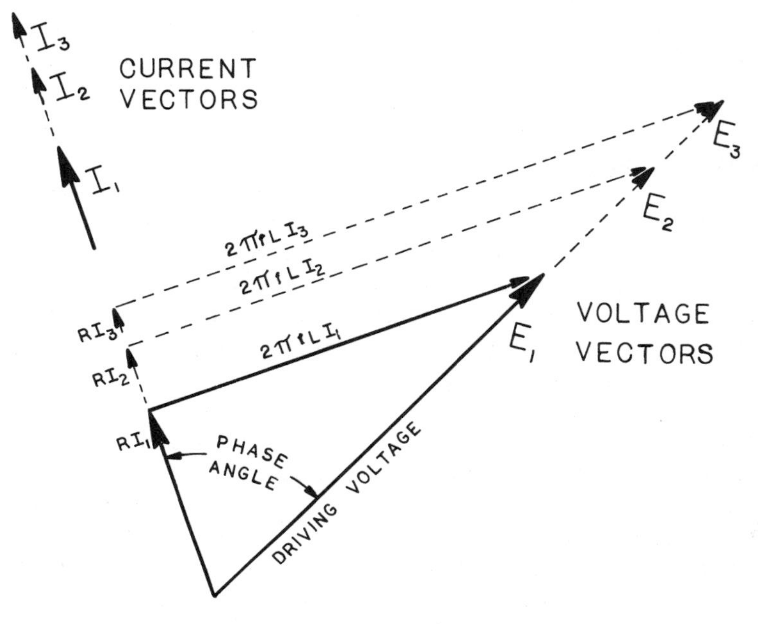


Fig. 11.6. Inductive and resistive vectors: The current in a circuit containing both inductive and resistive elements lags an alternating driving voltage.

## 11.9 PHASE ANGLE

If two time-varying functions cycle at the same rate but their minimums and maximums occur at different times, this temporal difference can be expressed in terms of a *PHASE ANGLE.*

Figure 10.3 presents the concept of a coil of wire rotating in a magnetic field at a constant speed. The voltage generated by this coil rotation follows a sinusoidal pattern in which the instantaneous voltage is determined by the instantaneous rate of change of magnetic flux passing through the coil. In the configuration of the particular instant shown, the flux is directed at right angles to the coil axis. In this position, none of the flux is intercepted by the coil, and the instantaneous rate of change is zero.

When the coil is rotated by 90 degrees from the position shown, both the flux through the coil and the rate of change of that flux are maximum. The same is true when the coil is rotated by 270 degrees. In the 270-degree position, however, the flux passes through the coil in a direction opposite to that of the 90-degree position. As a consequence, although the voltages generated in these two positions are equal in value, they are opposite in sign.

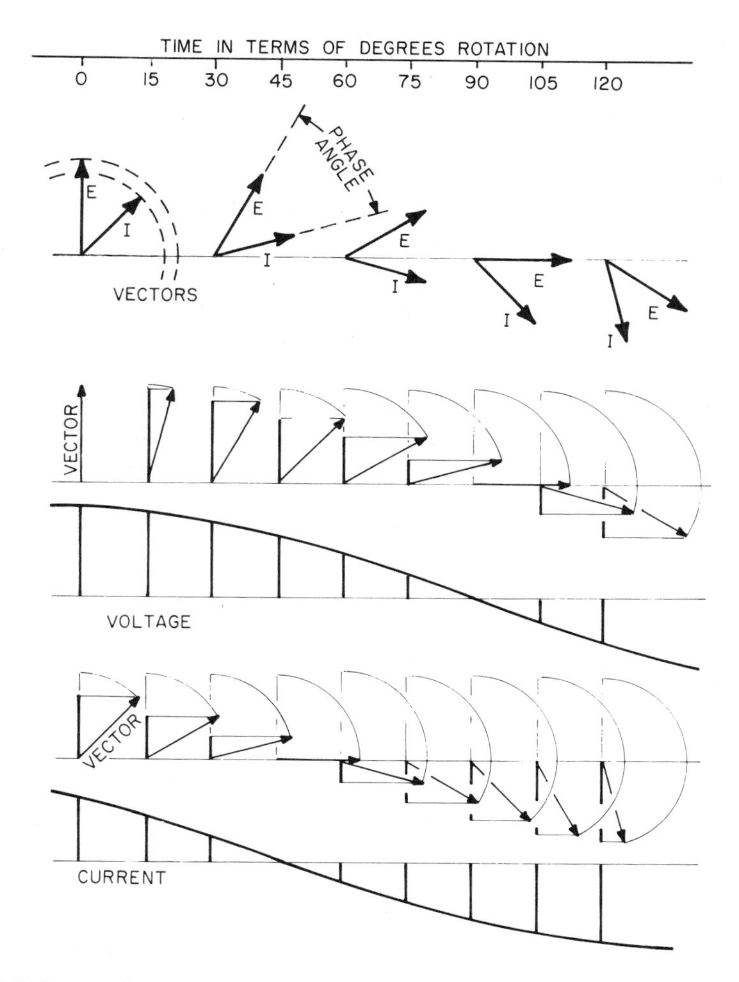


Fig. 11.7. Voltage and current vectors: The angular difference between the voltage and the current in a given circuit and at a given frequency remains the same no matter what the time or what the voltage magnitude.

A vector diagram is a convenient means of illustrating the electrical effects of a coil's rotation in a magnetic field. Fig. 11.7 shows a "voltage" vector that duplicates the coil's rotating activities. The projection of that vector on the vertical axis at any particular time represents the magnitude of the voltage generated at that time.

Since the coil, and hence the vector, rotates at constant speed, it takes the same length of time to traverse each complete rotation or equal interval of rotation. Consequently, time can be measured in terms of degrees of rotation as well as the elapsed time unit it takes to complete an interval represented by degrees of rotation. When it is measured in this way, one vector diagram can be applied to the rotating experience of any coil, in any field, and at any speed.

Since other voltage sources follow sinusoidal variations analogous to that of a rotating coil, the same logic can be applied to their performance. In fact, the concept of a Fourier integral shows that any time-varying function can be broken down into the vector sum of a number of sinusoidal variations of different frequencies and different magnitudes. The basic concepts displayed in Fig. 11.7 then assume a universality that can be applied to any circuit situation.

As shown in this figure, the voltage vector changes its direction with the passage of time. A particular direction is sometimes called a "phase," and the angle between two phases is called a "phase angle." The phase angle of a single rotating vector is constantly changing, but the phase angle between two rotating vectors remains the same if they both rotate at the same speed.

Figure 11.7 also shows a voltage vector and the resulting current vector of a current driven by that voltage. The current revolves at the same speed as the driving voltage. In the illustrated circumstances, the current is a response to a capacitive reactance so that the current vector leads the voltage vector. In a circuit characterized by inductive reactance, the current would lag the voltage.

Figure 11.6 shows the phase angle between a voltage $E$ and a current $I$ that results from a circuit consisting of a resistance in series with an inductance. Here the phase angle between the voltage and the current can be expressed as follows:

$$\phi = \tan^{-1} 2\pi f(L/R) \qquad \text{(Eq. 11.7)}$$

where $\phi$ is the phase angle between the voltage and the current; $f$, the frequency of voltage oscillation; $L$, the inductance; $R$, the resistance; and $\pi$, the universal constant "3.14." The symbol $\tan^{-1}$ may be translated as "an angle whose tangent is."

The time constant of Eq. 11.4 and the phase angle of Eq. 11.7 both represent the same phenomenon viewed in two different ways. In Fig. 11.5, a current is shown to follow a stimulating voltage by a time constant. In Fig. 11.7 a current is shown to follow a stimulating voltage by a phase angle. In one circumstance, the time delay is described by $L/R$, and, in the other, by an angle whose tangent is $kf(L/R)$.

## 11.10 ELECTRIC ENERGY

Energy is consumed when electric currents flow in electric ciruits, as follows:

$$J = EI \cos \phi \qquad \text{(Eq. 11.8)}$$

where $J$ is the energy consumed by the circuit; the product $EI$ is commonly referenced in terms of volt-amperes; $I$ is the current that flows; $E$ is the voltage driving that current; and $\phi$ is the phase angle between voltage and current.

As shown by Fig. 11.1, the dynamic energy of a changing magnetic field can be used to establish potential energy in a secondary circuit. No energy is actually consumed in the secondary circuit, however, as a result of this potential energy. On the other hand, as shown by Fig. 11.2, this potential energy can be converted into kinetic energy, and thereby consumed, if the induced voltage is allowed to drive a secondary current. It may then be deduced that, although the changes of one magnetic field provide the mechanism, energy is transported through space only as an interaction between two changing magnetic fields—the fields of the inducing (primary) and the induced (secondary) currents.

In electromagnetic induction, as a consequence, a changing magnetic field associated with a changing electric current in one circuit induces a changing electric current in a second circuit. The second-circuit current changes create a second changing magnetic field in turn, and this interacts with the changing magnetic field of the first circuit. The energy absorbed by the current flow in the secondary circuit is abstracted from, or reflected back into, the first circuit.

Conversely, if the second circuit is open, no secondary current can flow, and no secondary magnetic field is created to interact with the primary magnetic field. Under these circumstances, no energy can be consumed in the secondary circuit; no energy is reflected back into the primary circuit; and the primary circuit is not affected by whatever else takes place in the secondary circuit.

# 12
# REFLECTED IMPEDANCE

When energy expended in a sensing coil is used to illuminate conducting targets, some of that energy is reflected from these targets back into the sensing coil. Energies so reflected are detected as apparent changes in sensing-coil impedance.

Such energy reflections are derived from eddy currents induced in the conducting targets. The magnitude, the phase, and the variations of both phase and magnitude with frequency of these energy reflections can be used to deduce various target characteristics.

## 12.1 REFLECTED SIGNAL CURRENT

When a flow of current in a sensing coil is used to induce a voltage in a target coil, the target current that flows in response to the target voltage induces a signal voltage in the sensing coil. Signal currents then flow in the sensing coil in response to the signal voltages. These reflected signal currents are added to the original sensing currents to create new sensing currents.

Figure 12.1 shows a conducting target in the form of a ring and exposed to the changing field of a sensing coil. This configuration could be described equally well by the target coils of Fig. 12.6 or even the transformer of Fig. 13.5. Figure 12.2 offers a vector diagram describing the electric circumstances of a general condition that embraces the examples of all these figures.

In the absence of a target, a sensing coil responds to a driving voltage with a lagging current $I_0$. Here the phase angle $\phi_0$ between the driving voltage and the lagging current is characteristic of sensing coil impedance—its resistance and its inductance. In the presence of a target, a sensing coil responds to the same driving voltage with a new lagging current, $I_s$. Here the phase angle $\phi_0$ between driving voltage and lagging current has been affected by the current induced in the target. This phase angle assumes a new $\phi_s$ in the presence of a target.

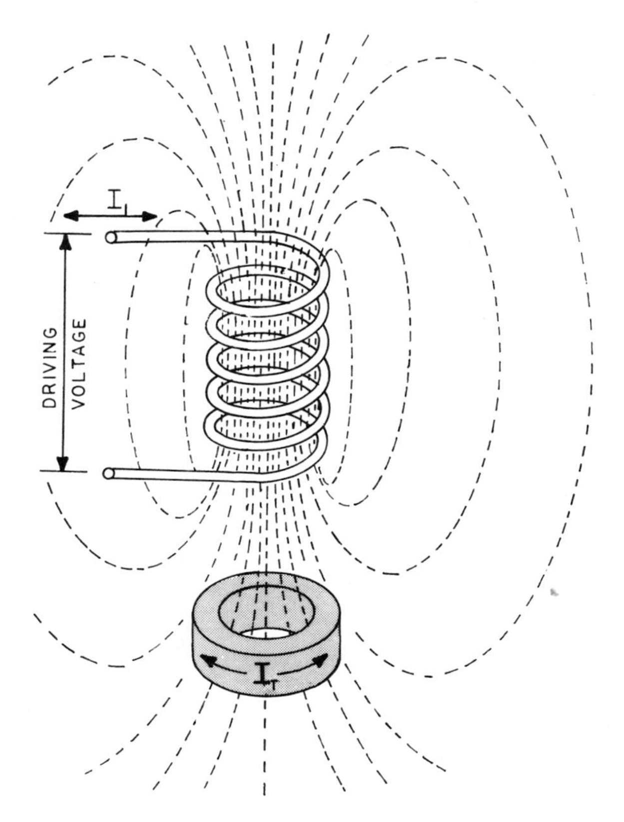


Fig. 12.1. Current induced in a target ring by a sensing coil: The impedance of the coil is increased when its changing magnetic field is used to induce eddy currents in a nearby ring acting as a target.

Following Ampere's Law, an in-phase field accompanies a sensing-coil current. In the figures here, targets are exposed to this changing magnetic field. Following Faraday's Law, a voltage $E_t$ is induced in a target that lags the inducing field (and hence the sensing-coil current $I_s$) by 90 degrees. A target current $I_t$ then flows in response to the target voltage $E_t$. This target current lags the target voltage by some phase angle $\phi_A$, which depends on both target inductance and target resistance.

Following Ampere's Law, an in-phase changing magnetic field accompanies a target current. A sensing coil is exposed to this changing magnetic field. Following Faraday's Law, a signal voltage $E_x$ (reflected voltage) is induced in a sensing coil that lags the inducing field (and hence the target current $I_t$) by 90 degrees. A signal current $I_x$ (reflected current) then flows

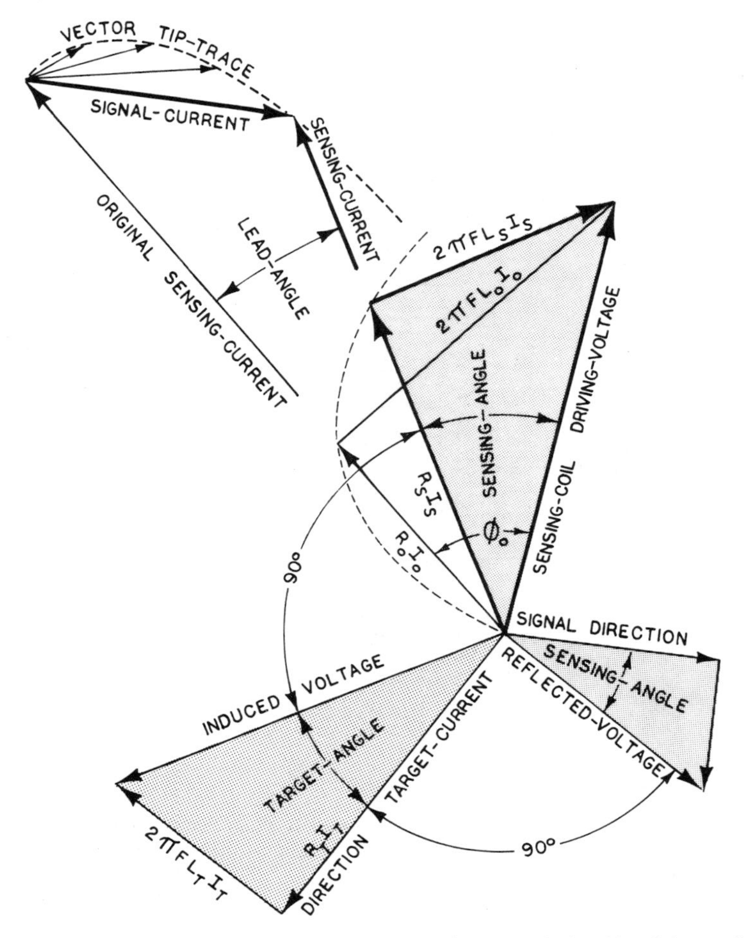


Fig. 12.2. Sensing coil and target vectors: The voltage/current relationship of the configuration shown in Fig. 12.1 may be described by a vector diagram. Here current flows in a sensing coil in response to a driving voltage and lags the driving voltage by some phase angle. This current further induces a voltage in any nearby conductor acting as a target. This target voltage then drives a current in the target that in turn induces a voltage back in the sensing coil that changes the sensing coil current by some phase angle and magnitude.

in a sensing coil in response to the signal voltage $E_x$. This signal current lags the signal voltage by the same phase angle $\phi_s$ by which the sensing coil current $I_s$ lags the driving voltage.

A signal current $I_x$ (in response to a conducting target) always leads the original driving current $I_0$. Hence, $I_s$ always leads $I_0$ under these conditions. As a result of the energy reflected back into a sensing coil from a conductive target, the new current $I_s$ that flows is the vector sum of the original sensing-coil current $I_0$ and the signal current $I_x$.

If the original sensing coil current $I_0$ is used as a reference, the angle by which the new sensing-coil current $I_s$ leads $I_0$ is an indication of either the size or the proximity of a conducting target; that is, although the phase relation between $I_x$ and $I_s$ is always the same for a given target, the magnitude of the signal current $I_x$ that affects the phase angle between $I_0$ and $I_s$ is variable.

## 12.2 MAGNITUDE OF REFLECTED ENERGY

The magnitude of energy interchanged between two coils (as indicated by the magnitude of the signal current $I_x$) is maximum when the two coils share the same axis. The relative strength of a magnetic field along a coil's axis is mathematically stated as follows:

$$B = kINr^2/d^3 \qquad \text{(Eq. 12.1)}$$

As long as $d$ is much larger than $r$, $B$ is the strength of the magnetic field along the centerline of a coil of radius $r$ and number of turns $N$ at a distance $d$ from the coil's center when a current $I$ flows through the coil. In this case, k is a rather complex constant that depends on the relative orientation and location of the two coils.

Combining all relevant features of both Ampere's and Faraday's Laws, the magnitude of the signal current induced in a sensing coil as a result of energy reflected from a target coil is mathematically stated as follows:

$$I_x/I_s = kZ_s^{-1}(N_s r_s^2)^2 Z_t^{-1}(N_t r_t^2)^2 d^{-6} f^2 \qquad \text{(Eq. 12.2)}$$

where a sensing coil's driving current $I_s$ of frequency $f$ experiences a change $I_x$ as a result of energy reflected back from the current induced in the target coil; $N_s$ is the number of turns of the sensing coil wound on radius $r_s$, separated by a distance $d$ from a target coil with the number of turns $N_t$ wound on radius $r_t$; $Z_s$ is the impedance of the sensing coil; and $Z_t$ is the impedance of the target coil. Here, again, k is a rather complex constant that depends on the relative target-coil location and orientation with respect to the sensing coil.

## 12.3 TARGET ORIENTATION

As illustrated by Fig. 12.3, a coil of wire can be rotated in a magnetic field. The voltage generated by such rotation follows a sinusoidal pattern

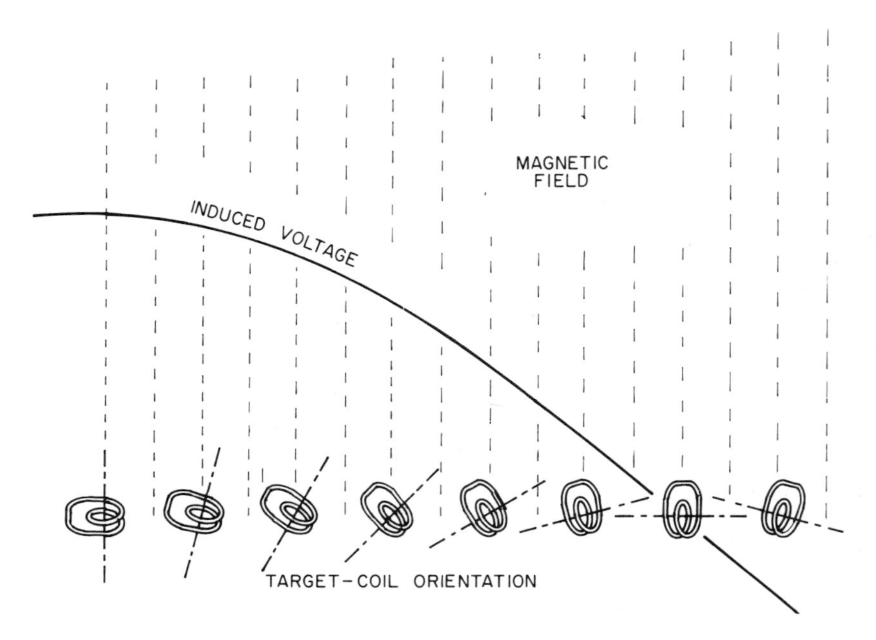


Fig. 12.3. Target orientation: The voltage induced in a rotating coil of wire is proportional to the instantaneous rate-of-change of magnetic flux through the coil.

in which the instantaneous generated voltage is determined by the instantaneous rate of change of magnetic flux passing through the coil. If flux is directed at right angles to the coil's axis, none of it is intercepted by the coil, and the instantaneous rate of change through the coil is zero. On the other hand, when a rotating coil's axis is parallel to a magnetic field, both the flux through the coil and its rate of change are maximum.

The relationship between two coils when the target coil is fixed in space and illuminated by a sensing coil also fixed in space is analogous to the circumstances of a rotating coil. In the latter, the time rate of change is provided by coil rotation; in the former, by the alternations of sensing-coil current.

In both sets of circumstances, the magnitude of the induced voltage follows the same pattern in relation to the target-coil orientation. As shown by Fig. 12.3, an induced voltage is equal to a maximum induced voltage times the cosine of the angle made by the intercept of the sensing-coil and target-coil axes. The maximum induced voltage, as determined by Eq. 12.2, must then be multiplied by the cosine of this orientation angle.

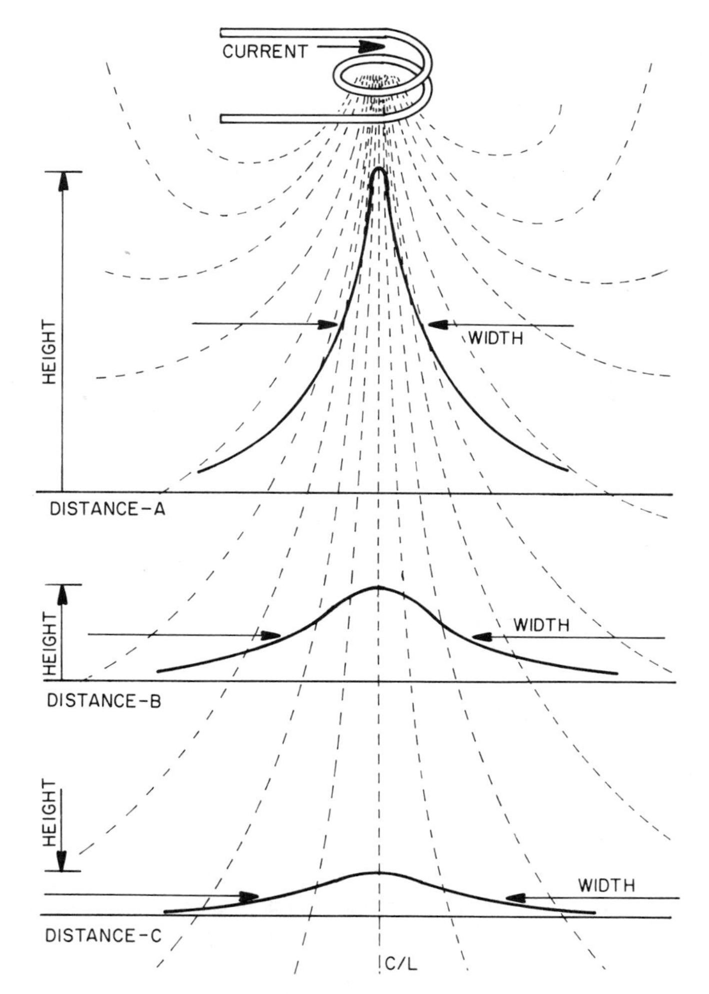


Fig. 12.4. Target location: The strength of a magnetic field surrounding a current-carrying coil of wire varies significantly in a spatial sense.

## 12.4 TARGET LOCATION

Sensing coils respond variously to the energy reflected from target coils located variously. A mathematical analysis of all possibilities is beyond the scope of this discussion. It should be pointed out, however, that the field strength encountered along a common centerline, as described by Eq. 12.2, is modified by the trends displayed in Fig. 12.4. In other words, if the target is not located along the sensing coil's centerline, the reflected voltage is less than it would be if it were on the centerline. This figure suggests a somewhat oblique solution to the general problem. Certainly, when a target coil is immersed in the toroidal magnetic field of a sensing coil, the strength of the magnetic field experienced by the target coil is a complex function of its position as well as its orientation within the toroidal field.

The most practical thing to do is to determine the magnitude of this field experimentally, but a general plot of it can be made by envisioning a number of planes located at right angles along the axis of the sensing coil. As shown by Fig. 12.4, it can be said that such a magnetic field is always strongest along the axis of the sensing coil and that the ratio of on-axis to off-axis voltage is much greater as the distance between target coil and sensing coil decreases.

The "sharpness" of a field plot—of sensitivity versus distance away from a sensing-coil axis—can be described in terms of a breadth/height aspect ratio. (There is a different aspect ratio in each plane located along a sensing-coil axis, and both the maximum field strength and the aspect ratio depend on the diameter of the sensing coil, the number of turns in it, and the frequency of the energy driving it.)

## 12.5 MULTIPLE TARGETS, MULTIPLE RESPONSES

When a sensing coil approaches a target coil representing inductive reactance, the phase angle of the sensing coil is reduced. Figure 12.5 shows three possible responses induced by a target coil in a sensing coil. The three responses shown are the phase angles $\phi_1$, $\phi_2$, and $\phi_3$. All three of these angles are progressively less than the original phase angle $\phi_0$. In fact, Fig. 12.5 shows how these three phase angles can be created by three different target coils each with different phase angles, $\phi_A$, $\phi_B$, and $\phi_C$.

The progression of the sensing-coil phase angle from $\phi_0$ to $\phi_3$ can be calibrated in terms of distance between the sensing coil and the target coil for a particular target coil.

A tip trace of the vector sum of the original sensing-coil current $I_0$ and the various possible signal currents $I_x$ is different for target coils with different phase angles. This trace then, is a characteristic of a target coil's unique phase angle, and it can be used for target-discrimination purposes.

As shown by Fig. 12.5, a target can either increase or decrease the magnitude of a sensing coil's current. As a general statement, however, the smaller the target coil's phase angle, the more that the energy reflected from that target tends to reduce the magnitude of the sensing-coil current.

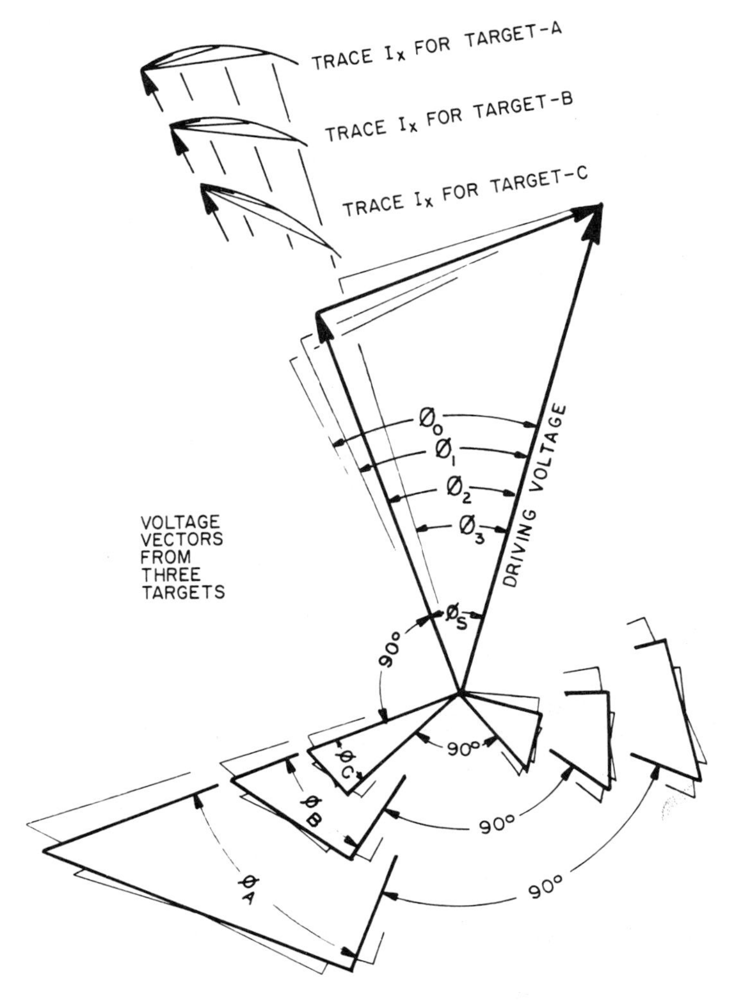


Fig. 12.5. Multiple targets, multiple reading vectors: Expanding the concepts of Fig 12.2, targets with differing phase angles ($\phi_A$, $\phi_B$, $\phi_C$) reflect voltages back into sensing coils that have different phase angles and different magnitudes. The trace of the tip of the sensing coil current vector when a sensing coil approaches a particular target is then characteristic of that target.

## 12.6 REFLECTED PHASE-ANGLE RANGE

Figure 12.6 shows two different target coils of differing phase angles and two positions with differing orientations for one of these target coils. All three of these variables—location, orientation and electric characteristic—affect the coupling between the sensing coil and the target coil. All

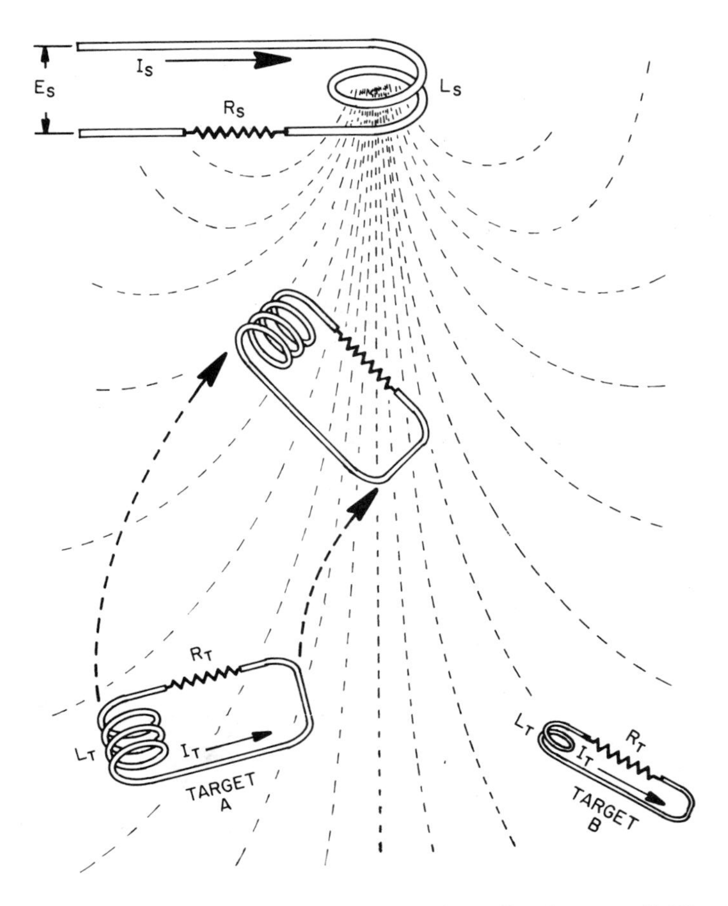


Fig. 12.6. Variables affecting the coupling between sensing coil and target coil: The signal voltage reflected back into a sensing coil from the current induced in a target depends on the impedance of the target and on both its location and its orientation in the sensing coil's field.

three of these variables affect the magnitude and direction of the reflected current $I_x$.

The current vectors of Fig. 12.5 show the current in a sensing coil as a result of the energy reflected back from various target coils. The vector sum of the original sensing current $I_0$ and the reflected current $I_x$ creates new sensing currents that always lead the original sensing current $I_0$. The amount of this lead depends on both the phase and the magnitude of the reflected current $I_x$.

The direction of the reflected current $I_x$ as described by the phase angle $\phi_x$ is determined both by the degree of coupling between the sensing coil

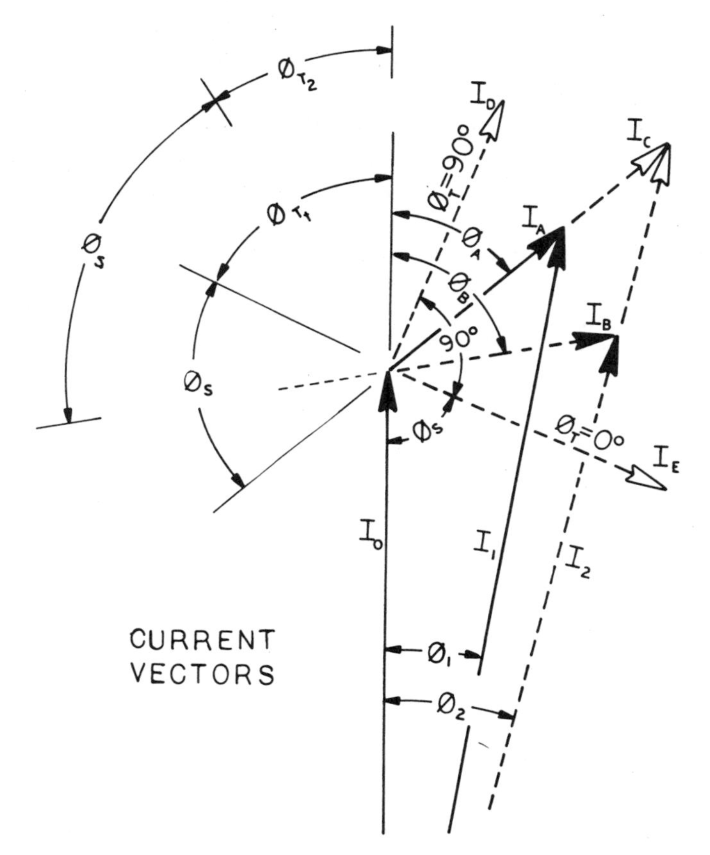


Fig. 12.7. Vector range of signal current: A target's phase angle can vary over a 90-degree range, from pure resistance to pure inductance.

and the target coil and by the electric characteristics of both the sensing coil and the target coil.

Figure 12.7 is an expansion of the current vectors in Fig. 12.2. This sketch is included to show the range of possibilities presented by sensing coils and target coils of various electric characteristics. The results of two different target coils driven by the same sensing coil are displayed. Here, the two reflected currents are indicated as $I_A$ at phase angle $\phi_A$ and $I_B$ at phase angle $\phi_B$.

For a given sensing coil whose phase angle is normally $\phi_0$, various targets might be encountered whose phase angles lie anywhere between pure resistance, where $\phi_t$ is equal to 0 degrees, and pure inductance, where $\phi_t$ is equal to 90 degrees. The reflected currents for these targets are shown as $I_E$ and $I_D$ in Fig. 12.7.

Including all of the theoretical 0-to-90-degree possibilities for the phase angle of both the sensing coil and the target coil, reflected currents can be encountered that lead the original sensing-coil current by any angle from 0 to 180 degrees. It is thus possible for a reflected current to be either in-phase or out-of-phase with the original sensing-coil current. Under all possible circumstances, however, the reflected current of a sensing coil characterized by inductive reactance, after encountering a target coil also characterized by inductive reactance, always leads the original sensing-coil current. On the other hand, if the target circuit is characterized by capacitive reactance, a new sensing-coil current can lag the original sensing-coil current.

Figure 12.7 also shows the possibility in which two targets whose electric characteristics are significantly different can cause an exposed sensing-coil current to flow at the same phase angle. In this case, the reflected current $I_C$ at phase angle $\phi_A$ and the reflected current $I_B$ at phase angle $\phi_B$ cause two different amplitudes of an exposed sensing-coil current $I_2$ to flow at the same phase angle, $\phi_2$. Under these circumstances, if $I_2$ is greater than $I_0$, $R_2$ must be smaller than $R_0$. It is then possible for both $R_2$ and $L_2$ to be either greater than or less than the values experienced under no-load conditions.

## 12.7 EDDY CURRENTS

Voltages are induced in all conductors exposed to changing magnetic fields regardless of the cause of such changes or of the configuration of such conductors. The Foucault currents of Fig. 10.5 flow in response to voltages induced when a conducting body is moved through a nonuniform magnetic field.

The transformer of Fig. 13.5, the target coils of Fig. 12.6, and the ring of Fig. 12.1 all illustrate various configurations of the same phenomenon that differ only in terms of degree. Each figure represents a "target coil" with unique electric characteristics. Figure 12.8 envisions the ring of Fig. 12.1 embedded in the surface of a conducting body. Here, too, the changing magnetic field of a sensing coil induces a current that circulates within a conducting body. Currents so induced are commonly called *EDDY CURRENTS*.

Circulating eddy currents reflect energy back into sensing coils in exactly the same manner as currents flowing in any type of target coil do.

## 12.8 SKIN EFFECT

In the *SKIN EFFECT*, alternating currents flow only near the surfaces of conductors. The depth to which such a current penetrates beneath a

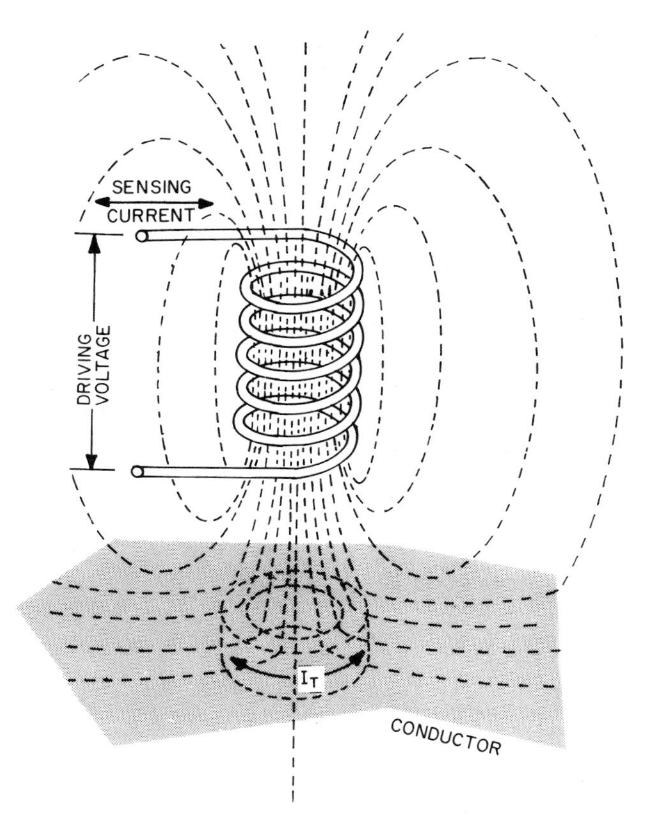


Fig. 12.8. Eddy currents induced in the surface of a conducting body: When eddy currents are induced in a conducting body by the changing magnetic field of a sensing coil, the relations between voltages and currents are the same as those described by the vector diagram of Fig. 12.2.

conductor's surface at 36.8 percent of its maximum (surface) value is mathematically expressed as follows:

$$d = (\mathrm{k}fpc)^{-1/2} \qquad \text{(Eq. 12.3)}$$

where $d$ is the depth of penetration; $f$, the frequency; $p$, the magnetic permeability; $c$, the electric conductivity; and k, a constant.

The higher the frequency, or the faster the current's rate of change, the more pronounced the effect. At microwave frequencies, currents penetrate below the surface of conductors by only a few molecular spacings.

Electrons flowing near the surface of a conductor—in the process of contributing their bit to an electric current—are exposed to less of the magnetic flux accompanying other moving electrons than are those that are more deeply imbedded in that conductor. The reason is that surface electrons are exposed to the effects of their fellow electrons on only one side whereas imbedded electrons are completely surrounded. Since the imbedded electrons of an alternating current are exposed to changing magnetic fields that are stronger, they experience greater Lenz forces. Or, to look at this phenomenon in another way, the imbedded electrons experience a greater "mutual inductance" with their fellow electrons than do those near the surface.

It is easier, then, for an electron to change its motion near the surface of a conductor than it is in the conductor's interior. Since carriers invariably seek the easiest path (a minimum energy condition), those carriers making up an alternating current are exposed to lateral Lenz forces that move them outward toward a position of minimum inductance, that is, toward the surface of a conductor.

The inductance-difference gradient within the body of a conductor establishes a condition in which the current phase angle varies over the cross section of that conductor. It is even possible for electrons in two parts of the same conductor to be traveling in opposite directions instantaneously.

The Skin Effect is opposed by the electrostatic repulsion between carrying electrons.

Combining the concepts of the Skin Effect and eddy currents, it is possible to control how far below the surface of a conductor an eddy current flows by adjusting the frequency of the driving voltage applied to a sensing coil.

Similar action can be observed in the *PROXIMITY EFFECT*, in which there is a redistribution of alternating current carriers when two conductors are moved into each other's vicinity. The carriers flowing in one conductor exert forces upon those of the nearby conductor. The carriers of each conductor then move to a position of minimum "mutual inductance."

Both the Skin Effect and the Proximity Effect cause current carriers to flow through the equivalent of a reduced conductor area. This has the result of increasing the resistance of a conductor so that the higher the frequency, the greater the resistance.

## 12.9 TARGET PROFILE

A target resistance $R_t$ as seen by an eddy current $I_t$ is inversely proportional both to the depth of penetration and the conductivity of the target material. By substituting the conclusions of Eq. 12.3 for depth, the target resistance may be determined as follows:

$$R_t = \mathrm{k} f^{1/2} c^{-1/2} p^{1/2} \qquad \text{(Eq. 12.4)}$$

where $R_t$ is the target resistance; $f$, the driving frequency; $c$, the conductivity of the target material; $p$, the permeability of the target material, and k, a constant.

By using this value of target resistance in Eq. 12.2, ignoring permeability, and assuming that the target inductance does not change significantly, the change in resistance to which eddy currents are exposed as a result of changes in frequency can be illustrated by the following relationship:

$$I_x/I_s \propto f^2(\mathrm{k}'c^{-1}f + \mathrm{k}''f^2)^{-1/2} \qquad \text{(Eq. 12.5)}$$

where $I_x/I_s$ is the ratio of current flow at a particular frequency to the current flow at some high frequency; $C$, the conductivity of the target material; $f$, the driving frequency; and $k'$ and $k''$, constants.

This equation has a peak that occurs at some frequency that depends only on the conductivity of the target material. Here, the greater the conductivity, the lower the frequency at which this peak occurs. Each type of conducting material can thus be identified by a characteristic peak that occurs at a frequency dictated by this equation. More than one peak in a frequency spectrum establishes the presence of more than one different

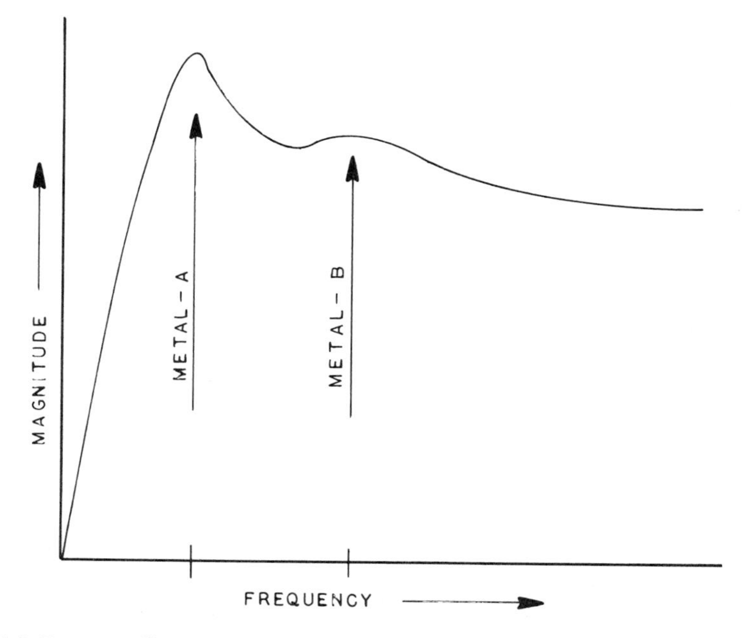


Fig. 12.9. Target profile: The changing resistance to which eddy currents are exposed during a swept-frequency cycle peaks at a frequency that is characteristic of a particular material.

type of target. A particular aluminum alloy has a characteristic peak in the range of 300 Hz.

Figure 12.9 represents a solution of Eq. 12.5 in which two different types of metallic elements are present in the changing magnetic field of one sensing coil. This figure shows how it is possible to discriminate between targets of different types of material by sweeping through a range of driving-voltage frequencies. In fact, if a target is constructed from a number of different materials, a sweep through the frequency spectrum over which peaks for each of these materials might occur will yield a spectral signature, or *TARGET PROFILE*, that is characteristic of that particular composite target.

## 12.10 BEAT FREQUENCY OSCILLATOR

If two resonant circuits oscillate at almost the same frequency, a third frequency is created that is equivalent to the difference between the two primary frequencies. This derived frequency is called a "beat" frequency.

A combination of two primary oscillators creating a beat frequency is called a *BEAT FREQUENCY OSCILLATOR*. One of the two primary oscillators can be used as a reference if it is stabilized at constant frequency. If the frequency of the other primary oscillator is changed by changing some circuit parameter, the resulting "difference," or "beat," frequency can be a very sensitive measurement of parametric change.

If a changing magnetic field associated with a changing current in a sensing coil is used to illuminate a target, the energy reflected back from that target is detected as a change in sensing-coil impedance. A beat-frequency oscillator provides one means of detecting the impedance change.

When a sensing coil is exposed to a target, the characteristics of the sensing-coil current are subject to changes in phase, changes in magnitude, or changes in both. In all circumstances, changes of this nature are electrically equivalent to changes in sensing-coil impedance. In some circumstances, such changes represent changes in sensing-coil inductance.

When exposed to a permeable target, as shown by the vector diagram of Fig. 13.3, the driving current of a sensing coil lags its driving voltage by an increased angle. This increase represents an increase in the sensing-coil inductance. When exposed to a reflective target, on the other hand, as shown by the vector diagram of Fig. 12.2, the current of a sensing-coil lags its driving voltage by a decreased angle. This decrease represents a decrease in the sensing-coil inductance.

A "change" in lagging phase angle is of interest in these circumstances in contrast to the absolute value of the angle itself. As a means of maximizing the effects of such change, the magnitude and phase of the sensing-coil current when it is not exposed to a target is used as the reference rather than the sensing-coil driving voltage. Figure 14.2 combines the effects of Fig. 13.3 and 12.2 as these vector diagrams relate to zero-loaded sensing-coil current. When viewed in this manner, permeable targets cause lagging phase angles, whereas reflective targets cause leading angles.

If a sensing coil is used as the inductive, or L, part of the L/C tank circuit of a primary oscillator, the sensing coil resonates with a fixed capacitor C at a particular frequency. As the sensing-coil inductance is changed by target exposure, the frequency of the oscillator changes. When this oscillator is beat against a reference oscillator, the resulting beat frequency can be calibrated in terms of inductance change. Such beat frequencies can then be related to the combination of target characteristics—distance, orientation, location, and electric nature. If all variables are kept constant except the one of interest, the beat frequencies can be calibrated in terms of the latter.

In such a mensuration technique, the frequency stability of the parametric oscillator and an ability to resolve small parametric changes are opposing objectives since the parametric oscillator frequency must change with the changing parameter of interest. If this frequency is easy to change with the variations of any particular factor, however, it is also subject to change as a result of extraneous signals.

Phase-lock presents one such problem. This is a tendency for two resonant circuits tuned to almost the same frequency to oscillate at the same frequency, a mode, of course, that eliminates the beat frequency completely. Then, too, the relationship between a sensing coil and its environment represents capacitance as well as inductance. If there is a change in this capacitance, the parametric frequency will change, thus introducing error in the beat frequency. This effect is relatively easy to minimize with a *FARADAY SCREEN*, however, which represents both an electrostatic shield and a fixed capacitance.

## 12.11 REFLECTED RESISTANCE THERMOMETER

Consider a ring constructed from some material whose resistance changes significantly as its temperature is changed. The change in sensing-coil current, which represents the change in sensing-coil impedance caused by the change in target resistance, can then be calibrated in terms of ring temperature. See Fig. 12.10.

The vector diagram of Fig. 12.11 (a variant of Fig. 12.2) illustrates the effects encountered when the resistance of a target is changed without changing any other circuit factor.

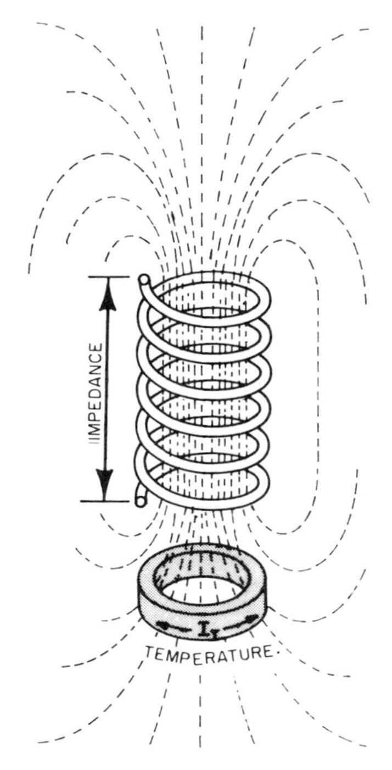


Fig. 12.10. Reflected resistance thermometer: The impedance of a sensing coil, which generates eddy currents in a target, changes as the resistance of the target changes with temperature.

## 12.12 REFLECTED IMPEDANCE GAGE

Differences in the apparent inductance and apparent resistance of a sensing coil as indicated by differences in the phase and amplitude of its current can be used to measure a number of different parameters. This doesn't mean that several measurements can be made simultaneously, but if all variables are standardized except the one of interest, differences in the latter can be monitored by this technique.

Such a gage responds to several different phenomena or combinations thereof. In the first, increases in the reluctance of the circuit seen by the sensing coil cause the sensing-coil currents to lag; in the second, increases in target capacitance also cause the sensing-coil currents to lag; in the third, increases in target resistance cause the sensing-coil currents to lead; and in the fourth, increases in target inductance also cause sensing-coil currents to lead.

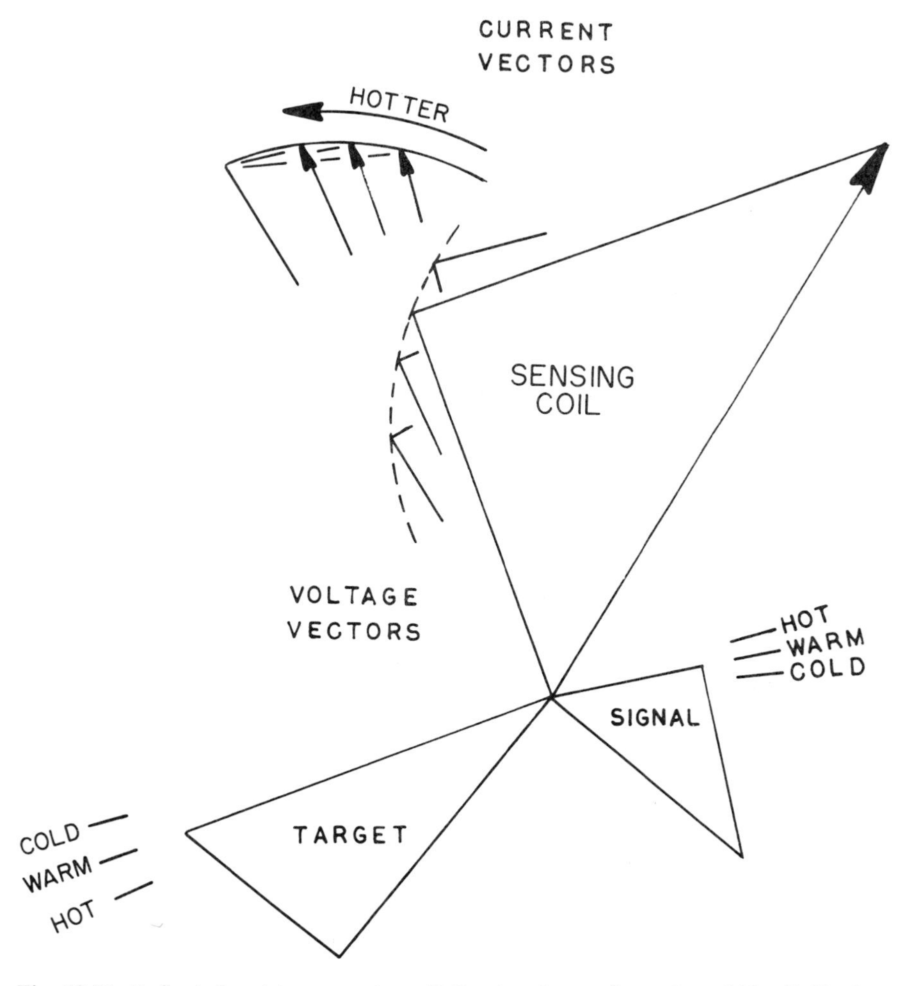


Fig. 12.11. Reflected resistance vectors: Following the configuration of Fig 12.10, the impedance of a sensing coil that generates eddy currents in a target changes as the resistance of the target changes with temperature. The trace made by the tip of the sensing-coil current vector can then be calibrated in terms of target temperature.

As shown in Fig. 12.5, a trace of the signal-current vector with different signal magnitudes detected in a sensing coil provides an indication of what type of variable—reluctance, capacitance, resistance, or inductance—that is affecting the sensing-coil current. In the configuration of Fig. 12.12, if the distance between a sensing coil and the surface of a conductor is standardized, the difference in the resistivity of various conductors that

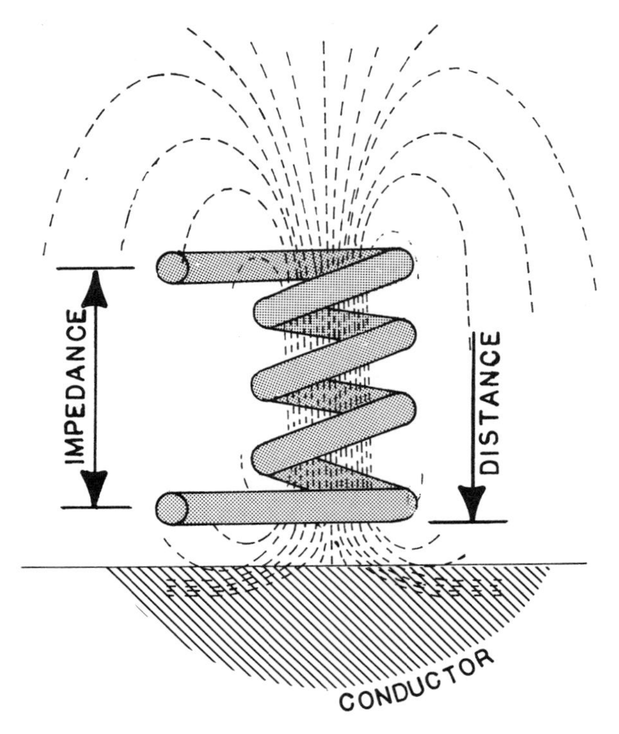


Fig. 12.12. Reflected impedance gage: The impedance of a sensing coil depends both on the distance between the coil and target and on the resistivity of the latter. If one of these two factors is held constant, coil impedance can be calibrated in terms of the other.

might be exposed to that sensing coil will affect both the phase and the magnitude of the induced eddy currents. These differences can be used to discriminate between various types of conductors.

Figure 12.11 shows how a vector diagram can be used to develop a calibration scale for one type of phenomenon—in this case, temperature. The same concept can be used to calibrate for other phenomena—conductor types, for instance. Similarly, if a coating and a base conductor have different conductivities, if both the magnetic and the electric characteristics of the base material are reasonably uniform, and if the sensing coil is kept in contact with the top surface of the coating, the magnitude of the induced eddy currents will depend to a significant degree on the thickness of that coating. The change in the phase angle of the sensing-coil current can then be calibrated in terms of coating thickness.

## 12.13 REFLECTED IMPEDANCE FLAW DETECTOR

If a cylindrical conductor is placed within the center of a sensing coil, eddy currents can be induced to flow around the circumference of the cylinder. If the cylinder is moved through the sensing coil at a constant velocity, the impedance changed experienced by these circumferential eddy currents can be monitored by the changes reflected back into the current of the sensing coil.

The ability to measure two variables—the phase and the amplitude of the sensing coil's current—makes it possible to measure more than one phenomenon simultaneously. Such multiple measurements are based on separation between the changes in resistance or inductance and the changes in the rate of change of either of them.

As an example, if a metal tube is passed through a coil of wire at high speed, it is possible to separate gradual changes from rapid ones. A localized flaw in a tube wall can be separated from a change in tube-wall thickness by differentiating circuits working from the output of the one sensing coil. One such circuit can detect the shift in amplitude and phase of the reflected eddy currents caused by changes in wall thickness, whereas

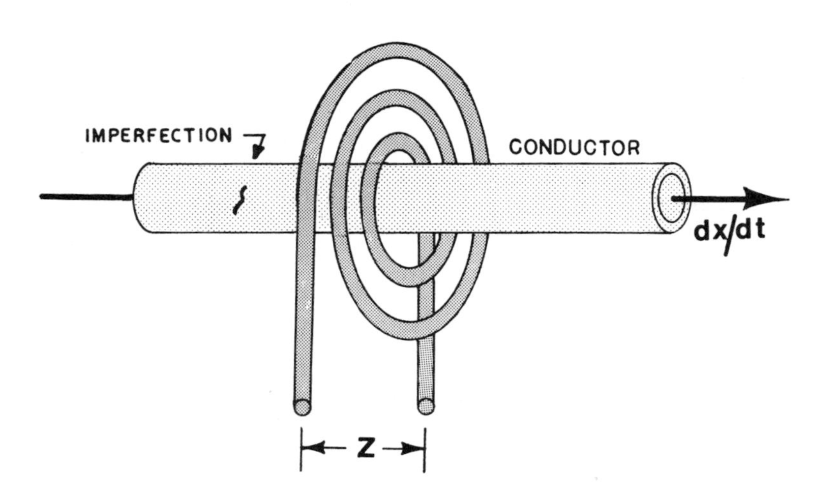


Fig. 12.13. Reflected impedance flaw detector: If the eddy currents in a conducting target are interrupted by a flaw in the target, that fact is indicated by a shift in sensing-coil impedance.

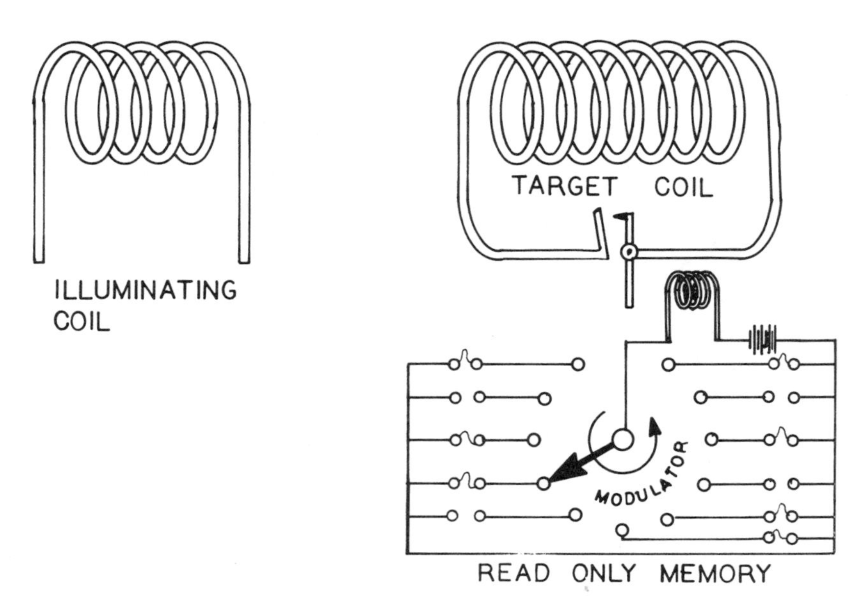


Fig. 12.14. Reflectivity modulation: If the impedance of a target changes in a time-varying pattern, this fact is detected by a sensing coil whose impedance changes in the same pattern.

the other circuit can detect the sudden rate of change caused by a flaw in that wall. (See Fig. 12.13.)

## 12.14 REFLECTIVITY MODULATION

In electromagnetic induction, energy from a transmitting coil illuminates a target; the target reflects a part of the impinging energy; and a receiving coil detects some of the reflected energy. As described in other sections,

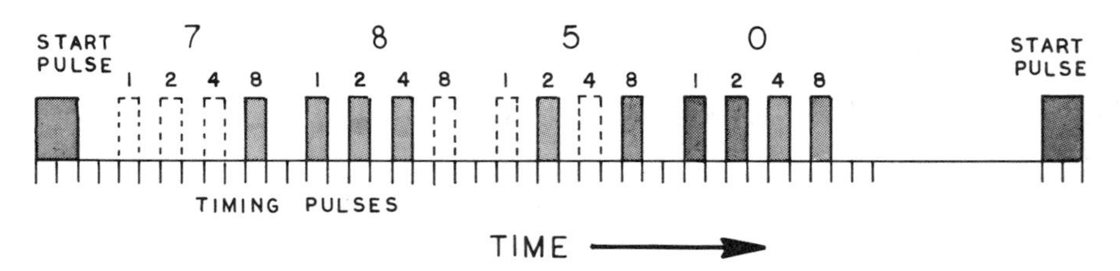


Fig. 12.15. Pulse code modulation: The impedance of a target can be varied according to the dictates of a pulse-code-modulated message.

one coil can be used as both transmitter and receiver, or individual coils can be used to perform each of these functions. In all circumstances, the amount and character of the reflected energy detected by the receiving coil depends on the amount and character of the current induced in the target. This current, of course, is a function of target impedance.

In a given configurational relationship between transmitting/receiving coil assembly and target, the reflected energy changes if the target impedance is changed. A change in target impedance can then be detected by the receiving coil as a change in the magnitude and phase of the reflected energy. In Sec. 12.11, impedance is a function of temperature, whereas in Sec. 12.13 impedance is a function of cross-sectional variation.

If the induced target current is modulated by changing the target impedance according to some time-varying pattern, the receiving coil current will track that modulation by changes in both amplitude and phase. In one concept, a switch is used either to interrupt or to activate a target coil's current by opening and closing according to a pattern of pulse-code modulation. This form of "reflectivity modulation," which determines target-current modulation, can be detected in a receiving coil.

Figure 12.14 illustrates the essential features of a circuit capable of impedance modulation according to some pulse code, whereas Fig. 12.15 schematizes the current induced in a receiving coil as a result of such target-impedance modulation.

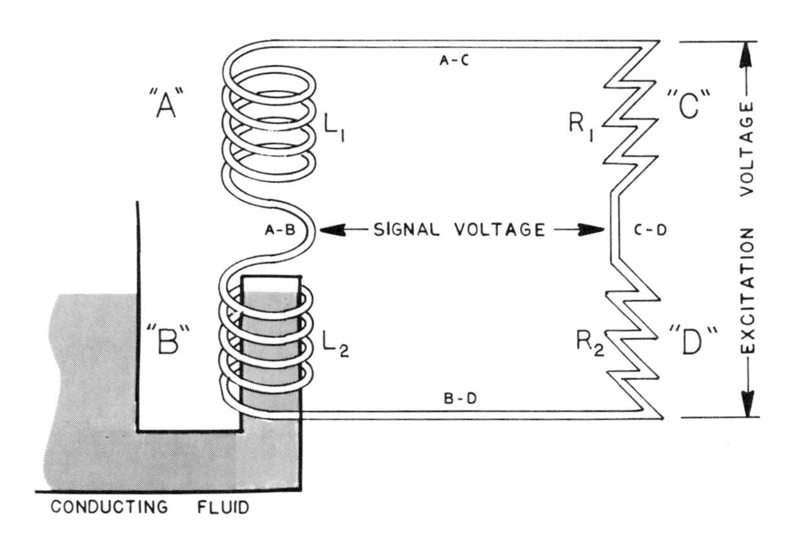


Fig. 12.16. *L/R* bridge circuit: An *L/R* bridge can be unbalanced by the intrusion of a conducting fluid into one of the bridge arms acting as a sensing coil.

## 12.15 BRIDGE CIRCUIT

A bridge circuit provides one means for significantly improving the character of transductive transfer functions. By using transductive elements as bridge components, it is possible to straighten out nonlinear responses, to improve output signal stabilities, to increase sensitivities, and to accomplish zero supressions.

As shown by Fig. 12.16, a bridge circuit consists of four elements connected as a loop by four nodes. Two of the alternate nodes are used to energize the loop. Each of the four elements is chosen to have those characteristics that will produce a common potential across the other two alternate nodes when all elements are in a quiescent state. Or, to look at this configuration from another point of view, a bridge consists of two pairs of elements driven from a common power supply. One member of each pair is connected to the high side of the power supply, whereas the other pair members are connected to the low side. If the elements are labeled A, B, C, and D, A and B constitute one pair and C and D the other. Power is then supplied to nodes A–C and B–D with the bridge output detected between nodes A–B and C–D.

A bridge can have one, two, or four active elements. In a one-element bridge, position B is active. If the ratio A/B is identical to C/D, the voltage detected between A–B and C–D is zero. Therefore, zero supression has been accomplished.

There are two possible two-element bridge configurations. If A and B are both active, linearization is possible. In this case, the transductive element A is made to increase in value as element B decreases. If elements A and B are otherwise equivalent, the nonlinear variations of one are used to compensate for the nonlinear variations of the other. The bridge output signal between nodes A–B and C–D will then follow a reasonably linear traverse. If B and D are both active, stabilizing factors of various kinds can be introduced. For instance, if transductive element B has a significant temperature coefficient, a temperature sensitive element D can be used in compensation.

In a four-element bridge, any function that affects all four elements equally will not affect the bridge output signals. If all transductive elements are the same, and if A and D are made to decrease in value as B and C increase, all bridge benefits will be achieved.

Fig. 12.16 illustrates a possible one-element bridge used to detect the height of a conducting fluid in a surrounding coil. The intrusion of the fluid into the coil unbalances the bridge, and the magnitude of the signal voltage indicates the degree of intrusion.

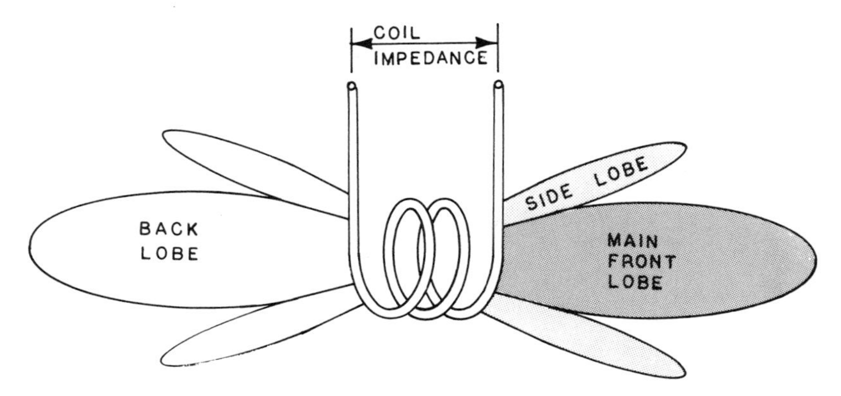


Fig. 12.17. Antenna pattern: The changing magnetic field surrounding all coils (and the electromagnetic fields surrounding all antennas) is indicated by the multiplicity of lobes in a pattern. The number of these and the shape of each depend both on the frequency of the driving energy and the configuration of the coil.

## 12.16 ANTENNA PATTERNS

Coils of wire whose currents support magnetic fields in space function as antennas radiating electromagnetic energy. Although concern may well be with magnetic-field strengths alone at low frequencies, basic principles apply for all configurations over the entire frequency spectrum. As a result, the flux-line representations accompanying the coil sketches in this text, such as those of Figs. 11.4, 12.4, and 14.3, are correct for the one condition where direct current flows through these coils. As the coils are used with alternating currents, the sketches provide only a first approximation of the flux-line distributions.

Coils exposed to time-varying electric currents establish magnetic fields where field strengths are distributed throughout three-dimensional space in much more complex patterns. These patterns depend both on coil configuration and driving-current frequency. A plot of field-strength distribution is called an *ANTENNA PATTERN*. Such a plot consists of a number of field-strength maxima called *LOBES*.

All antenna patterns consist of multiple lobes. Although antennas can be designed so that most of the energy is radiated by one main lobe, some energy will always be found in side lobes and back lobes that have only been reduced to an acceptable minimum. Lobes are characterized by a maximum, a width at the half-power point, a direction, and a phase. It should be remembered that adjacent lobes can be operating in opposite phases. Fig. 12.17 illustrates one possibility.

# 13 RELUCTANCE VARIATIONS

The inductance of a coil is changed whenever the reluctance of the magnetic path associated with that coil is changed. A change in coil inductance is detected as a change in the phase and/or the magnitude of a current flowing in a coil in response to an unchanging driving voltage.

Coil inductance is proportional to the number of flux lines passing through a coil divided by the magnitude of the flux-supporting current flowing in the turns of that coil. Since the presence of magnetic material in flux paths influences flux density, a coil's inductance is affected whenever accompanying flux paths are exposed to magnetic materials. The inductance of a sensing coil thus increases as it approaches a permeable target.

*RELUCTANCE* is a term used to describe those characteristics of flux paths that determine the efficacy of flux passage. The less the reluctance, the greater the flux passage. The inductance of a particular coil is inversely proportional to the reluctance of the paths seen by the flux associated with the coil.

As indicated by Eq. 2.5, if the magnetomotive force driving a magnetic circuit is held constant and if the reluctance of this circuit is changed by any means, the flux through the circuit will change as a function of reluctance change.

Equation 2.6 indicates that reluctance is a function of the length and cross-sectional area of a magnetic circuit and the permeability of the material from which the circuit is constructed. If any one of these three possible variables is altered in any way, the flux through the circuit will change in concert.

Reluctive changes brought about by permeability changes are discussed in other chapters of this book depending on the force that brings about the permeability changes. These include Chemical (3), Hysteretic (4), Magnetostriction (16), Magnetic Resonance (8), and Flux Quantization (25). This chapter covers those means other than permeability changes by which reluctance changes are brought about and detected.

The literature is somewhat confused by the lack of commonly accepted meanings for the terms "variable inductance," "variable reluctance," "variable permeability," and "variable mhu" phenomena. In this discussion, the terms "variable permeability" and "variable mhu" are considered to be synonymous, and relevant details will be discussed elsewhere. The first two terms are defined here as follows:

In *VARIABLE RELUCTANCE* transduction, the magnetic characteristics of a magnetic path are changed by means other than permeability changes. In *VARIABLE INDUCTANCE* transduction, the physical characteristics of a magnetic body are unchanged, but the location of that body within a coil's magnetic field is changed.

## 13.1 FERROMAGNETIC CORES

The minimum possible reluctance for a magnetic circuit is achieved when all of the flux is contained within a ferromagnetic material. If, then, a coil is designed to include a continuous ferromagnetic path linking all turns of a coil, the self-inductance of that coil will be greater than it would with an air-core. Fig. 13.1 illustrates such a design. Under these conditions, the magnitude of coil inductance is determined both by the number of turns of wire and by the reluctance experienced by the flux.

The inductance of a coil with a ferromagnetic core is stated mathematically as follows:

$$L = kN^2pa/l \quad \text{(Eq. 13.1)}$$

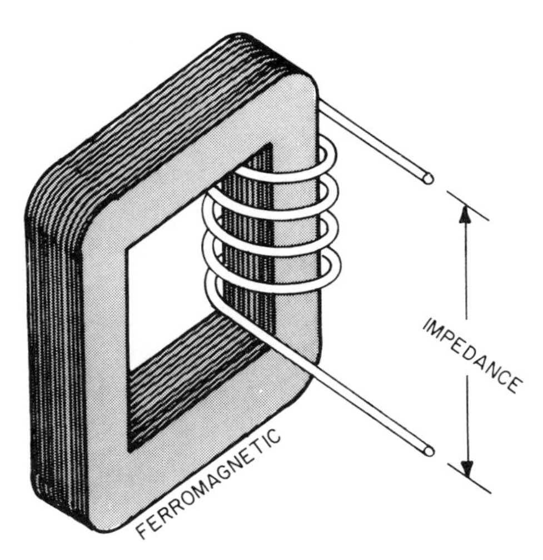


Fig. 13.1. Ferromagnetic inductance: The impedance of a coil is increased significantly by the presence of a body of ferromagnetic material within the coil's magnetic field.

where $L$ is the inductance when $N$ is the number of turns: $a$, the cross-sectional area of the core; $l$, the length of core; $p$, the permeability of the core material; and k, a constant.

Although this equation is mathematically precise, a number of other factors enter into the consideration of core design. For instance, ferromagnetic cores driven with alternating flux will encounter all of the characteristics of the hysteresis loops described in Chapter 4. The incremental permeability of a magnetic material of any kind becomes less as the flux through that material increases. In fact, at saturation, the incremental permeability of a material is the same as the permeability of empty space. In addition, exercising hysteresis loops with alternating flux generates losses that are defined by the area enclosed within a loop. The minimization of these losses requires the use of materials with the minimum possible coercivity.

Further losses are derived from the eddy currents induced in core materials. These increase as the frequency of flux alternations increase. At low frequencies, these losses are minimized by using ferromagnetic materials of high resistivity and by laminating cores as shown in Fig. 13.1; at medium frequencies, by powdering cores; at high frequencies, by using nonconducting ferrites; and at still higher frequencies, by eliminating ferromagnetic materials from cores entirely.

Magnetostrictive losses are minimized by using materials with minimum magnetostrictive coefficients.

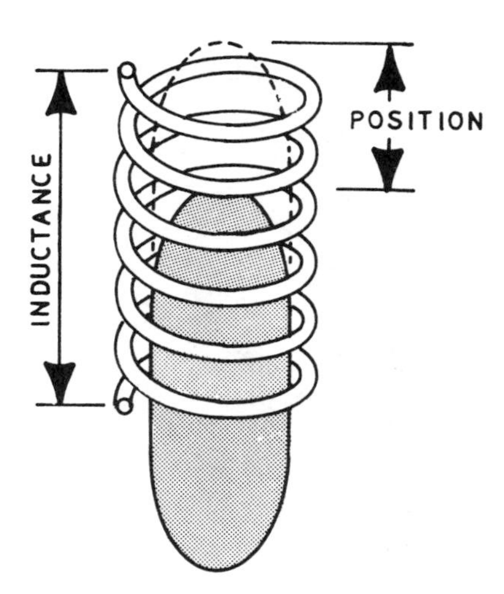


Fig. 13.2. Variable inductance: The inductance of a coil of wire depends on the position of a ferromagnetic body within that coil's magnetic field.

## 13.2 VARIABLE INDUCTANCE

In Fig. 13.2, a slug of ferromagnetic material is moved into or out of a coil's magnetic field. Under these circumstances, the inductance of the coil is maximum when the slug is centered in the coil and minimum when the slug is removed from the coil's vicinity. The position of this slug can then be deduced by measuring coil inductance, coil current, or the phase angle between driving voltage and coil current.

Fig. 13.3 is a vector diagram that illustrates the electric effects encountered in a resistive-inductive circuit when the inductance is changed without changing any other circuit factor. It schematizes electrically the

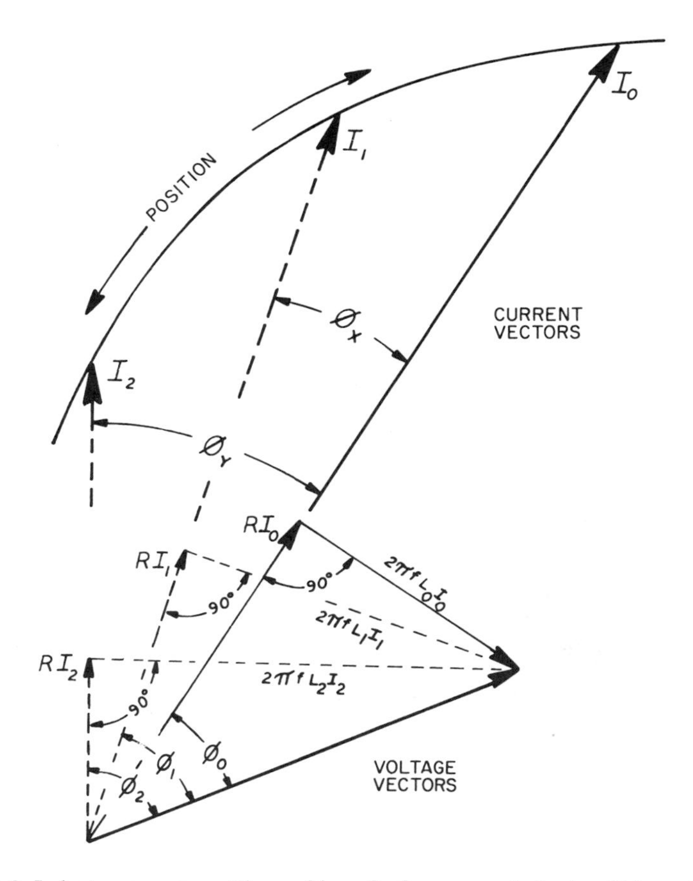


Fig. 13.3. Inductance vectors: The position of a ferromagnetic body within an illuminating coil's magnetic field can be tracked by noting both the magnitude and the phase angle of the current flowing in that coil.

physical circumstances of Fig. 13.2. Here the voltage across the resistive and inductive elements are always at right angles to each other and always add to duplicate the driving voltage. In addition, the current flowing in this circuit is always in phase with the voltage drop across the resistive element.

As shown, if the inductance is increased, the inductive reactance is increased, the product of current times inductive reactance is also increased, but the product of current times resistance is decreased. When the resistance is held constant, a decrease in the product of resistance and current indicates a decrease in current. In addition, the current flowing through such a circuit is always in-phase with the voltage across the resistive element. As a consequence, the position of the slug of Fig. 13.2 can be deduced by calibrating the path traversed by the tip of the current vector of Fig. 13.3 as this vector changes in phase and magnitude.

## 13.3 VARIABLE RELUCTANCE

From an electric standpoint, variable inductance and variable reluctance are the same phenomenon. In both, the inductance of a coil is changed

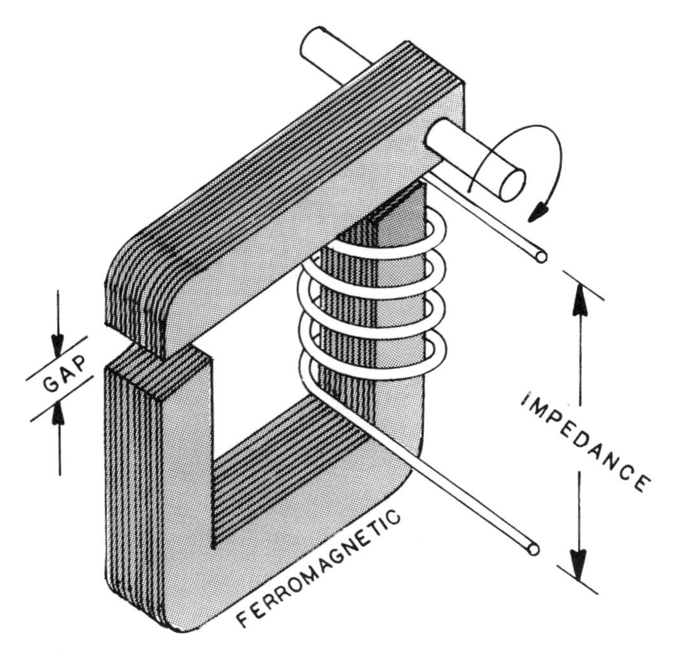


Fig. 13.4. Variable reluctance: In an association of magnetic and electric circuits, the impedance of a coil can be related to the reluctance of the former.

by changing the reluctance of the path seen by that coil's magnetic field. The vector diagram of Fig. 13.3 applies equally well, therefore, to both effects.

From a magnetic standpoint, however, these phenomena are two different means of accomplishing a change in path reluctance. In one, a magnetic path of constant reluctance is moved into or out of a coil's magnetic field. In the other, the reluctance of a magnetic path is somehow altered.

For instance, as shown in Fig. 13.4, if an otherwise continuous ferromagnetic path contains an air gap, the reluctance of that path can be a sensitive function of the dimensions of that gap. If one element of a magnetic circuit can be moved to open or close such a gap, the self-inductance of a ferromagnetically cored coil is a function of the gap length. This gap length, or the position of the movable magnetic element, can then be inferred by a measurement of coil inductance, coil current, or the phase angle between coil voltage and coil current. As before, the gap length can be calibrated in terms of the trace of the tip of the current vector shown in Fig. 13.3.

## 13.4 TWO COIL LINKAGE

Figure 13.5 shows expansion of the concept illustrated in Fig. 11.1. In this sketch, two electrically isolated but magnetically linked coils experience an interaction, known as *MUTUAL INDUCTANCE*, in which the voltage generated in a secondary coil as a result of changes in the current of a primary coil is determined as follows:

$$E_s = -M dI_p/dt \qquad \text{(Eq. 13.2)}$$

where $E_s$ is the voltage generated in one coil (the secondary) by the time rate of change of current $dI_p/dt$ in another coil (the primary); $I_p$, the primary current; and $M$, the *COEFFICIENT OF MUTUAL INDUCTANCE*.

## 13.5 TRANSFORMER

As an expansion of the concept illustrated in Fig. 11.2, the voltage generated in the secondary coil of Fig. 13.5 can be used to drive a secondary current. Such a two-coil arrangement is commonly called a *TRANSFORMER*, in which energy is transfered from a primary coil to a secondary coil by means of an interaction between two magnetic fields. The greater the flux linking two coils, the more efficient the transfer of energy. If the two coils are linked by a ferromagnetic circuit containing most, if not all, of the flux, the transfer will be maximized.

Figure 13.6 displays the essential features of a ferromagnetically cored

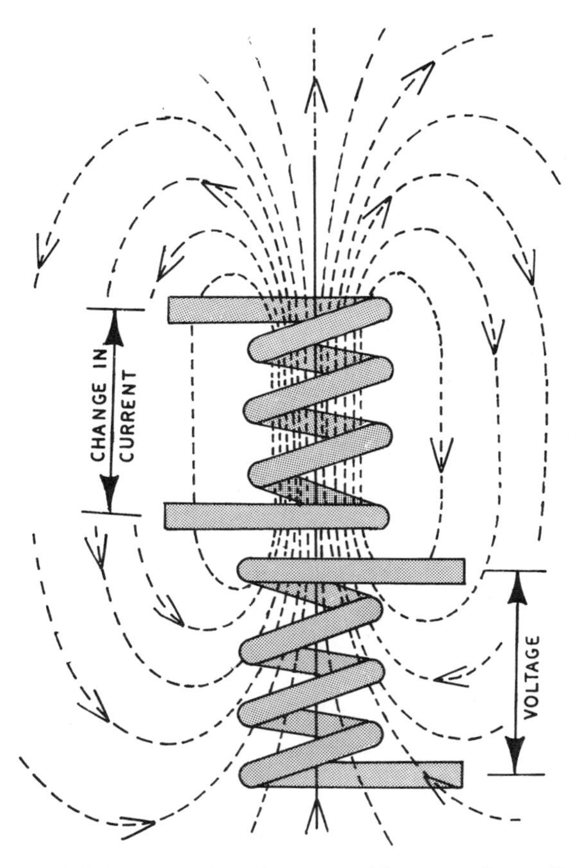


Fig. 13.5. Two-coil linkage: A voltage is generated in a secondary coil when the current is changed in a primary coil.

transformer. The ratio of energy consumed in the secondary coil to the energy expended in the primary coil in support of that consumption is known as the *TRANSFORMER EFFICIENCY*, as follows:

$$T_e = P_s/P_p \qquad \text{(Eq. 13.3)}$$

where $T_e$ is the transfer efficiency; $P_p$ the energy, or power, expended in the primary; and $P_s$, the energy consumed in the secondary.

In the process of using a transformer to transfer power, the secondary voltage may be made either greater than, equal to, or less than the primary voltage. In other words, energy is both transferred and transformed. If the transfer efficiency is very near to 100 percent, the primary and sec-

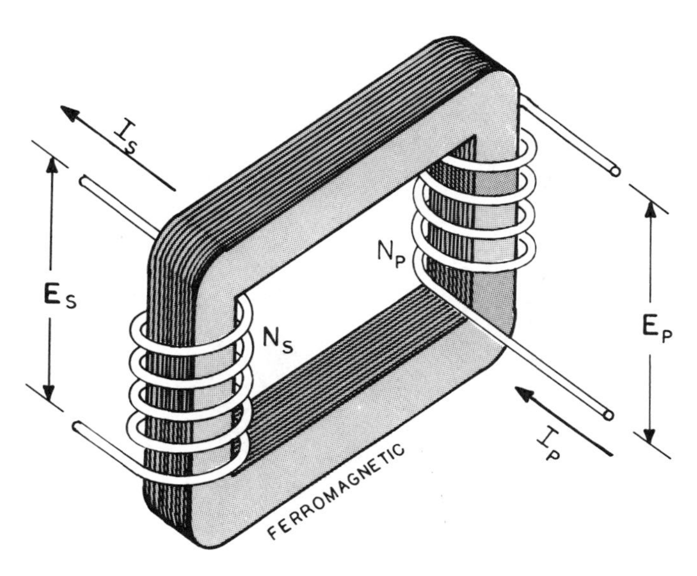


Fig. 13.6. Transformer: When two coils are linked by one ferromagnetic circuit, the energy supplied at one voltage level is reproduced in the secondary at some other voltage level depending on the efficiency of transference.

ondary voltages are related by the following equation:

$$E_s = E_p\ (N_S/N_p) \qquad \text{(Eq. 13.4)}$$

where $E_s$ is the secondary (transformed) voltage; $E_p$, the primary voltage; $N_s$, the number of secondary turns; $N_p$, the number of primary turns.

If the transfer efficiency is less than 100 percent, $E_s$ will have a value less than this equation will indicate.

## 13.6 CURRENT TRANSFORMER

Equation 13.2 indicates a direct proportionality between the secondary voltage of a transformer and the alternating primary current. The magnitude of this current can thus be deduced from a measurement of the secondary voltage.

The simplest possible current transformer consists of a coil of wire wrapped around a conductor carrying the primary current. The voltage generated in this coil is calibrated in terms of the primary current. The efficiency of such a device is enhanced by encompassing the conductor

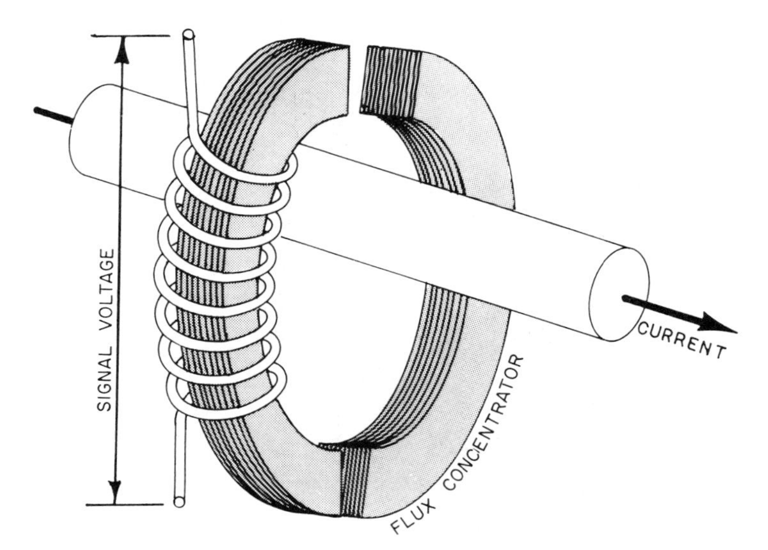


Fig. 13.7. Current transformer: The Biot-Savart field surrounding the changing current in a linear conductor can be used to generate a voltage in a detector coil that shares a magnetic circuit with that field.

with a ferromagnetic circuit, or *FLUX CONCENTRATOR*, and wrapping the secondary around the magnetic circuit instead of around the primary conductor directly.

Figure 13.7 illustrates a "clip-on" type of current transformer in which the magnetic circuit is divided into two parts that can be separated, placed around the primary conductor, and recombined for use.

## 13.7 DIRECTIONAL DETERMINATION

The transductive effects that occur in Figs. 13.2 and 13.4 generate electric signals whose linearity decreases markedly with increasing magnitudes. In fact, this nonlinearity is so pronounced that useful signals can be obtained only over a short range of possible values. The results include small signal currents that are difficult to interpret in the presence of large zero-base currents.

The combination of two more or less identical elements makes it possible to minimize zero-base effects. In Fig. 13.8, when a driving voltage is applied to points 1 and 3, the voltage between points 1 and 2 is the same as the voltage between points 2 and 3 as long as the magnetic characteristics of gap B are the same as those of gap A.

Changes in the A dimension can then be detected as the difference between voltages 1–2 and 2–3. In this configuration, gap B can be adjusted

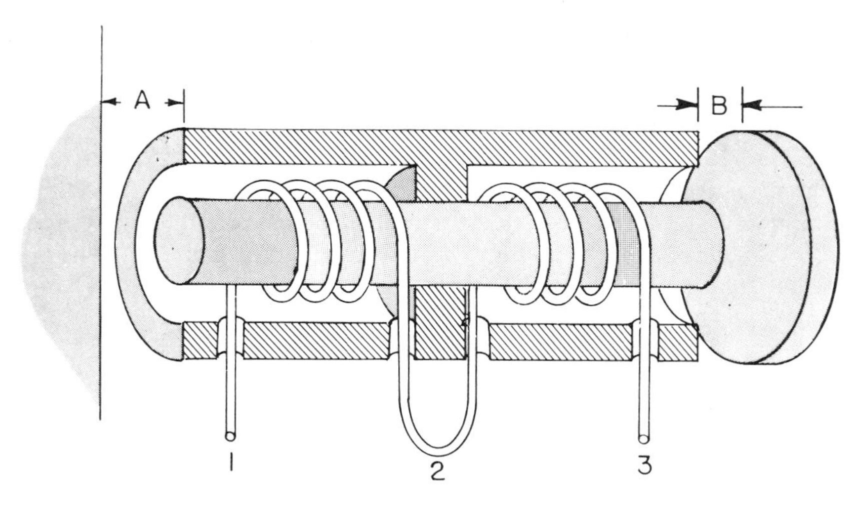


Fig. 13.8. Reluctance reference: A reluctance bridge can be brought more or less into balance by varying the known *B* dimension against the unknown *A* dimension.

to the mean of the variations of the desired gap A. The signal voltage is then a plus-minus variation about the null of the desired gap A.

## 13.8 NONLINEARITY COMPENSATION

In Fig. 13.8, gap B is held constant while gap A changes in response to a variable of interest. If gap B can be made to decrease by the same force

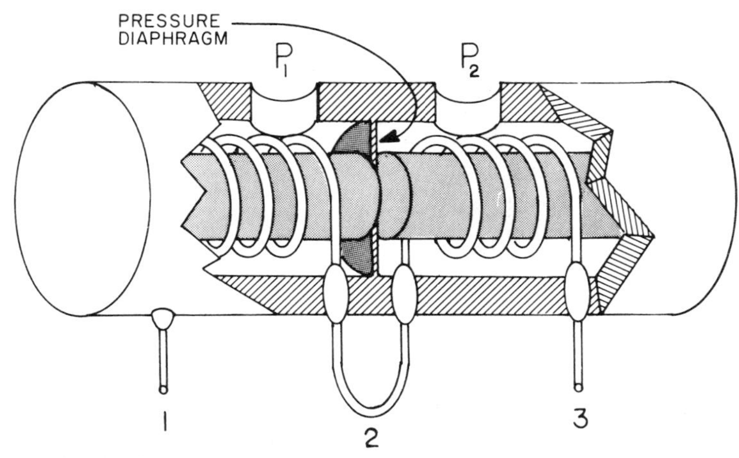


Fig. 13.9. Push-pull reluctance: As the position of a ferromagnetic pressure diaphragm moves in one direction or the other, it increases the reluctance seen by one bridge arm as it decreases that seen by the other.

that increases gap A (and vice versa), not only will the output signal be responsive to directional changes, but the useful range of the device will be extended as the linearity of the transfer function is significantly improved. The nonlinear deviations of A compensate for the nonlinear deviations of B.

Figure 13.9 illustrates a differential pressure transducer in which the position of a pressure sensitive diaphragm establishes both the gap A and gap B dimensions as well as the mechanism by which gap B decreases as gap A increases.

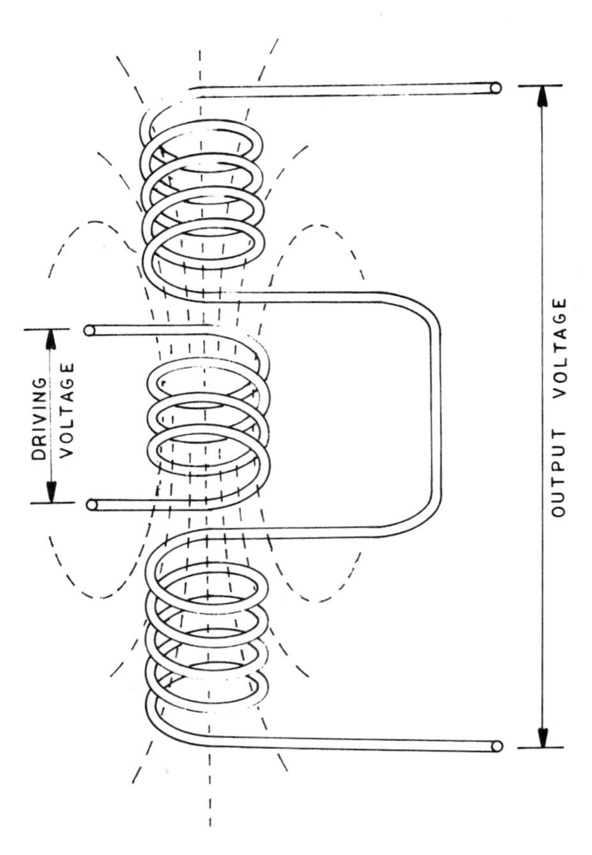


Fig. 13.10. Differential transformer: A single driving coil can be used to energize two out-of-phase secondary coils. As long as the two secondary circuits are completely symmetrical, there is no secondary voltage in response to any primary voltage.

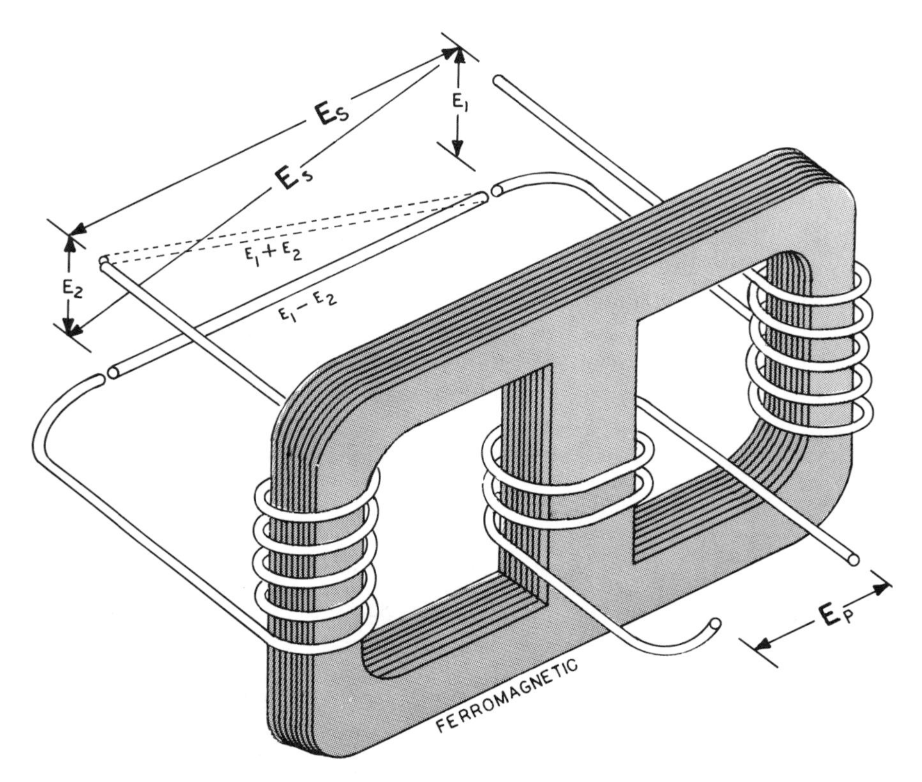


Fig. 13.11. Ferromagnetic differential transformer: When one primary coil is used to drive two symmetrical secondary coils, the two secondary voltages can be configured either to add or to subtract.

## 13.9 DIFFERENTIAL TRANSFORMER

An alternating current flowing in one primary coil can be used to induce alternating voltages in two secondary coils. If the two secondary coils are identical, and if the two magnetic paths about which they are wound are also identical, the two secondary voltages will be the same. A device configured in this manner is called a *DIFFERENTIAL TRANSFORMER*. A differential transformer may have either an air core, as shown in Fig. 13.10, or a magnetic core, as shown in Fig. 13.11.

As shown by Fig. 13.11, the two secondary windings can be connected either in-phase, in which case their two voltages are added to each other, or out-of-phase, in which case their two voltages are subtracted from each other. If the two windings are connected to subtract, and if the two voltages are actually identical, the combined secondary voltage will be zero. On the other hand, if the magnetic characteristics of one magnetic circuit in relation to the other magnetic circuit are deliberately altered, the two secondary voltages will no longer be equal, and their difference will not be zero.

Under the latter circumstances, the phase of the combined secondary voltage will indicate which magnetic path has the greater reluctance, whereas the magnitude of the combined secondary voltage will indicate the degree of the reluctance difference.

If one force is used both to increase the reluctance of one path and decrease the reluctance of the other path, the output voltage reflecting this force will be maximized, and the transfer function will be at its most linear.

Since neither of the two secondary windings nor of the two magnetic

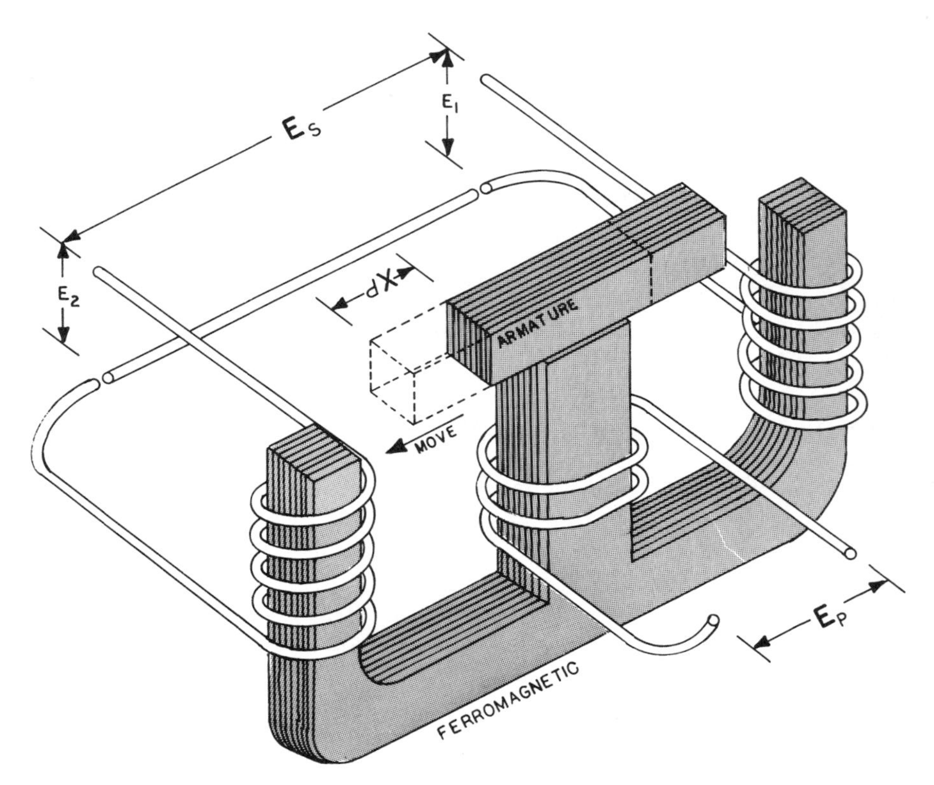


Fig. 13.12. Variable reluctance differential transformer: The couplings between one primary coil and two symmetrical secondary coils can be varied by varying the reluctances in the two magnetic circuits that couple primary to secondaries.

paths can be made exactly the same, a differential transformer always has some output voltage under zero stimulus. In addition, the magnetic characteristics of the magnetic circuits are nonlinear. Such nonlinearity creates even harmonics in the fundamental driving frequency that are not reducible to zero by any secondary-winding arrangement.

## 13.10 VARIABLE RELUCTANCE TRANSFORMER

The reluctance of a ferromagnetic circuit containing an air gap is a very nonlinear function of the length of that gap. As a consequence, the self-inductance of the coil (described in Fig. 13.4) that is wrapped around such a circuit is also a nonlinear function of gap length. On the other hand, if two more or less identical magnetic paths are provided, each with an air gap, and if the length of one gap is increased while that of the other is decreased, the difference in reluctance between the two paths can be made reasonably linear.

Figure 13.12 shows a differential transformer with an air gap in each arm. A movable armature opens one gap as it closes the other. The absolute difference in voltage between the two secondary voltages is a measure of how far the armature has been moved from a central, or null, position, and the phase of the difference voltage is an indication of movement direction.

## 13.11 TORSIONAL DIFFERENTIAL TRANSFORMER

The principles applicable to differential transformer designs are utilized in a number of different configurations for a number of different purposes. The possibilities are too numerous to list here. Figure 13.13 is included merely as one example of the many possibilities. It shows a configuration capable of measuring the torsional force on a drive shaft.

In this device, three cylinders fabricated from a magnetic material are mounted on a common nonmagnetic shaft. One cylinder is located at the center of the shaft, whereas the other two are threaded on the ends of the shaft. One end cylinder is subject to a left-hand thread and the other to a right-hand thread. Each threaded cylinder is spring-loaded to establish a zero-load position in which air gap 1 is magnetically equivalent to air gap 2.

If this assembly is subjected to torsional stress, one end cylinder will *unwind* against its spring, thereby expanding its associated air gap, while

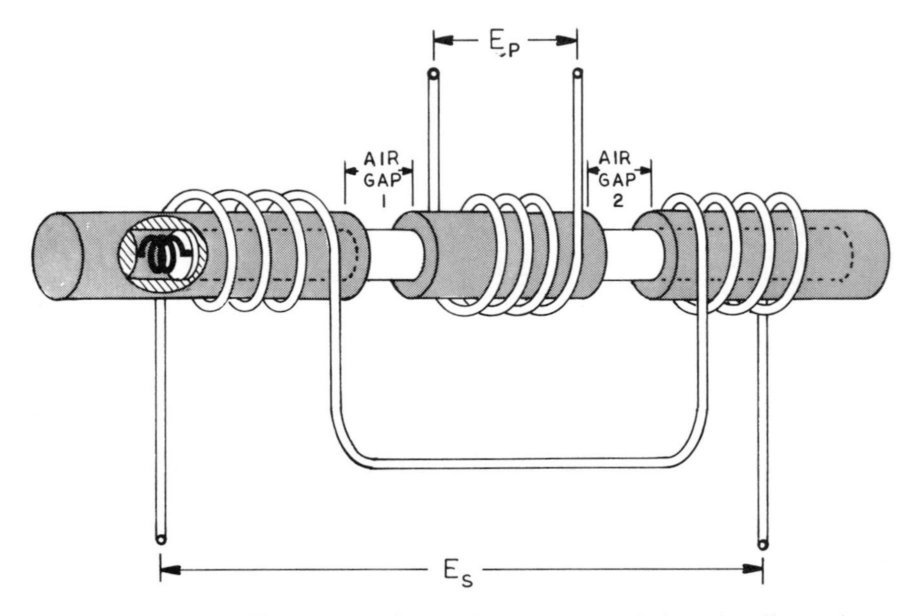


Fig. 13.13. Torsional differential transformer: The torque on a shaft can be offset against a spring force that controls the air gaps in two ferromagnetic circuits linking a primary coil with two symmetrical secondary coils.

the other will *wind* against its spring, thereby contracting the second air gap. Differential transformer windings around all three cylinders detect the changes in air gap dimensions.

## 13.12 SHADED POLE EFFECT

If a magnetic circuit is split into two parallel paths of equal length and equal area, and if a coil of very low impedance is placed around one of the two paths, both the magnitude and the phase of the flux in one path are altered in relation to those in the other path. This alteration is derived from the effects of the very large circulating currents induced in the shorted coil.

Referring to Fig. 13.14, if the shorted single turn is placed at the central position of the air gap, the same flux will flow in each arm of the differential transformer, and the two secondary windings may be connected in phase opposition to give a zero output signal. If the shorted turn is then slipped linearly in one direction, the flux will decrease in the corresponding arm and increase in the other arm. The phase and amplitude of the summed output signal will then indicate the direction and distance of the movement in a manner analogous to that described in Sec. 13.10.

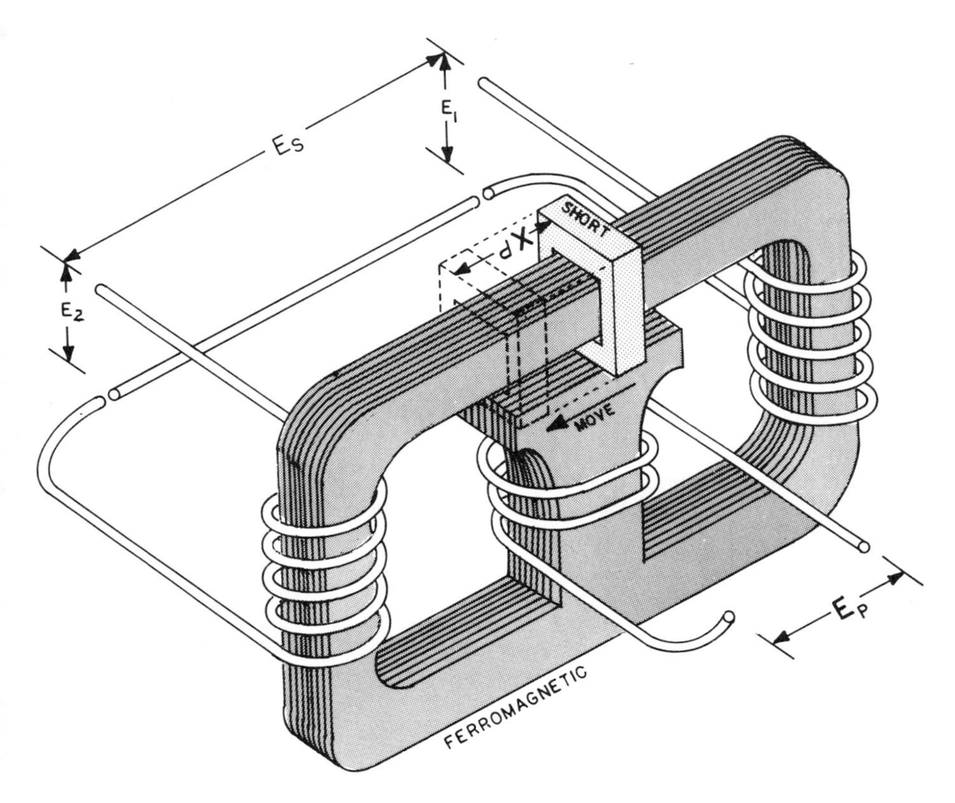


Fig. 13.14. Shaded pole effect: The effects of eddy currents induced in a shorted ring are added to the effects of current flowing in a primary coil, and the two of these together drive two symmetrical secondary coils through two symmetrical magnetic circuits. The effects of the eddy current in the shorted ring are distributed between the two secondary coils in a nonsymmetrical fashion by varying the ring's location ($dx$) in relation to the two otherwise symmetrical magnetic circuits.

## 13.13 LINEAR DIFFERENTIAL TRANSFORMER

The inductance of a coil containing a ferromagnetic slug is a very nonlinear function of the position of that slug within, or adjacent to, that coil. As a consequence, the self-inductance of the coil described in Sec. 13.2 is also a nonlinear function of the position of the slug it contains. On the other hand, if such a slug comprises the ferromagnetic circuit of a differential transformer, the secondary difference voltage can be a reasonably linear indication of slug position.

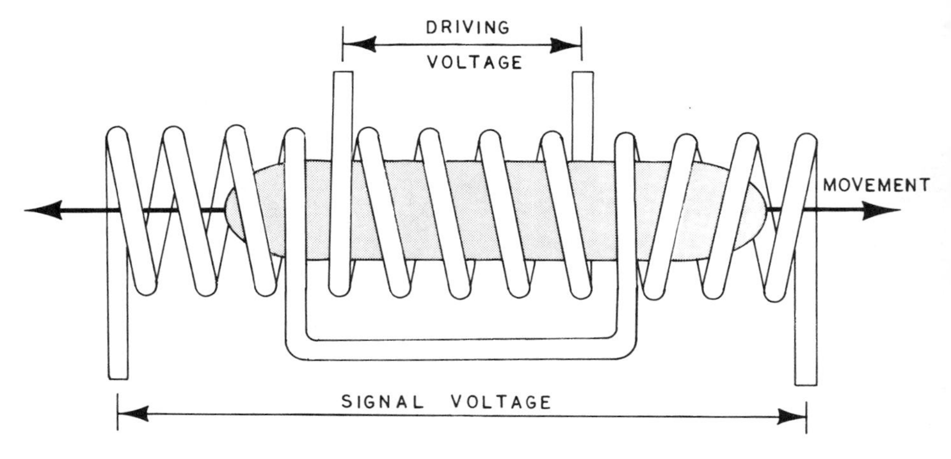


Fig. 13.15. Linear differential transformer: The position of a symmetrical ferromagnetic core mounted in a differential transformer configuration can be deduced by noting the phase and magnitude of the secondary voltage.

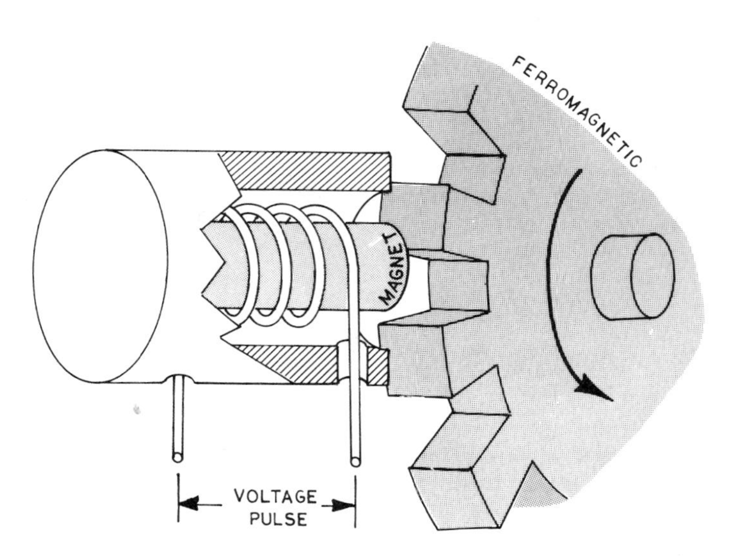


Fig. 13.16. Dynamic variable reluctance: The velocity of a shaft can be measured by the rate of pulse generation in a pick-up coil as the reluctance of the path seen by that coil is varied.

Figure 13.15 shows a differential transformer with a ferromagnetic slug that extends equally into each of two secondary windings. A movement of this slug increases its extension into one secondary coil and decreases its extension into the other. The absolute difference in voltage between the two secondary voltages is then a measure of how far the slug has been moved from a central, or null, position, and the phase of the difference voltage is an indication of movement direction.

## 13.14 DYNAMIC VARIABLE RELUCTANCE

A permanent magnet can be used to drive a ferromagnetic circuit interrupted by an air gap. If the length of this air gap is changed, the changing reluctance of the magnetic circuit will cause the flux through the circuit to change. Such air-gap dimensional changes modify the loading experienced by a permanent magnet, with two effects. In the first, a change in loading directly affects the flux density in the magnetic circuit. In the second, a change in loading changes the operating point of the magnet, which further changes the flux density.

If the ferromagnetic core encloses a coil, a voltage will be generated in this coil while the flux is changing. The voltage so generated is a direct function of the time rate of change of the flux.

Time rate of change is the first derivative of change. The change encountered in this case is a change in the length of the air gap. Since a change in length as a function of time is a mathematical expression of velocity, transducers utilizing this principle are commonly called *velocity transducers*.

In the particular configuration of Fig. 13.16, a voltage pulse is generated each time the passage of a gear tooth changes the air gap between the permanent magnet and the ferromagnetic core.

## 13.15 DYNAMIC RELUCTANCE BRIDGE

Figure 13.17 illustrates an assembly in which the air-gap dimensional changes $dx$ are accomplished by rotating an armature mounted on a pivot. The voltage generated in a coil wrapped around this armature is proportional to the angular rate-of-change, $d\Theta/dt$.

## 13.16 CHOKE

Inductors impede the flow of time-varying electric currents. Although they provide a minimum of impedance for constant-flow currents, their imped-

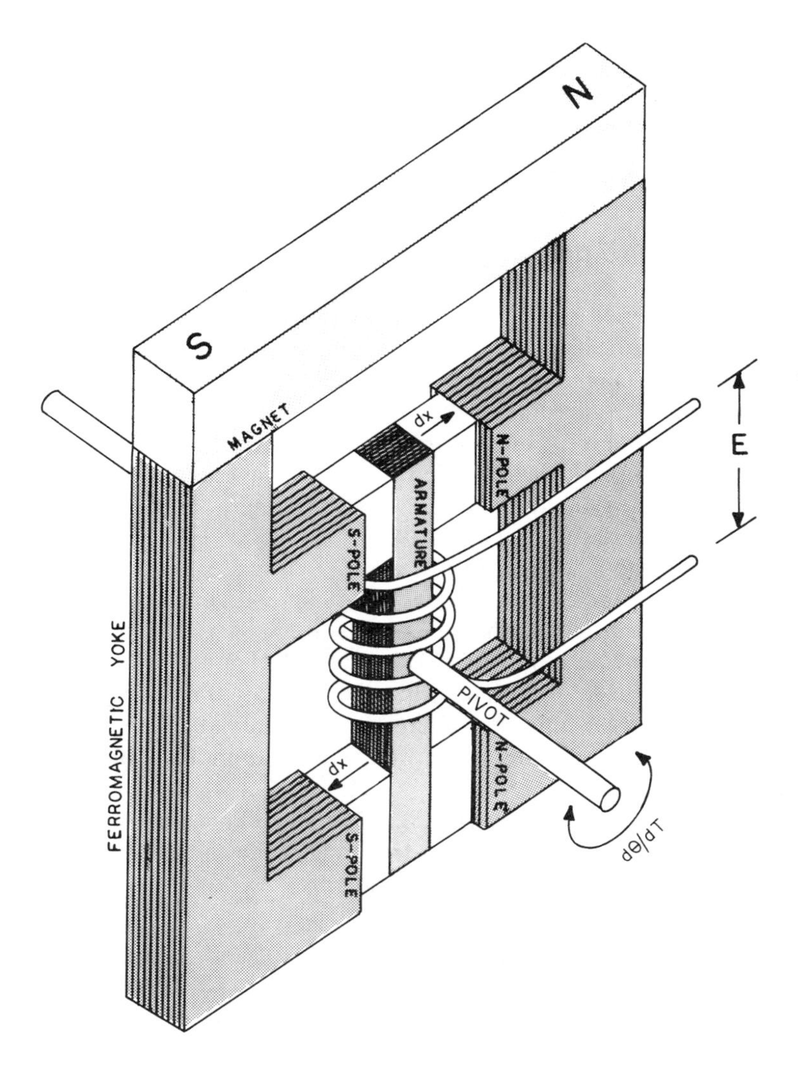


Fig. 13.17. Dynamic reluctance bridge: No flux passes through the pick-up coil when the armature is in the null position. As the armature is oscillated, the reluctance of the two paths driving it are alternately increased and decreased, thus driving flux through it (and hence the coil) in an oscillating mode.

ance increases with time rate of change as expressed by the frequency in Eq. 11.5.

Inductors can be either air-core, as shown in Fig. 11.14, or cored with a ferromagnetic material, as shown in Fig. 13.1. As indicated by Eq. 13.2, ferromagnetic cores enhance a conductor's ability to impede current flow at low frequencies, but eddy-current, hysteretic, and magnetostrictive losses preclude their use at high frequencies.

An inductor used to pass direct current but block alternating currents is commonly called a *CHOKE*. A capacitor is thus a device that operates in a mode inverse to that of an inductor since it passes alternating currents but blocks the flow of direct current.

# 14
# COMPOSITE TARGETS

When a sensing coil is exposed to a conducting, ferromagnetic target, the sensing current responds both to the increased self-inductance caused by the presence of ferromagnetic material and to the energy reflected back from induced eddy currents. The general circumstances of composite targets are shown in Fig. 14.1. The current that flows in response to a composite target can be determined by combining the reflectivity phenomena of Chap. 12 with the reluctance phenomena of Chap. 13. (This combination is schematized in Fig. 14.2 as the addition of the vector diagrams of Figs. 12.2 and 13.3.)

## 14.1 SENSING-COIL RESPONSE

The various circumstances that might be encountered when a sensing coil is exposed to targets of significantly different electric/magnetic characteristics include the following:

*Permeable Targets with Low Conductivities*:

1. Ferrites combine high permeability with low conductivity.
2. A ferrite's low conductivity minimizes the induction of eddy currents to a point at which the effects of reflection can be ignored.
3. A ferrite's high permeability makes it possible for the same coil current to drive more magnetic flux. This increase in the flux/current ratio increases a sensing coil's self-inductance.
4. When a sensing coil approaches such a target, the coil's self-inductance increases, its sensing current decreases, and the changed sensing current lags the original sensing current by an increasing angle.
5. The vector diagram of Fig. 13.3 indicates these changes.
6. The information available from these measurements is limited to the fact that the sensing current is both diminished and lagging. Both these changes are affected by target size and target proximity.

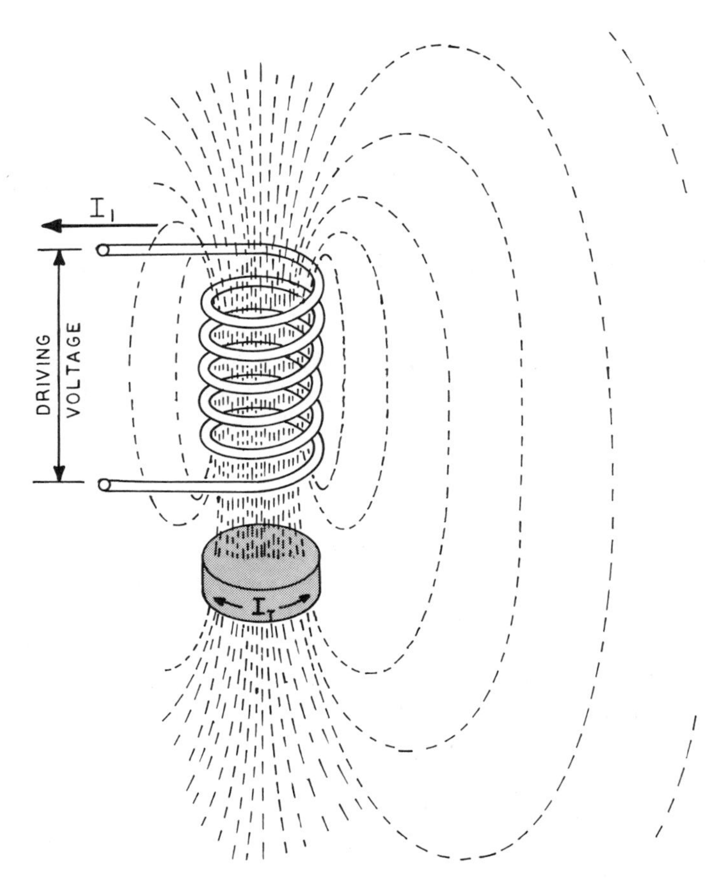


Fig. 14.1. Composite targets: The impedance of a coil is changed both by the presence of magnetic dipoles in a target and by the eddy currents that are induced to flow in that target.

*Conductive Targets with Low Permeabilities*:

1. Nonferromagnetic conductors combine high conductivity with low permeability.
2. Eddy currents are induced in bodies constructed from nonferromagnetic materials. These eddy currents reflect signal currents back into sensing coils. For a given sensing coil, these signal currents always lead the original sensing current.
3. The vector diagrams of Figs. 12.2 and 12.11 indicate the variables involved.

*Inductive Targets with Low Permeabilities*:

1. Coils of wire fabricated from nonferromagnetic, conductive materials combine high inductivity with low permeability.
2. Eddy currents are induced in target coils of inductance $L$ and resistance $R$. For a geven sensing coil, the reflected signal current always leads the original sensing current by an angle derived from the ratio $2\pi fl/R$. The vector diagrams of Fig. 12.2 apply to targets of this nature.
3. The information available from these measurements includes the fact of a changed and leading sensing current and a characteristic "trace" of the signal current vector (see Figs. 12.5 and 14.2). The latter can be used to interpret the target's electric characteristics.

*Resonant Circuits*:

1. This condition is represented by a circuit consisting of an inductance $L$, a capacitance $C$, and a resistance $R$ connected in series. Currents in the sensing coils induce voltages in the inductance. Currents then flow in all circuit elements in response to the induced voltages.
2. At resonant frequency, the induced current is in phase with the induced voltage; below resonant frequency, the current lags the induced voltage; above resonant frequency, the induced current leads the induced voltage.
3. As a sensing coil approaches such a target, the changed sensing current can either lag or lead the original sensing current, and the magnitude of the changed current can either increase or decrease, depending on the frequency of the driving current in relation to the resonant frequency of the target.

*Capacitive Targets with Low Permeabilities*:

1. Tuned circuits that operate well below resonant frequency and whose coils are fabricated from nonferromagnetic materials combine high capacitive reactance with low permeability.

*Targets That Are Both Permeable and Conductive*:

1. Ferromagnetic conductors combine high permeability with high conductivity.
2. The availability of permeable paths increases a sensing coil's self-inductance.

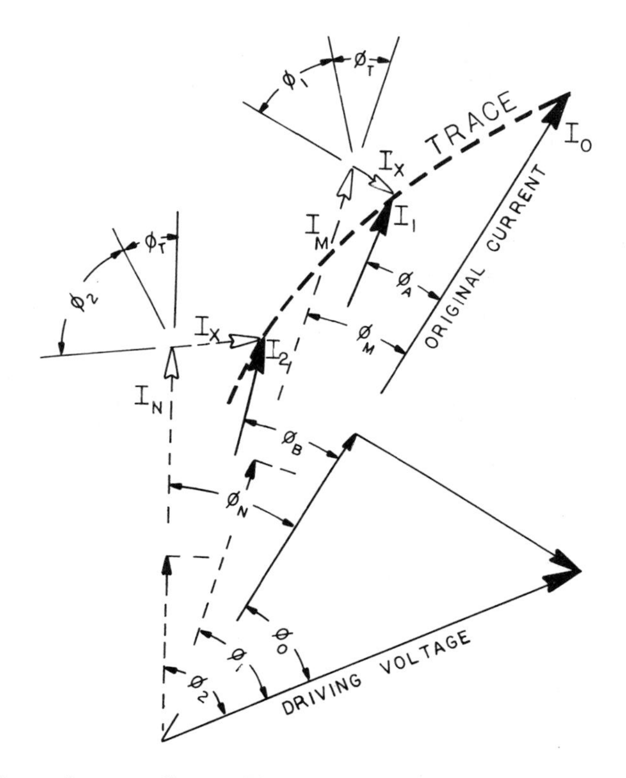


Fig. 14.2. Composite vector diagram: The phase angle between voltage and current is modified by both magnetons and eddy currents in composite targets.

3. Induced eddy currents reflect signal currents back into sensing coils.
4. A sensing coil exposed to such targets combines the effects of permeability with the effects of the energy reflected from eddy currents. This combination is illustrated by the vector diagram of Fig. 14.2.

## 14.2 TRANSMIT-RECEIVE CONFIGURATION

In a crossed field, or transmit–receive arrangement, a transmitting coil is used to illuminate a target and a receiving coil detects the energy reflected back from that target.

As outlined by Eq. 13.2, two coils share a mutual inductance when the changing magnetic field of one coil, associated with its current changes, is intercepted by the other coil. The stronger the shared field, the greater the mutual inductance.

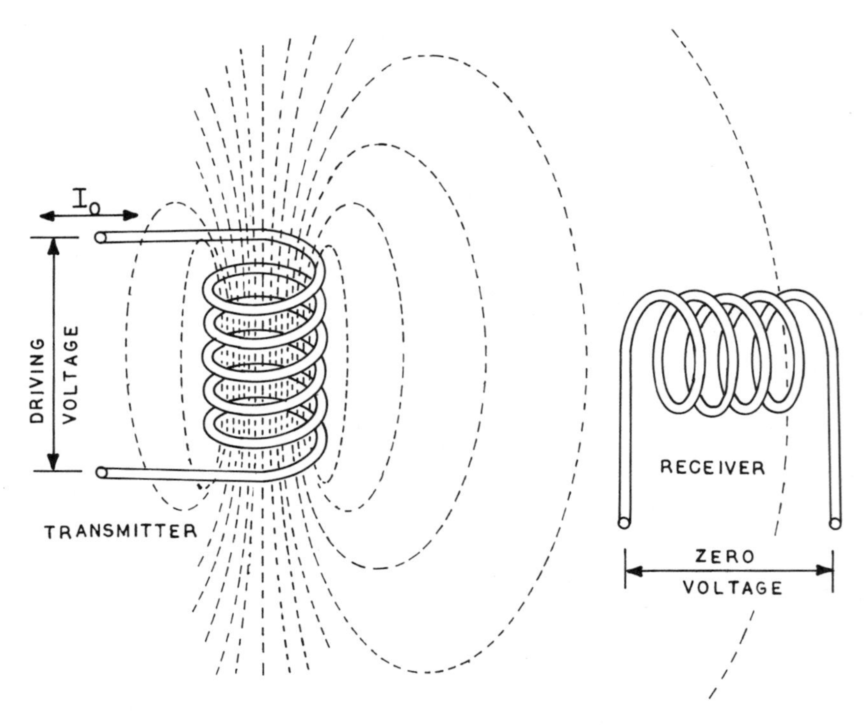


Fig. 14.3. T/R configuration: Two coils can be configured in such a way that no energy is transferred between transmitting and receiving coils.

Since magnetic fields are characterized by direction as well as by strength, mutual inductance is a function of relative coil orientation and location as well as distance of separation. In fact, as shown by Fig. 12.13, mutual inductance is zero if a receiving coil is oriented with the axis of its turns at right angles to the direction of the field produced by a transmitting coil.

Because of the curvature of toroidal magnetic fields, zero mutual inductance cannot be achieved if two coils are physically close or if even one of the two is wound with excessive axial length. As shown in Fig. 14.3, however, if both coils are relatively flat, separated by a considerable distance, and oriented with their axes at right angles to each other, with the axis of one lying in the plane of the other's windings, zero mutual inductance can be approached.

As shown in Fig. 14.4, the conditions of zero energy interchange are altered if either a permeable or conducting object enters the space shared by the field patterns of both coils. If the intruding object is permeable, the field pattern surrounding a receiving coil is distorted, and energy passes

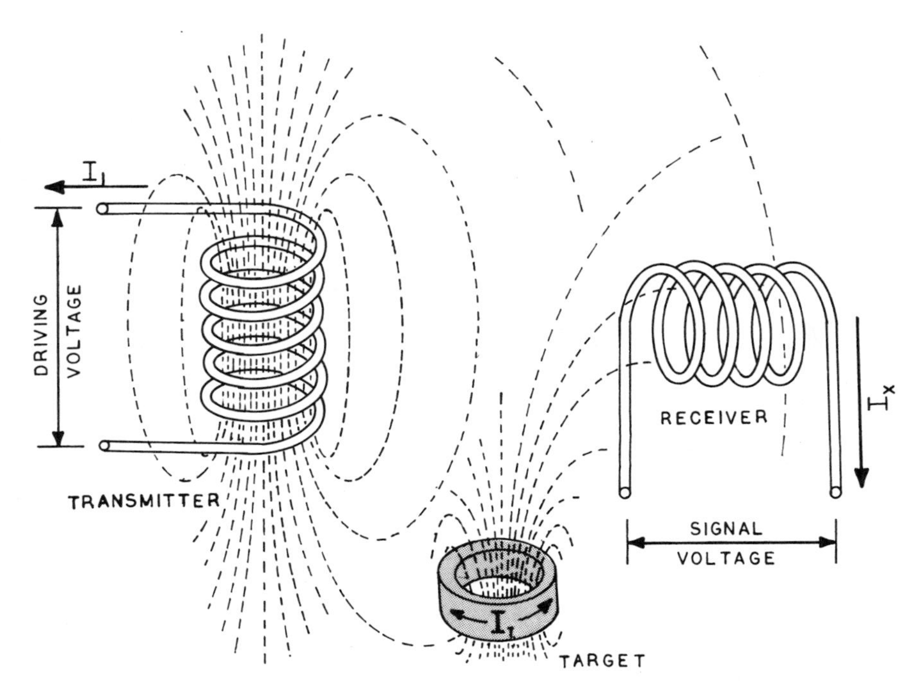


Fig. 14.4. T/R configuration with target: When a target intrudes into the changing magnetic field of a transmitting coil, energy is passed from transmitter to target and from target to receiver coil.

directly from the transmitter coil to the receiver coil through the distorted field. The vector diagram of Fig. 13.3 illustrates these circumstances. If a target is conducting, the induced eddy currents create fields that interact with those of the receiver coil, and energy is transferred from the transmitter coil to the eddy currents and thence to the receiver coil. In other words, transmitter coils illuminate targets that reflect energy into receiver coils.

The vector diagram of Fig. 12.2 shows the voltage vectors involved in this process. Here the signal current $I_x$ is induced directly into a receiving coil in contrast to being induced back into a sensing coil as it is in a one-coil system. It is then possible to monitor the signal current directly rather than indirectly as required in a one-coil system.

Because of the relatively large amount of power that can be generated by the transmitting coil without affecting the receiving coil, the transmit–receive configuration is commonly used for the detection of distant targets.

In the configuration shown in Fig. 14.5 the rotation of a target affects

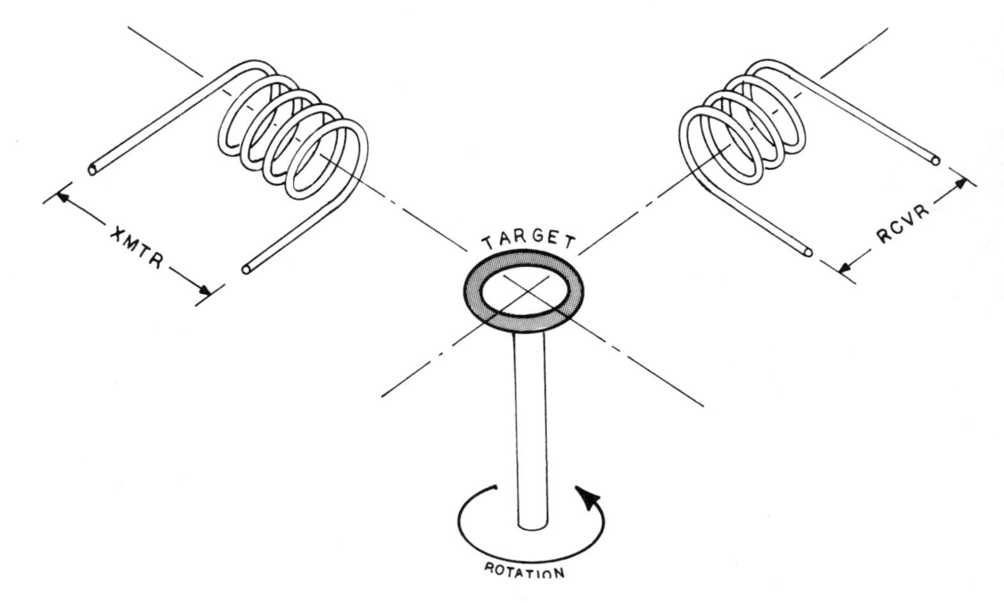


Fig. 14.5. T/R configuration with rotating target: When a rotatable target intrudes into the field of a transmitting coil, the amount of energy passing from transmitter to target and from target to receiver depends on the orientation of the target in relation to the transmitter and receiver coil configuration.

the energy transferred between transmitter coil and receiver coil. The energy detected in the receiver coil can be calibrated in terms of degrees of rotation.

## 14.3 INDUCTION BALANCE

In an *INDUCTION BALANCE*, an activated pair of coils is used to illuminate a third coil. Each member of this pair is oriented so that the effects of their magnetic fields on the third coil are of equal magnitude and in phase opposition. Under these circumstances, if the two coils constituting the pair are identical, the energy transferred from the pair to the third coil is zero. No voltage is generated in the third coil.

Now, if one member of the pair is exposed to a target, as shown in Fig. 14.6, its electric characteristics are no longer the same as those of the other member of the pair. Energy is then transmitted from the primary to the secondary to a degree, and in a phase, appropriate to the asymmetry.

The voltage vectors of Fig. 12.2 show the signal current $I_x$ as a secondary current. It is then possible to monitor this secondary signal current directly, as is the case in the two-coil "crossed-field" configuration. This is in

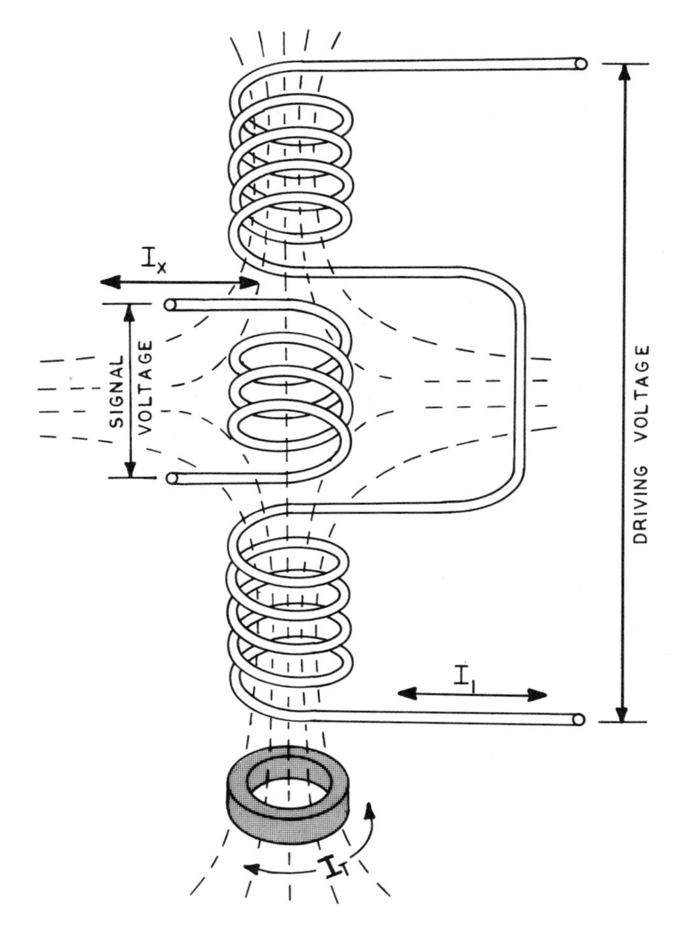


Fig. 14.6. Induction balance: If the magnetic fields of two primary coils are symmetrical but out-of-phase, their effects are cancelled in the secondary. If a target is introduced into the field of one primary, however, the effects are no longer symmetrical, and energy is passed to the secondary.

contrast to the one-coil system, where the change in the sensing-coil current is caused by the signal current.

This three-coil arrangement is symmetrical in that the energy transmitted from a single coil to an out-of-phase pair is also zero. If two out-of-phase coils are used to illuminate a single coil, however, no voltage is generated in the single coil under null conditions. On the other hand, if one coil drives two out-of-phase coils, voltages are generated separately in the latter but, since these two voltages are out-of-phase, no current flows under null conditions.

As shown in Fig. 14.6 if two out-of-phase coils are used to drive a third, the configuration is called an *induction balance*. As shown in Fig. 13.10, if one coil is used to drive two out-of-phase coils, the system is called a *differential transformer*.

Configurations with more than three coils can be used to create specialized field patterns with maximum sizes and maximum strengths in variously shaped sensitized lobes.

## 14.4 GEOPHYSICAL TARGETS

Minute electric currents induced in the earth can be monitored with two-coil, crossed-field configurations. If the electric conductivity of the earth is reasonably consistent, the presence of buried anomalies that have either greater or lesser conductivities than the earth's mean can be detected. If such a detection system is driven at low frequency with high power, the detectable anomalies include masses of stone, tunnels, significant chemical differences, or, in fact, any phenomenon that localizes distinct conductivity differences.

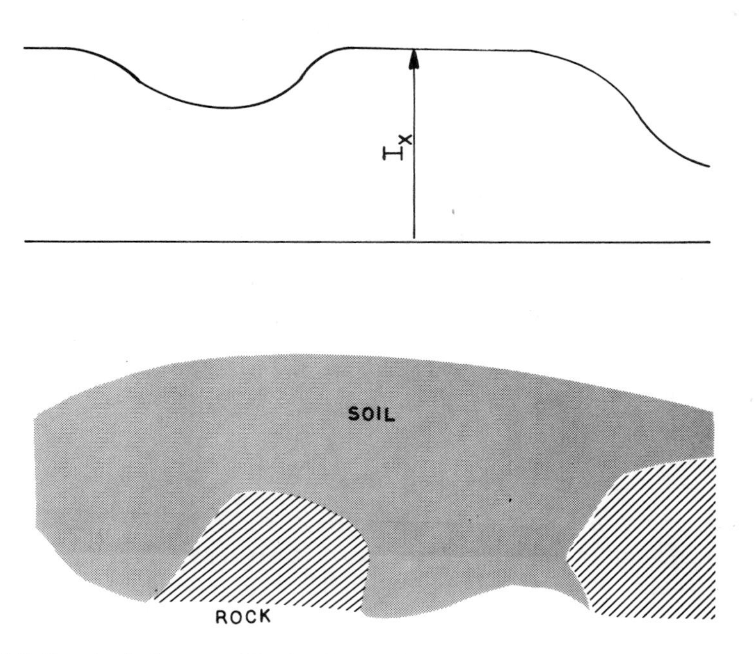


Fig. 14.7. Geophysical targets: To the degree that earth is conductive, eddy currents are induced by a sensing coil. The earth's conductivity varies and is reflected as changes in sensing coil impedance.

Since the earth's mean field is subject to considerable spatial variations in its own right, the presence of an anomaly is detected as a change in conductivity rather than as an absolute measure of conductivity when a sensing instrument is moved through a raster over the earth's surface. The spatial pattern of conductivity change revealed by this raster will then indicate the location of a buried anomaly. Figure 14.7 illustrates the signal-magnitude variations that might be encountered during a pass of such a raster. Here the signal current $I_x$ is reduced whenever a relatively non-conducting stone passes under the receiver coil instead of relatively more conducting soil.

Systems utilizing high frequencies similarly detect small, but still discrete, anomalies near the earth's surface. The operation of some non-magnetic mine detectors is based on this technique. At high frequencies, however, permittivity variations must be added to the variables of permeability and reflected energy, as previously discussed.

# 15
# MOTOR PHENOMENA

As discussed in Chap. 10, a voltage is generated between the two ends of a conducting wire when that wire moves in a magnetic field. This is the *generator function*, in which mechanical movement is converted into electric energy.

The Lorentz forces of Eq. 2.4 are activated when conduction electrons are moved as part of a conductor's motion. The force on the electrons is exerted at right angles to the direction of both the magnetic field and the wire movement.

In a reciprocal mode, a conducting wire carrying an electric current experiences a lateral force when it is immersed in a magnetic field. The Lorentz forces of Eq. 2.4 are activated when conducting electrons are moved as part of an electric current. This is the *motor function*, in which electric energy is converted into mechanical energy. Once again, the force on the electrons, and hence the force on a wire carrying these electrons, is exerted at right angles to the direction of both the magnetic field and the current flow.

## 15.1 STRING GALVANOMETER

A conducting wire carrying an electric current experiences a lateral force when it is immersed in a magnetic field. If the wire is constrained at both ends, but free to move in between, the wire's displacement assumes the form of a catenary curve. This configuration, called a *STRING GALVANOMETER*, is shown in Fig. 15.1.

The amount of catenary displacement is a function of the strength of the magnetic field, the strength of the electric current, and the longitudinal tension on the wire. When the string galvanometer is driven by an alternating current, which does not significantly affect wire tension, the movement of the wire's center reproduces the waveform of the electric current.

An *EINTHOVEN GALVANOMETER* is a string galvanometer that is made extra sensitive by the use of a strong magnetic field and a fine gold-plated quartz thread.

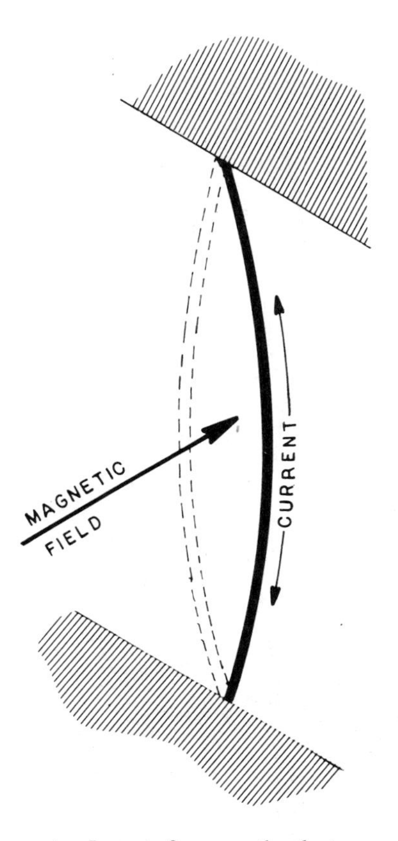


Fig. 15.1. String galvanometer: Lorentz forces on the electron carriers in a conducting filament cause it to deflect in the presence of a magnetic field.

## 15.2 VIBRATRON

A uniform, flexible wire constrained at both ends, subjected to tension, and caused to vibrate transversely tends to oscillate at certain discrete frequencies, each known as a *STRING RESONANCE*, determined as follows:

$$F = kNL^{-1}(T/m)^{1/2} \qquad \text{(Eq. 15.1)}$$

where $F$ is the resonant frequency for an integer mode $N$; $L$, the length of wire; $T$, the tension on the wire; m, the wire's mass per unit length; and k, a constant.

Such a wire immersed in a magnetic field is caused to vibrate while it

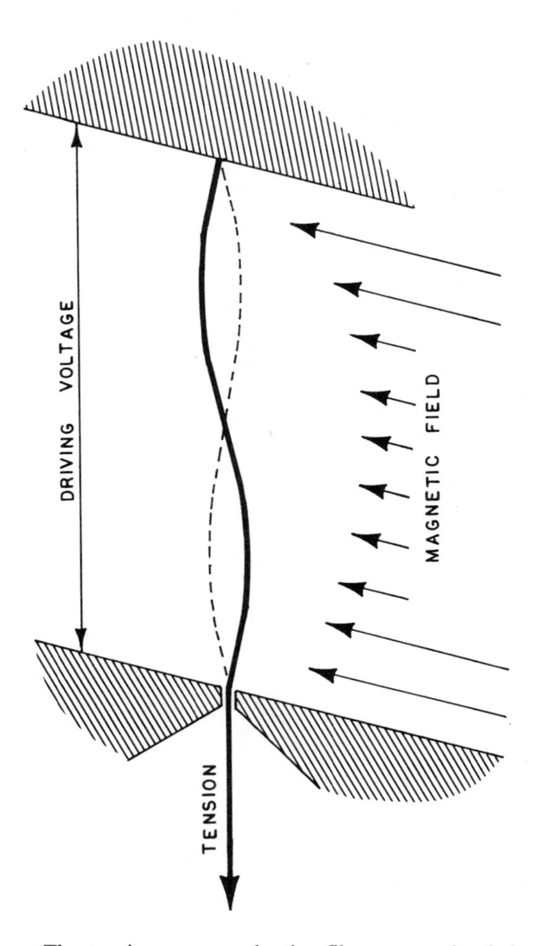


Fig. 15.2. Vibratron: The tension on a conducting filament can be deduced by noting the frequency of a driving voltage that sets it in mechanical resonance.

is conducting an oscillating current. The impedance experienced by this current is minimized by the oscillations that occur at resonant frequencies. While such a wire is vibrating, moreover, an alternating voltage is generated between its two ends because of its movement in the magnetic field. Using conventional servo techniques, the voltage generated at minimum impedance can be used to adjust the frequency of the oscillating current to that of the wire's resonant frequency.

In this device, called a *VIBRATRON* (see Fig. 15.2), the frequency of generated voltage is a measure of wire tension.

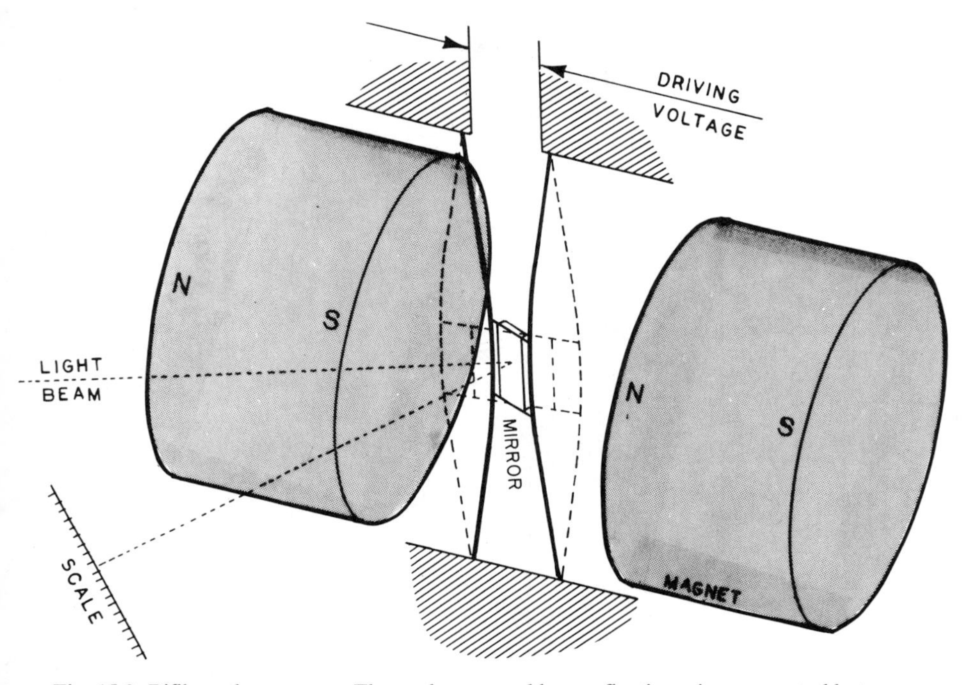


Fig. 15.3. Bifilar galvanometer: The angle assumed by a reflecting mirror mounted between two filaments carrying the same current in opposite directions depends on the magnitude of this current if all other factors are held constant.

## 15.3 BIFILAR GALVANOMETER

If a mirror is suspended between two string galvanometers and the current through the two "strings" is passed in opposite directions, the magnitude of mirror rotation can be calibrated in terms of current intensity. When such an assembly, called a *BIFILAR GALVANOMETER*, is driven by an alternating current, which does not significantly affect wire tension, the rotation of the mirror reproduces the waveform of the electric current.

Figure 15.3 illustrates the essential elements of the bifilar galvanometer.

## 15.4 D'ARSONVAL GALVANOMETER

A *D'ARSONVAL GALVANOMETER* consists of a coil of wire suspended in a magnetic field in a way that allows the coil to rotate around an axis

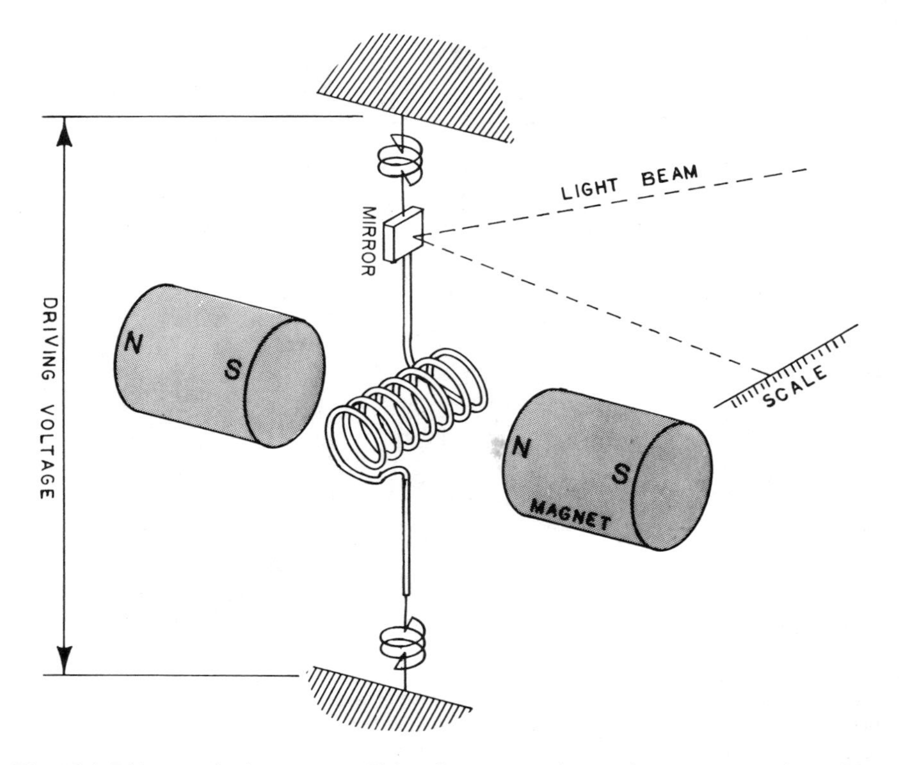


Fig. 15.4. D'Arsonval galvanometer: If the alignment tendency of a current-carrying coil in a magnetic field is offset against a spring-restoring force, the angle of rotation is a function of driving voltage.

that is maintained at right angles to a plane defined by the coil's centerline and the magnetic field vector. As shown by Fig. 15.4, a spring arrangement tends to orient the coil's centerline at right angles to the magnetic field vector, whereas an electric current passing through the coil tends to rotate the coil's centerline toward the direction of the magnetic field vector.

If a mirror is attached to the coil, the rotation of the mirror reproduces the waveform of the electric current flowing through the coil. If the electric current is standardized, rotation is a function of magnetic field strength. On the other hand, if the field strength is standardized, rotation is a measure of current magnitude.

## 15.5 GAUSSIAN GALVANOMETER

In the *GAUSSIAN GALVANOMETER*, which has a configuration inverse to that of the D'Arsonval galvanometer, a permanent magnet is suspended

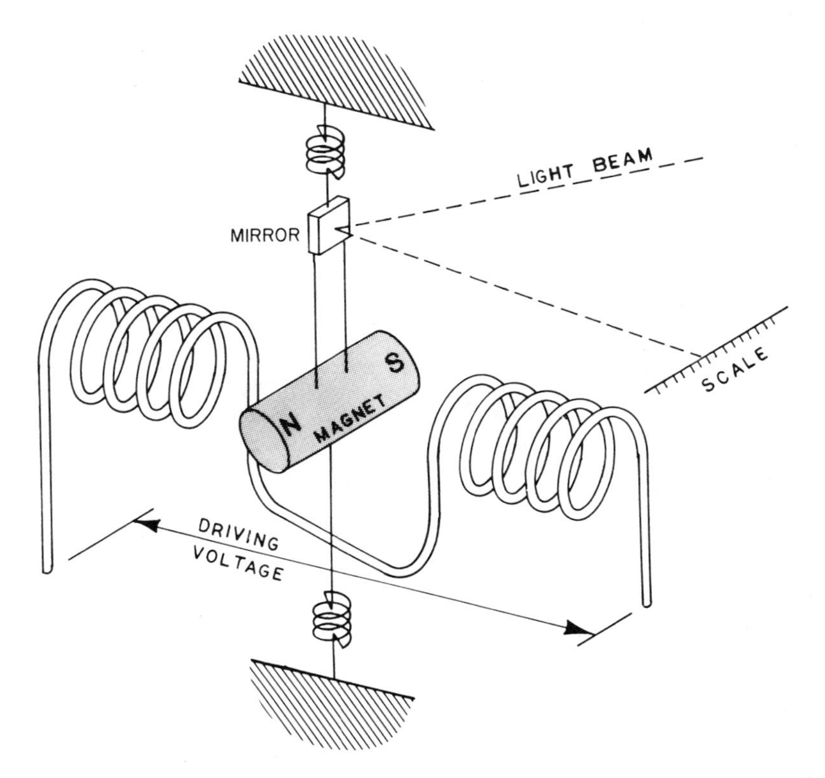


Fig. 15.5. Gaussian galvanometer: If the alignment tendency of a suspended magnet is offset by a spring-restoring force, the angle of rotation is a function of the driving current.

in a magnetic field accompanying a current flowing in a coil of wire. The suspension of this body allows it to rotate around an axis that is maintained at right angles to a plane defined by the coil's centerline and the vector of the magnetic field produced by the body.

As shown by Fig. 15.5, a spring arrangement tends to orient the vector of the magnetic body at right angles to the coil's centerline whereas an electric current passing through the coil tends to rotate this vector in the same direction as the coil's centerline.

If a mirror is attached to this magnetic body, the rotation of the mirror will reproduce the waveform of the electric current.

## 15.6 EDDY CURRENT MULTIPLIER

Consider the configuration of Fig. 15.6, in which eddy currents from two different sources are induced to flow in one disk. If both these eddy currents flow at the same frequency, with one induced to flow around the

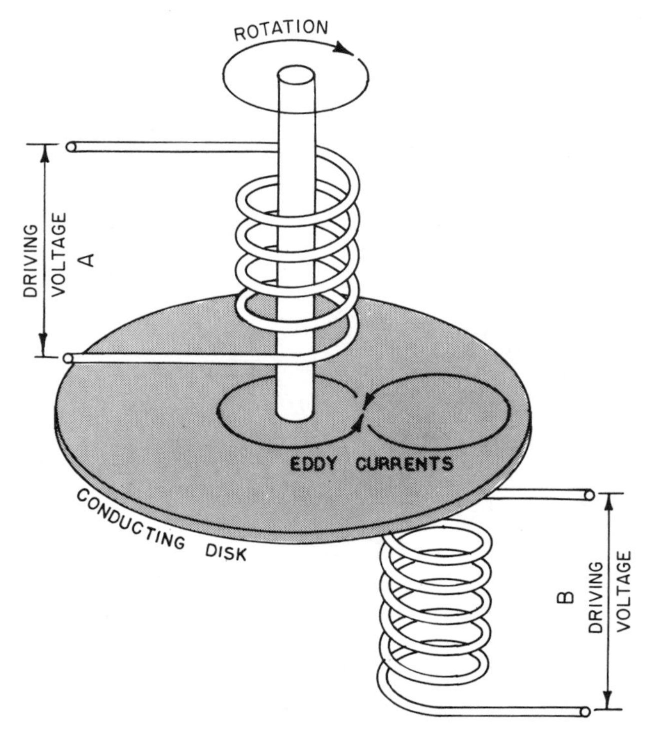


Fig. 15.6. Eddy current multiplier: Eddy currents can be induced to flow from two different sources, and their interaction can be used to rotate a conducting disk.

center of the disk and the other directed to oppose the first, the two of them will interact in such a way as to induce a torsional force on the disk.

Under such circumstances, this torsional force is proportional to the product of the strengths of the two eddy currents. If both of them are in-phase, the torsional force is maximum. Otherwise, the torsional force is reduced.

When the central coil is driven by a current proportional to the voltage applied to a load and the peripheral coil is driven by the current flowing through that load, the summed revolutions of the disk represent the power consumed by that load.

The next three parts discuss what happens magnetically to moving magnetons: Part III, where the magnetons move through solid structures when there is an interchange of energy between the moving magnetons and the structures; Part V, where there are no moving-magneton/structure interactions; and Part IV, where magnetons move free of all such structures.

# PART III

# MAGNETONS MOVING UNDER TIGHT CONSTRAINTS, AS IN A SOLID OR LIQUID

When an electrically charged particle finds itself surrounded by a dense population of other charged particles, its posture and motion are profoundly influenced by the postures and movements of its neighbors. Its magnetic field may be forced to align with, or against, those of its neighbors. If it tries to move, it may collide with, and if it doesn't move, it may well be swatted by one of its neighbors. Such an environment is characterized by a seething mass of particles moving at different velocities and different directions, with axes oriented in different directions, and all of these factors in a constant state of change.

At first glance, the situation might appear to be one of sheer chaos. In actuality, all of this activity is busily obeying basic laws that tend to minimize both the per-unit-volume electrostatic charge and the per-unit-volume magnetic field. And, of course, it is the forces that tend to minimize these values that provide the glue that holds all of these particles together.

Part III discusses a number of the magnetic consequences that may be expected when magnetons move in solids or liquids exhibiting various characteristics.

# 16 MAGNETOSTRICTION

The term *MAGNETOSTRICTION* refers both to the dimensional changes that occur in ferromagnetic materials in the presence of imposed magnetic fields and to the magnetization changes that occur in ferromagnetic materials exposed to mechanical stress.

## 16.1 MAGNETOMECHANICAL HYSTERESIS LOOP

The mechanical consequences of magnetostriction can be illustrated by adding a magnetomechanical hysteresis loop to the magnetic hysteresis diagram of Fig. 4.2. Figure 16.1 shows the relation of mechanical strain, as a function of magnetic field, to the magnetization cycle of a ferromagnetic material.

Interaction energy between elastic strain and magnetization is the fundamental basis for magnetostriction. The *MAGNETOELASTIC COUPLING CONSTANTS* represent coefficients in expressions relating these two energy forms.

Although the dimensional changes that occur under magnetostricion are larger than those associated with paramagnetic materials (as shown in Fig. 6.1), they are still very small. For instance, magnetic saturation in nickel causes a change in length of only about thirty parts per million.

As does hysteresis, magnetostriction depends on the formation of domains. The domains represent magnetic saturation along easy crystal axes. Their orientation and individual volumes are determined by both magnetic fields and by crystal electrostatic forces. Since lattice spacings are changed slightly by mechanical strain, the intermolecular self-aligning forces that set up domains are altered under strain. The magnetization that results from domain configuration is then changed.

Materials change in dimension and exhibit magnetostriction primarily during the process when many of the individual domains are forced from their preferred crystal axes in the upper part of the magnetization cycle. The domains tend to repel each other when they are lined up in the same

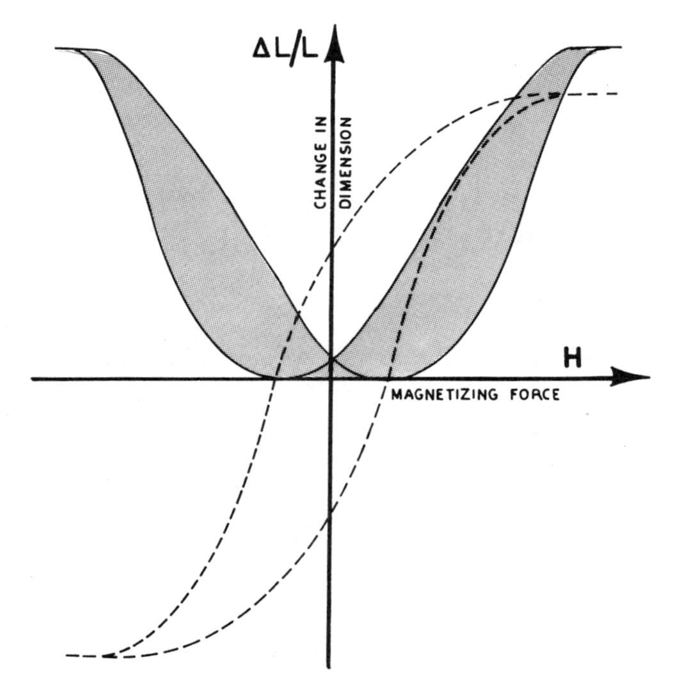


Fig. 16.1. Magnetomechanical hysteresis loop: The dimensions of a body fabricated from ferromagnetic material are a function of magnetization, and a dimensional hysteresis loop is thus derived from the magnetic loop.

general direction. A material containing them expands a bit in the direction of this repulsion.

A magnetic bias field is used in magnetostrictive transducers to position the central point of their operational minor loop in that area of the major hysteresis loop that gives the maximum magnetostrictive response in conjunction with the greatest linearity of transduction. The needed bias can be provided either by an externally applied magnetizing force or by the retentivity of the material itself.

Since magnetostriction depends directly on ferromagnetism, all of the peculiarities of ferromagnetism have an effect on magnetostriction. As might be expected, magnetstriction is completely lost above the Curie temperature. It is not possible, however, to predict any regular relationship between magnetizing force and magnetostriction in a particular ferromagnetic material. This can be established only by test. Ferromagnetic materials show wide variations in their manifestation of magnetostriction, both in magnitude and in sign. Perhaps a clue to the reason for this variation is given by Fig. 3.5. Here, if the columns are closer together than the rows, the sign will be opposite to what it would be were the rows closer than the columns.

All magnetostrictive effects exhibit complete reciprocity; that is, changes in dimension produce changes in magnetization and vice versa.

## 16.2 VILLARI EFFECT

The *VILLARI EFFECT* is a change in magnetization that occurs in the direction of mechanical strain. Magnetostrictive materials have either positive or negative Villari Effects; that is, magnetization can either increase or decrease with increasing strain. A "positive" Villari Effect is here defined as causing a decrease in permeability when a material is exposed to compression. This is accompanied by a permeability increase in the tension mode. Transducers making use of this effect are sometimes called "variable permeability" or "variable mhu."

In some circumstances, permeability increases with strain in weak magnetic fields but decreases in strong magnetic fields. The *VILLARI REVERSAL* is that point on a magnetization curve where magnetization is unaffected by strain.

The *TRANSVERSE VILLARI EFFECT* is a change in magnetization that occurs in a direction transverse to that of mechanical strain.

As in all magnetostrictive effects, if changes in dimension produce

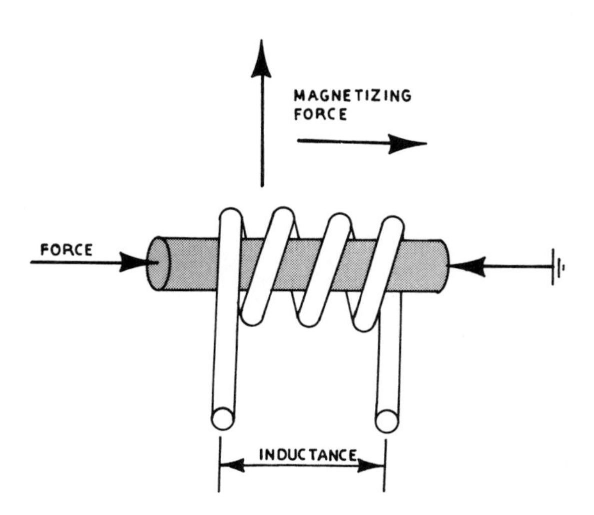


Fig. 16.2. Villari and Transverse Villari Effects: The inductance of a coil surrounding a ferromagnetic member is a function both of the axial force on that member and the immersion magnetic field.

changes in magnetization, then changes in magnetization are associated with changes in dimension. The reciprocals of the various Villari Effects are called *JOULE EFFECTS*.

Figure 16.2 shows a configuration in which the Villari and Transverse Villari Effects can be observed.

## 16.3 GUILLEMEN EFFECT

When a magnetic material is immersed in a magnetic field, the magnetons that are a part of that material tend to align themselves either with, or against, the immersion field. As shown in Fig. 3.5, there is more than one pattern of alignment possible. Once the magnetons are aligned, they exercise a mutual effect on each other, and, as a result, they affect the dimensions of the material of which they are a part. If the configuration is such that adjacent magnetons repell each other, the material expands slightly in the direction of repulsion. If, on the other hand, the adjacent magnetons are attracted toward each other, the material contracts slightly in that direction. The result is either magnetostriction as discussed in this chapter or the paramagnetic strain discussed in Sec. 6.4.

Magnetostriction takes place in a material in which exchange forces have already aligned magnetons into some kind of domain pattern. Under these circumstances, an imposed field tries to move the magnetons into a somewhat different orientation in which their mutal attraction/repulsion for each other will be slightly altered and the dimensions of the material thereby changed.

The mechanical processes of fabrication for a magnetic rod and the magnetic field to which the rod is exposed during fabrication tend to align the crystallographic axes and the associated domain axes in one direction. If such a rod is bent and placed in an axial magnetic field, there will be a force on those domains that are no longer directed along the original rod axis to reform around that axis. The result is a force that tries to straighten the bent rod.

In the GUILLEMEN EFFECT, a bent ferromagnetic rod tends to straighten in the presence of a longitudinal magnetic field.

## 16.4 VILLARI DIFFERENTIAL TRANSFORMER

Changes in magnetization are detected as changes in the self-inductance of a ferromagnetically cored coil; opposite changes in core magnetization for two arms of a differential transformer; or imbalance in a four-armed magnetic bridge circuit.

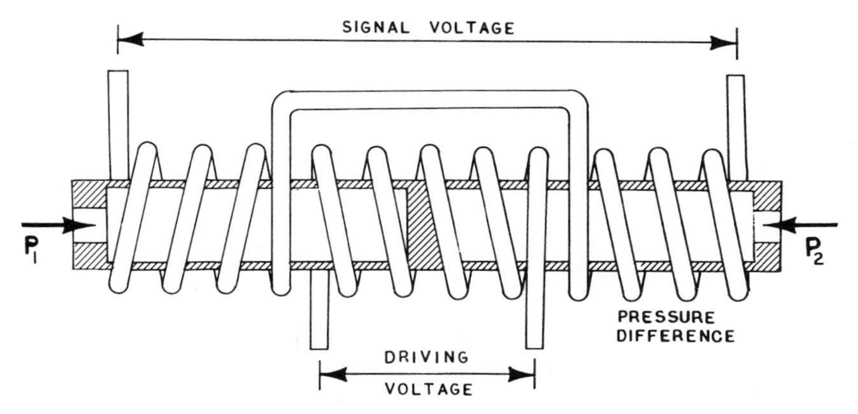


Fig. 16.3. Villari differential transformer: Two symetrical chambers within two secondary windings of a differential transformer can be used to unbalance that transformer as a function of the differential chamber pressures.

Figure 13.15 illustrates the significant features of a differential transformer. Utilizing the Villari Effect in the same general configuration, the differential magnetization of a transformer core is altered by strain rather than by a physical movement of core elements. As shown in Fig. 16.3, if a core constructed of magnetostrictive material contains two symmetrical cavities, differential pressures between these cavities produce shifts in magnetization that result in transformer unbalance.

## 16.5 VILLARI TORQUE DIFFERENTIAL TRANSFORMER

Loading a shaft in torsion induces shear stresses that are oriented at 45 degrees to the shaft axis. Here the principle tensile and compressive stresses are displaced from one another by 90 degrees and are directed as shown in Fig. 16.4. Reversing the direction of applied torque reverses the sign of the induced stress. This changes tensive forces to compressive forces or vice versa, but it does not change angular orientations.

In the Villari Effect, magnetization is effected by the presence of strain. It is, therefore, possible to utilize the Villari Effect in a differential transformer to measure torque in which the compressive strain influences one flux path and the tensive strain the other. In the configuration of Fig. 16.4, the output signal is zero under no-load conditions. A signal voltage increases with torque whether or not the shaft is rotating.

A number of such transformers, whose output signals are added, are commonly mounted around the circumference of a rotating shaft. This

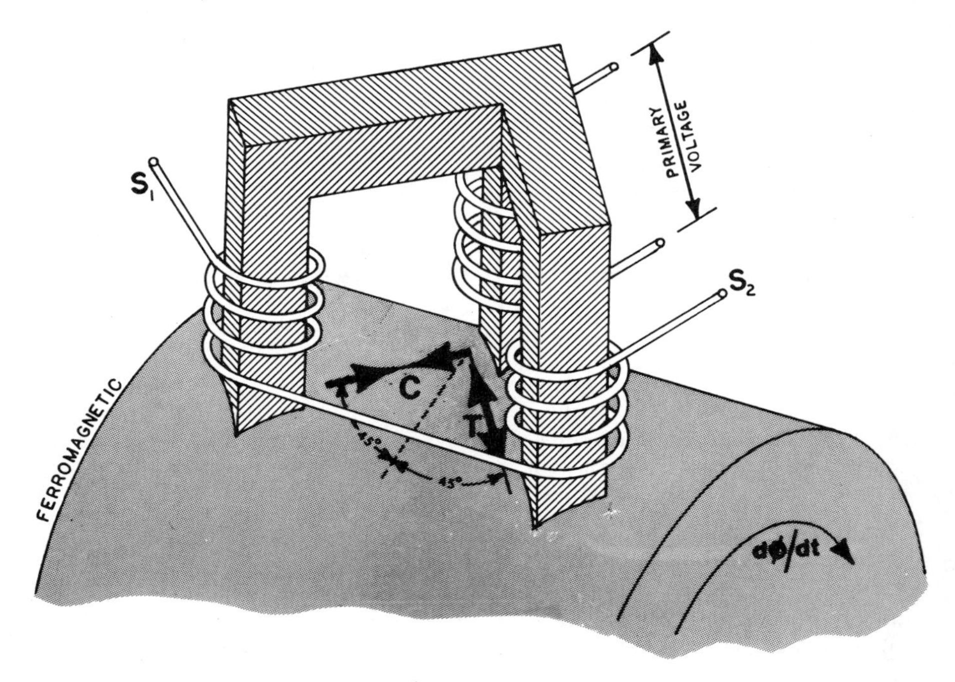


Fig. 16.4. Villari torque differential transformer: The degree of unbalance in the two arms of this transformer may be used to measure torque without physically contacting the shaft.

arrangement compensates both for minor shaft eccentricities and for variations in the magnetic properties of shaft materials.

## 16.6 VILLARI BRIDGE

*PRESSDUCTOR* is a trade name for devices utilizing the Villari principle in the construction of four-armed, magnetic-bridge circuits. Such an arrangement can be considered a "misdesigned" transformer in which the primary and secondary windings are oriented to produce zero mutual inductance (or coupling) when the unit is unstrained.

The magnetic flux accompanying the electric current in the primary winding of such a device flows symmetrically through the four arms (A–D and B–C) of the bridge—shown in Fig. 16.5 as "circular" flux loops. These circular loops intercept the secondary winding in a cancellation mode, and no energy is transferred from the primary to the secondary winding.

When the unit is subjected to compressive strain, magnetization is reduced in the direction of strain, and at least two of the bridge arms are magnetically altered. The flux then deviates in the transverse direction

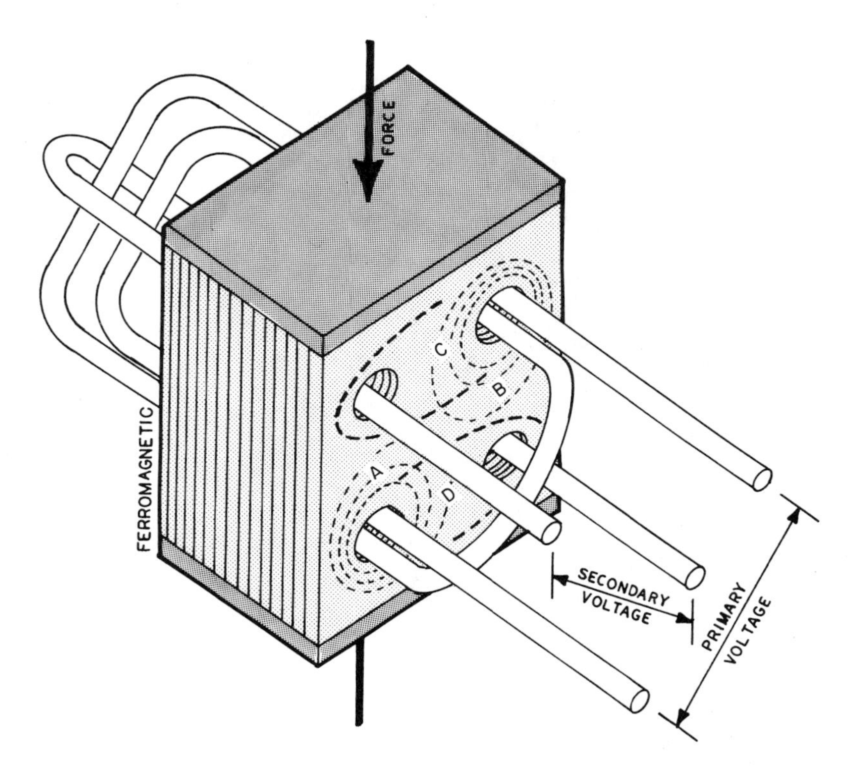


Fig. 16.5. Villari bridge: The amount of energy passed between primary and secondary windings is a function of the strain experienced by two of the four arms of a bridge.

by an amount proportional to the impressed force. This transverse flux intercepts the secondary winding to some degree—shown in the figure as "oval" flux loops. If the primary winding is powered by an alternating current, an ac voltmeter connected to the secondary winding reads proportional to the impressed force.

## 16.7 VILLARI STEEL-ROLLING SHAPE GAGE

The term "shape" here refers to the flatness of flat-rolled steel products. It can be defined in terms of a differential elongation along the length of a strip, as measured across the width of that strip. If the edges of a strip are compressed more than the center, they are longer than the center, and the edges become "wavy." Conversely, if the center is compressed more than the edges, the sheet "ripples" in the middle.

A Villari bridge can be used to sense the difference in permeability in the direction of ferromagnetic sheet movement as this relates to the

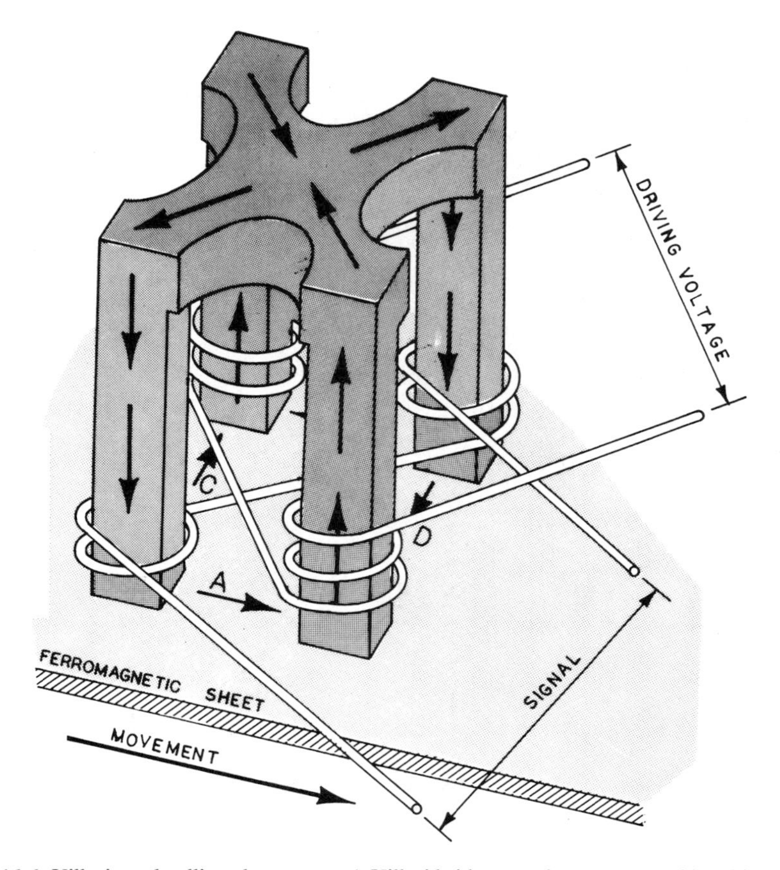


Fig. 16.6. Villari steel-rolling shape gage: A Villari bridge may be constructed in which each of its four arms (A, B, C, and D,) contain a portion of a sheet of ferromagnetic material. Two arms (A and B) are oriented in the direction in which the sheet is being rolled and the other two arms (C and D) in the direction transverse to the roll direction. The difference in strain experienced between these two directions as a result of rolling the sheet to a new thickness can then be monitored in terms of bridge output voltage.

permeability at right angles to such movement. Since permeability is a function of strain, the difference detected by a Villari bridge is a measure of strain distribution across the width of a rolled sheet. Such a measurement can be made at the exit of the rollers without touching the metal. The signals so obtained can be used to control roller position.

Figure 16.6 shows a bridge whose four arms include the rolled strip. Two arms are in the rolling, and two in the transverse to rolling, directions. Four arms are used to compensate the output signal for minor variations in the four air gaps.

Steel strip, after rolling, normally displays a higher permeability in the transverse direction than in the rolling direction. The output signal of a transducer designed as a symmetrical differential transformer would then have other than zero output when there is no stress on the sheet. The addition of a bias winding compensates for this anisotropy. Current through the bias winding adjusts the transducer output to zero for zero strain.

## 16.8 WERTHEIM EFFECTS

When a ferromagnetic rod is immersed in a longitudinal magnetic field and twisted, its longitudinally directed magnetons are changed to helical directions and a transient voltage is generated between its two ends. The generation of this voltage is called the *WERTHEIM EFFECT* (see Fig. 16.7).

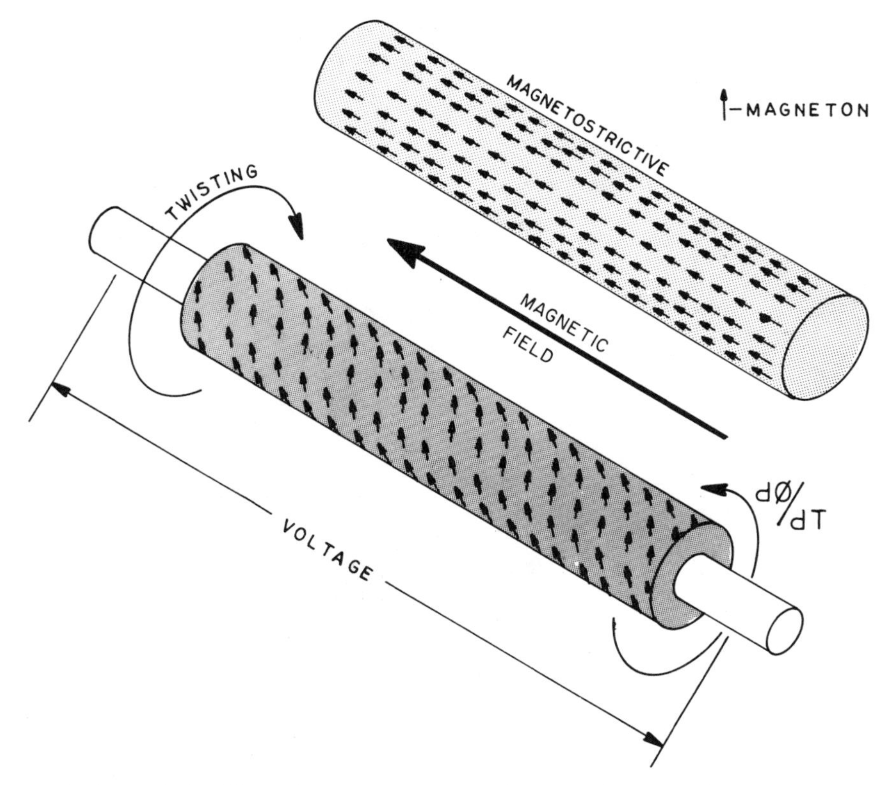


Fig. 16.7. Wertheim Effect: A voltage is generated between the two ends of a magnetized magnetostrictive rod while it is being twisted.

When the rod is placed in an axial magnetic field, more of its domain magnetons are forced to lie in the axial direction than in any other direction. In the presence of torsional strain, the alignments of some of these domain magnetons are diverted from predominantly axial directions to helical directions.

The same rod in the same magnetic field twists when a voltage is applied to its two ends. This twist is called the *INVERSE WERTHEIM EFFECT*.

Both of these effects are a result of imposing circular magnetization upon linear magnetization directed by longitudinal magnetic fields. In the Wertheim Effect, circular magnetization is supplied by the circular component of helical strain, whereas in the Inverse Wertheim Effect, circular magnetization accompanies the electric current that flows in response to an applied voltage.

## 16.9 WIEDEMANN EFFECTS

In Weidemann Effects, circularly directed magnetons are changed to helical directions as a result of twist induced in ferromagnetic rods.

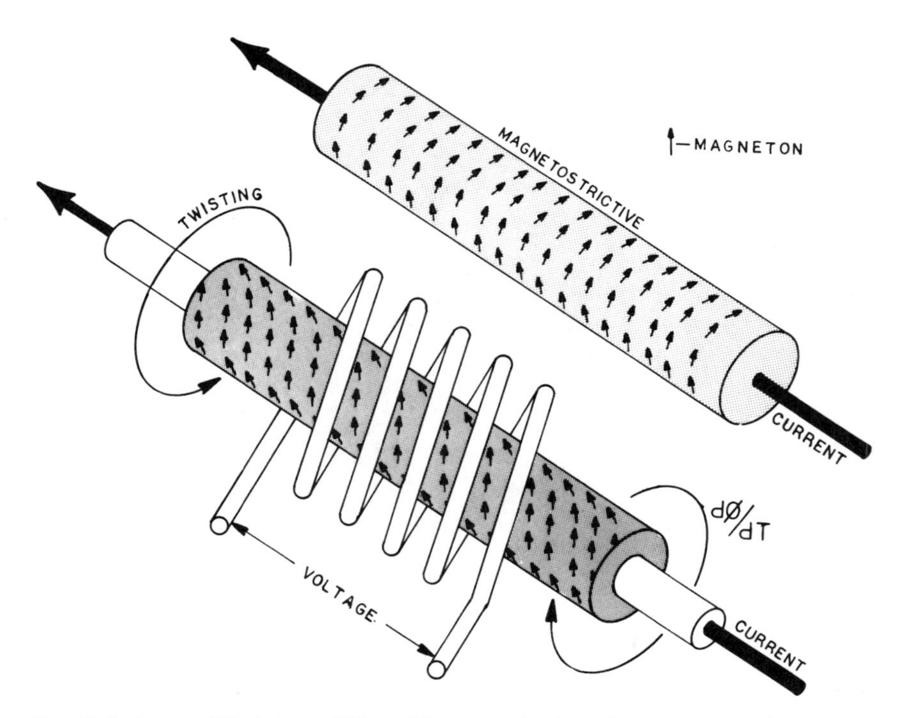


Fig. 16.8. Inverse Wiedemann Effect: The magnetization of a current-carrying magnetostrictive rod is changed when it is twisted.

A current-carrying ferromagnetic rod is axially magnetized when it is twisted. This axially directed magnetization is called the *INVERSE WIEDEMANN EFFECT*. A voltage will be generated in a coil surrounding such a rod while it is being twisted. See Fig. 16.8.

When a ferromagnetic rod carries an electric current, more of its domain magnetons are forced to lie in the circumferential direction than in any other direction, conforming to the direction of the Biot-Savart magnetic field that accompanies an electric current. In the presence of torsional strain, the alignments of some of these domain magnetons are diverted from predominantly circular directions to helical directions.

Conversely, the *WIEDEMANN EFFECT* is the twist experienced by a ferromagnetic wire carrying an electric current when the axially magnetic exposure is changed.

Both of these effects are a result of imposing linear magnetization upon the circular magnetization that accompanies a longitudinal electric current. In the Inverse Wiedemann Effect, linear magnetization is supplied by the linear component of helical strain, whereas in the Wiedemann Effect, linear magnetization is supplied by a change in the axial magnetic field.

## 16.10 NAGAOKA-HONDA EFFECT

According to the *NAGAOKA-HONDA EFFECT*, shown in Fig. 16.9, there is a change in the magnetization of a ferromagnetic material if, for any

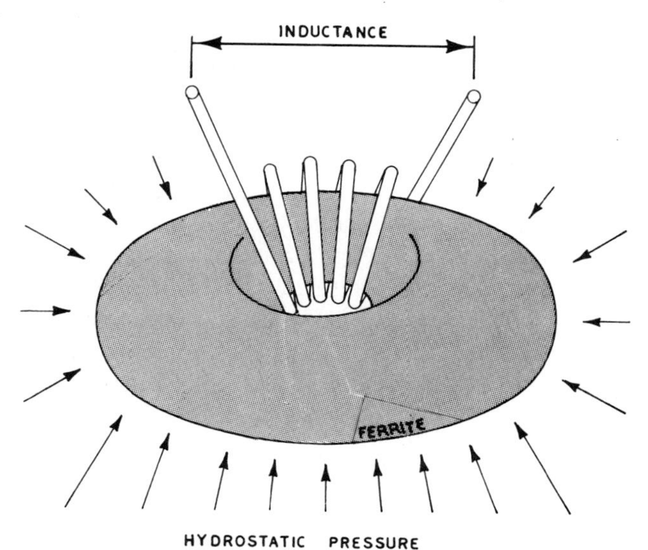


Fig. 16.9. Nagaoka-Honda Effect: The magnetization of a body constructed from a magnetostrictive material is changed when the body is subjected to hydrostatic pressure.

reason, there is a net change in the volume of a body fashioned from such a material. The converse of this effect, a change in volume for a change in magnetization, is called the *BARRETT EFFECT*.

Some ferrites with poor cohesion and a large Barrett Effect can be driven to the point of self-destruction by an imposed magnetic field.

## 16.11 MAGNETOSTRICTIVE ACOUSTICS

Magnetostrictive dimensional changes result from changes in magnetization. If a magnetization change is rapid, the resulting mechanical transient generates a sound wave.

The *PAGE EFFECT* is the "click" heard when ferromagnetic materials are either magnetized or demagnetized. In the same vein, the humming noises that emanate from ac transformers result partially from the dimensional changes of core materials.

In fact, mechanical resonance can result from magnetostriction because of the mechanical deformation of a material and the velocity of sound within that material. The effective $Q$ of a transductive element and the change in the velocity of sound with a change in temperature can significantly affect the performance of practical transducers.

## 16.12 MAGNETOSTRICTIVE/PIEZOELECTRIC EFFECT

In magnetostrictive materials, dimensions are functions of magnetic field intensity. In piezoelectric materials, output voltages are functions of dimension.

Where a magnetostrictive material is combined in a "sandwich" with a piezoelectric material and the mechanical coupling between the two materials is maximized, the voltage across the piezoelectric material is proportional to the magnetic field strength. Such a combination is here called the *MAGNETOSTRICTIVE/PIEZOELECTRIC EFFECT*. Very small devices based on this principle have been utilized in the reading heads for magnetic tape recorders and eliminate the need for relative motion between the recording medium and the reading head. (See Fig. 16.10.)

## 16.13 MAGNETOSTRICTIVE DETECTION EFFECT

The reduced dimensions of a ferromagnetic body exposed to a steady magnetic field are expanded slightly when that body is exposed to an alternating field of a frequency that stimulates the Larmor resonances of

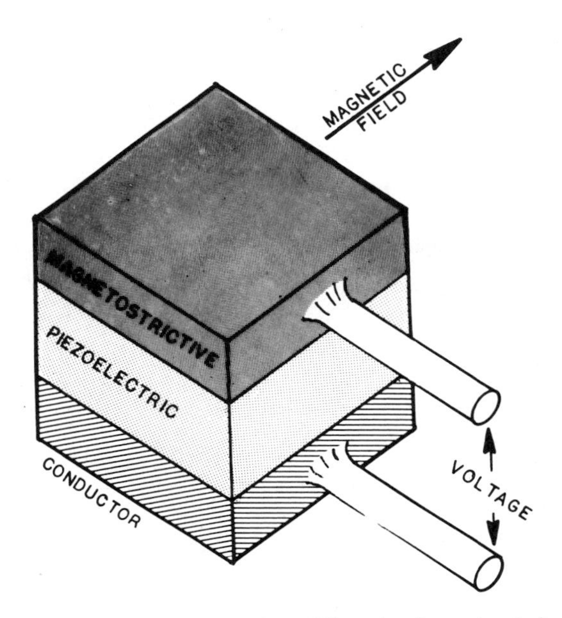


Fig. 16.10. Magnetostrictive/piezoelectric effect: When the dimensional changes of a magnetostrictive material induced by a magnetic field are imposed on a piezoelectric material, magnetic field strength is then related to piezoelectric voltage.

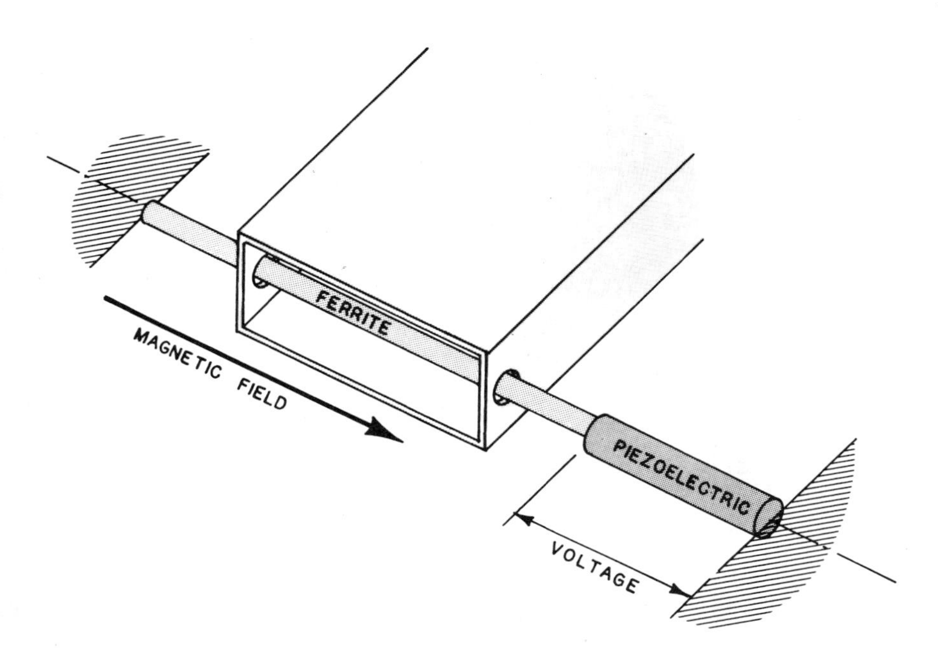


Fig. 16.11. Magnetostrictive detection effect: The voltage of a piezoelectric element activated by a change in rod length caused by precessing magnetons is related to the strength of the tuned electromagnetic radiation.

the constituent magnetons. This phenomenon is here called the *MAGNETOSTRICTIVE DETECTION EFFECT.*

As discussed in Sec. 8.17, the magnetization of a ferromagnetic material in response to a steady imposed magnetic field is reduced by Larmor stimulation. Since magnetostriction is a consequence of magnetization, the dimensional changes imposed by a steady field are relaxed by an alternating field of Larmor frequency. In the assembly of Fig. 16.11, the output voltage of the piezoelectric element is a function of the Larmor activities within the ferrite rod. If an electromagnetic wave traveling in the waveguide stimulates Larmor precession, the presence of that wave can be detected by the piezoelectric voltage.

## 16.14 GARSHELIS EFFECT

In the *GARSHELIS EFFECT*, the magnetization of a wire spring in a spiral, coiled configuration is a function of coil deflection. If the magnetization is a result of a direct current flowing through the spring, the

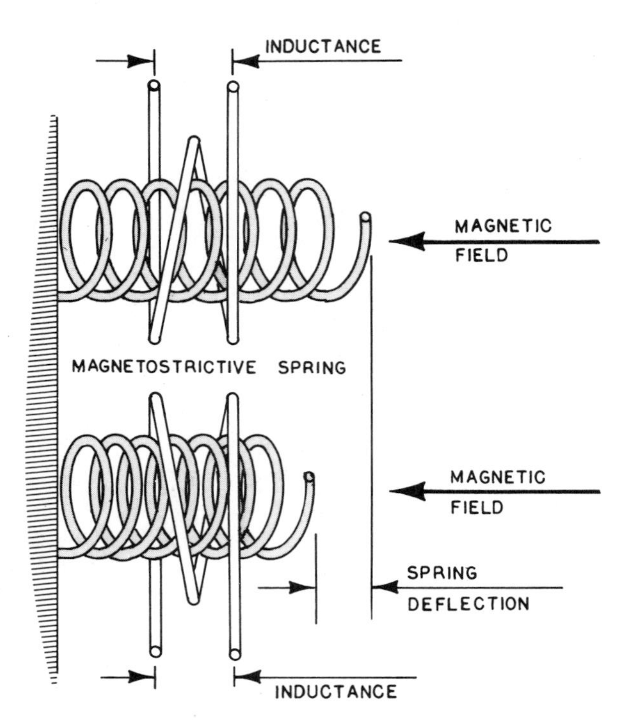


Fig. 16.12. Garshelis Effect: The magnetization of a spring fashioned from a magnetostrictive material immersed in a magnetic field of constant strength changes with spring deflection.

Garshelis Effect is an extrapolation of the Inverse Wiedemann Effect. If a remanent magnetization is directed along the wire axis, the Garshelis Effect is an extrapolation of the Wertheim Effect. As shown in Fig. 16.12, magnetization can also result from a magnetic field directed along the coil axis.

In all circumstances, linear spring deflection induces torsional stress in the wire from which the spring is constructed. Such torsional stress changes those magneton directions that are otherwise induced by both a current flowing through the coil and a magnetic field into which the coil is immersed. As a result of these magneton directional changes, the magnetization of the spring is modified.

Since magnetization is modified by either spring tension or spring compression, a device functioning in a Garshelis mode is somewhat analogous to a differential transformer. Changes in the magnetization of a material can be detected as changes in the inductance of a coil surrounding a body of that material.

## 16.15 MAGNETOSTRICTIVE RESONATOR

In the Joule Effect, changes in magnetization cause changes in mechanical strain. As a result of the Joule Effect, a ferromagnetic rod placed in a coil

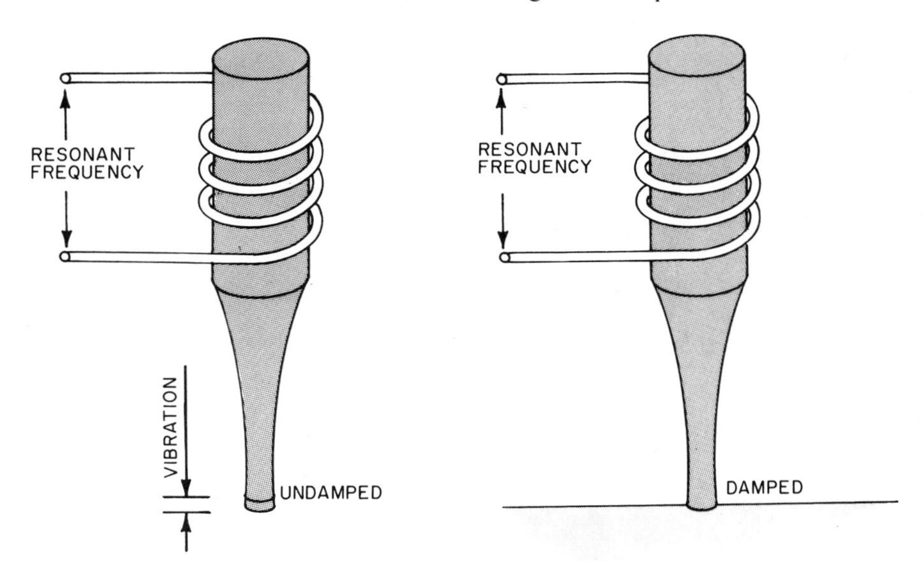


Fig. 16.13. Damped magnetostrictive oscillator: A body fashioned from magnetostrictive material may be driven at its frequency of acoustical resonance. The impedance of the driving coil is then dependent on the degree of acoustical damping.

driven by an alternating current is maintained in longitudinal vibration at the driving frequency. Such a rod exhibits a mechanical resonance derived from the velocity of sound within the rod material. The current driving the coil thus experiences a maximum impedance when it drives the rod at its resonant frequency.

In the Villari Effect, changes in mechanical strain cause changes in magnetization. As a result of the Villari Effect, changes in magnetization are maximum when a ferromagnetic rod is driven at its resonant frequency. This magnetization change accentuates the impedance maximum sensed by the driving current at the resonant frequency of the rod. Using conventional servo techniques, an oscillator can be constructed to oscillate at the rod's resonant frequency. Such a mechanism can be put to various useful tasks. In one of these (shown in Fig. 16.13), the presence of a fluid is detected as a sharp reduction in resonance caused by the damping action of fluid viscosity.

## 16.16 MAGNETOSTRICTIVE AMPLIFIER

A magnetostrictive rod exposed to an oscillating magnetic field experiences periodic changes in its length. This dimensional change is maximum when the rod is driven at its acoustical resonant frequency. If one of its ends is shaped into an acoustic horn as shown in Fig. 16.14, the sound wave

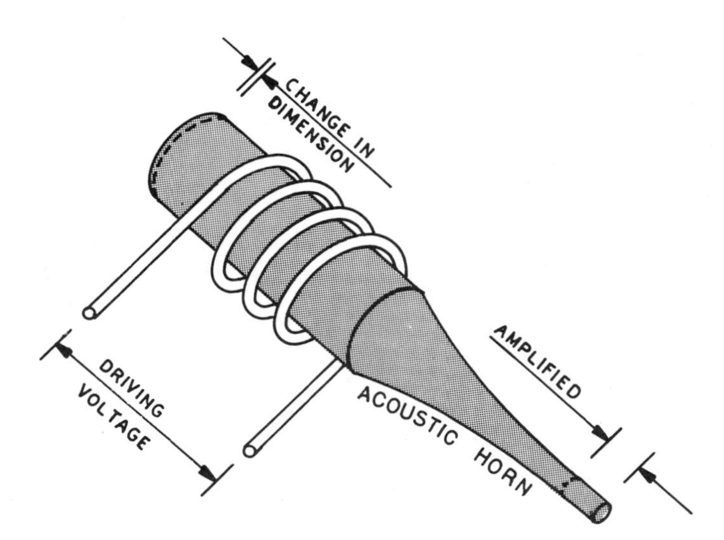


Fig. 16.14. Magnetostrictive amplifier: A rod fashioned from magnetostrictive material may be driven at its frequency of acoustical resonance at which there is an oscillating change in dimension. If an acoustical horn is added to the rod, the dimensional change is increased.

oscillating within the magnetostrictive material is amplified within the horn, and the tip of the horn moves over a greater distance than its butt. In other words, by adding the amplifying effect of an acoustic horn to that of magnetostriction, the total length change for a rod with an acoustic horn is significantly greater than for a rod without one.

### 16.17 MAGNETOSTRICTIVE MODULATOR

In a vacuum-tube amplifier, voltage on a control grid controls the flow of electrons from cathode to anode. The degree of control exercised by such a grid is a function both of the voltages on the various elements and of their spatial configuration. If the spacing between cathode and control grid is maintained by a magnetostrictive element, the control exercised by that grid is a function of magnetostrictive activity.

### 16.18 SONIC DELAY LINE

Sound propagates in a material at a velocity that is a characteristic of that material. A sonic pulse will then travel along a rod at some more or less constant velocity. This velocity depends on the material used to fabricate the rod.

The magnetization of magnetostrictive films plated on the surface of a rod will couple with the dimensional changes associated with sound travel through that rod. Such a magnetostrictive film can be used either to inject a sonic pulse into a rod or to detect such a pulse passing through a rod.

If magnetostrictive films are plated at two different locations on one rod, as shown in Fig. 16.15, a pulse introduced at one location will be detected at the other at some later time. The amount of time depends both on the velocity of sound and on the distance between locations.

### 16.19 DOMAIN FLIP

Consider a rod constructed from a ferromagnetic material characterized by positive magnetostriction and very little chemical anisotropy. If this rod is placed under tension (Villari Effect), most of the constituent domains will be aligned in the axis, or, alternatively, about one-half of them will lie in one direction, and the other half in the opposite direction.

When a saturation field is applied axially to this tensed rod, half of the domains undergo a 180 degree reversal. All of the domains then lie in the direction of saturation. Since domains that are directed both parallel and antiparallel to magnetic fields possess stability, and since tension en-

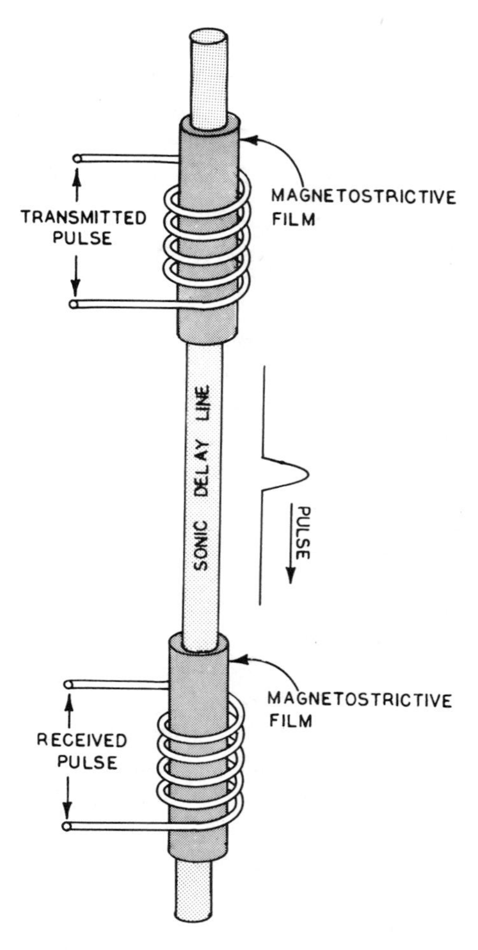


Fig. 16.15. Sonic delay line: Magnetostrictive films plated on a rod-shaped delay line will both create and detect sonic pulses.

courages this configuration, the rod will remain saturated even after the externally applied field has been removed! If a nucleus of magnetization opposed to the remnant state is introduced at some point along the rod, however, this reverse region immediately influences adjacent regions, and a wave of reversal is propagated in both directions down the rod until the original half-and-half condition is restored.

# 17 GALVANOMAGNETIC EFFECTS

Various consequences of Lorentz forces that deflect negatively charged electrons moving through solid materials are called *GALVANOMAGNETIC EFFECTS*.

When an electric current is passing through a solid conductor immersed in a magnetic field, the electrons constituting that current are deflected in a direction at right angles both to the direction of current flow and the direction of magnetic field, as shown in Fig. 2.2. As shown in Fig. 2.3, the faster moving electrons are deflected more than the slower moving ones.

Electronic deflections establish voltage gradients at right angles to current flow. Deflections of fast moving electrons, more than of slow ones, establish thermal gradients that also extend at right angles to current flow. Galvanomagnetics thus consists of electronic and thermal phenomena. Since electrons are moved by electric forcing fields, thermal forcing fields, and chemical forcing fields, galvanomagnetic effects are classified not only according to the type of forcing field involved but according to its consequences, whether thermal or electrical.

The term "galvanomagnetic" applies only to certain phenomena discernible in solid materials in which the only type of particle that can move in any significant numbers is an electron functioning either as a "free" agent or in the creation of a "hole." Galvanomagnetic phenomena are then further classified according to the type of carrier involved—free election or hole.

As one manifestation of thermal energy, some electrons within all solid materials are constantly moving about in random directions and with random velocities. If these movements are completely random, the sum of their individual displacements is zero, and the consequences of their individual Lorentz deflections cannot be detected. On the other hand, an electric current is transported by a number of charged particles, or "carriers," moving in the same direction. In a solid material, an electric current is effected by imposing a general, unidirectional drift on the otherwise random electron movements. A part of an electron's activities is thus a random response to thermal energy, and a part may be a unidirectional response to current transport. Although all electron movements react to Lorentz forces, only those parts that contribute to current transport are involved in galvanomagnetic phenomena. In short, galvanomagnetics is one result of immersing a solid material in a magnetic field and imposing a unidirectional drift on its electrons' otherwise random activities.

The establishment of carrier population density gradients at right angles to unidirectional drifts is one result of such a combination of circumstances. Here the Lorentz forces tend to push all carriers toward one side of a conductor. Since the carriers are charged particles, such population density gradients also function as voltage gradients. These voltage gradients provide forces that balance the Lorentz forces and are capable of driving electric currents in their own right. In the presence of such currents, the Lorentz forces, the galvanomagnetic voltages, and the ohmic voltages reach a three-way balance.

An electron's random movements are maintained by thermal energy whose magnitude is described by the temperature of a particular material. The energy required to maintain a unidirectional drift must be supplied by some other source. This second source cannot come from within the body of material when it is in a state of equilibrium but from some part of the surrounding environment.

Galvanomagnetic transduction is thus concerned with the electric consequences of the carrier population density gradients, established in solid materials when those materials are immersed in magnetic fields and exposed to various environmental conditions that impose a general, unidirectional drift on the otherwise random carrier activities.

## 17.1 CLASSIFICATIONS

There are six basic galvanomagnetic effects, as follows:

1. HALL EFFECTS: Voltage gradients that result from carrier deflections when carriers are moved by electric forcing fields. Here holes and/or electrons are moved in opposite directions and are therefore deflected in the same direction.
2. NERNST EFFECTS: Voltage gradients that result from carrier deflections when carriers are moved by thermal forcing fields. Here holes and/or electrons are moved in the same direction and are therefore deflected in opposite directions.
3. PEM and MEM EFFECTS: Voltage gradients that result from carrier deflections when carriers are moved by chemical forcing fields (population density gradients). Here both holes and electrons, occurring

as pairs, are moved in the same direction and are therefore deflected in opposite directions.

4. ETTINGHAUSEN and RIGHI-LEDUC EFFECTS: Thermal gradients that result from carrier deflections when hot carriers are deflected more than cold carriers. Such thermal gradients are called Ettinghausen Effects when associated with Hall Effects and Righi-Leduc Effects when associated with Nernst Effects.
5. Additions to electric resistance that result from carrier deflections when carriers are moved by electric forcing fields. Here there is both a reduction in the effective cross-sectional area of a conductor, since carriers are crowded to one side, and a reduction in the carrier forward motion caused by an increase in the length of traverse, since the carriers are forced to follow curved, rather than linear, paths.
6. Additions to thermal resistance that result from conditions analogous to those outlined above.

There are two basic combination effects:

1. When an electric current is allowed to flow in response to one of the above listed voltage gradients.
2. When heat is allowed to flow in response to one of the above listed thermal gradients.

In addition, there are combination effects in which a galvanomagnetic effect is combined with one or more nongalvanomagnetic effects:

1. Thermal Effects:
   (a) The mobility of carriers is altered by changes in temperature.
   (b) The mobilities of electrons and holes change at different rates with changes in temperature.
   (c) The population density of carriers is altered by changes in temperature.
   (d) The population densities of electrons and holes may change at different rates with changes in temperature.
2. Anisotropic Effects: The anisotropic characteristics of crystalline materials distort the result that would otherwise exist under isotropism.
3. Thermoelectric Effects:
   (a) The establishment of thermal gradients caused by a separation of hot and cold carriers introduces thermoelectric effects.
   (b) Additions to thermoelectric effects that result from carrier deflections. Here the unit-volume chemical potential is changed by a

change in the carrier population density. Such changes are called Nerst Effects.

4. Ferromagnetic Effects: The mobility of carriers in ferromagnetic materials is a function both of magnetic field strength and of magnetic field direction as in the Gauss Effect.
5. Size Effects: If a body of material is large in relation to electron trajectories, the bulk characteristics of a material predominate in affecting electron activities. On the other hand, if the body of material is small according to this same criterion, surface effects can dominate.
6. Strong Fields: Galvanomagnetic phenomena are influenced by how far a carrier moves along its cyclotron circumference. In strong magnetic fields, carriers move significant distances.

There are said to be over two hundred individual galvanomagnetic effects, but, in actuality, all of these can be constructed from some combination of the above listed phenomena.

## 17.2 CRYSTAL LATTICE

An *electron* is a very small particle associated with exactly one unit of negative electric charge. A *proton* is a much larger, but still very small, particle associated with one unit of positive electric charge that is exactly equal to but opposite, that of the negative electron.

The nucleus of an atom consists of a single, positively charged particle constructed in part from some finite number of protons. A complete *atom*, which has a neutral electric charge, consists of a positively charged nucleus surrounded by a fixed number of circulating electrons. The number of electrons in circulation is the number required to make the sum of their negative charges equal to the positive charge of the nucleus.

In a simplistic view of this relationship, the electrons and nucleus are attracted to each other, and held together, by their equal but opposite electric charges. In a more sophisticated view, the energy level of the association of electrons and nucleus is less than the energy level would be were they separated. It is, then, the energy difference between these two levels that defines the cohesive, or bonding, force.

An atom missing one or more electrons is called an *ion*. This is a particle whose positive electric charge is equal to the negative charge of the missing electrons. A molecule missing one or more electrons is also called an ion. There are, then, both *atomic ions* and *molecular ions*.

A complete molecule, which has a neutral electric charge, consists of two or more positively charged nuclei surrounded by a fixed number of circulating electrons. Again, the number of electrons in circulation is the number required to make the sum of their negative charges equal to the sum of the positive nuclear charges. As shown in Fig. 3.2, some of the electrons circulate around only one nucleus, some around another nucleus, whereas still others circulate around two or more nuclei. Two nuclei associated as one molecule by sharing electrons represent less energy under some conditions than they would if they functioned as two separate atoms. The difference in energy content of these two conditions—separate or sharing—is a measure of the binding force provided by the shared electrons.

A body of solid material represents a three-dimensional continuum in which each unit-volume is somewhat analogous to a giant molecule. A very large number of ions are held together in a repetitive geometric relationship in which the negative electric charges of the electrons in each unit-volume cancel the positive charges of the ions in the same unit-volume. Once again, however, the energy level of a continuum consisting of neutrally charged unit-volumes is less than the energy level would be of the same number of separate, neutrally charged molecules or atoms.

When ions and electrons are associated as a continuum, the pattern of structural possibilities is strictly limited. Although each type of ion has a different atomic weight, all have one of a very few possible electric-charge configurations. Since the unit-volume neutralization of electric charge is an overriding consideration, the continuum must be assembled with this fact in mind. A repetitive geometric relationship of this kind is called a *CRYSTAL LATTICE.*

Figure 17.1 represents a very small section of a greatly simplified crystalline structure in which 12 positive ions are held in their relative positions, one to every other, by 20 pairs of bonding electrons. Another 32 pairs, not shown, would be needed to tie the illustrated section into a continuing structure. In the illustrated structure, each ion carries the equivalent of six positive charges, and is surrounded by 12 electrons that it shares with other ions for one-half of the time. This arrangement neutralizes the unit-volume electric charge.

The electron pairs shown between each two ions provide the binding forces that hold the structure together, but these are not rigid. Their flexibility is indicated by the stylized springs shown between each ion and each of its shared electron pairs. One ion in the illustration is shown vibrating within the constraints of these springs (bonds). In actuality, all particles, ions and electrons, are similarly vibrating under all conditions except those near a temperature of absolute zero. The higher the temperature, the more violent the vibrations. In fact, at all temperatures, a few ions may break loose from the surface of a material and escape into space. This process is called *sublimation*.

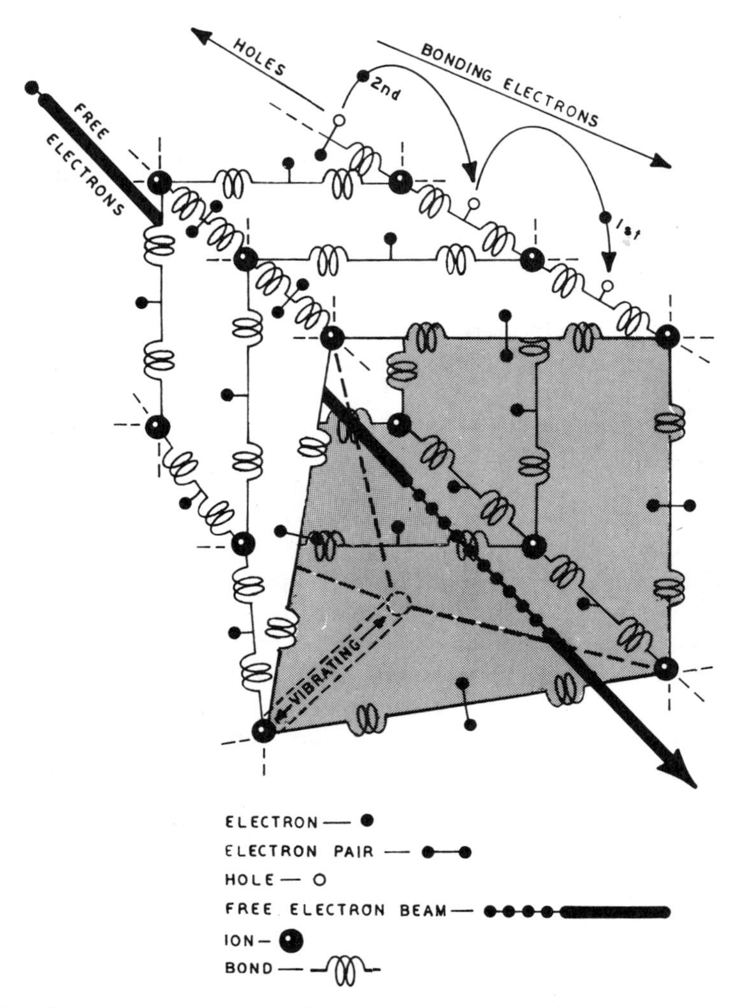


Fig. 17.1. Electron movements through a crystal lattice: In one of two mechanisms, bonded electrons hop from one lattice position to another; in the second, free electrons pass through the spaces between wildly swinging lattice elements.

At some particular temperature, all of the crystalline bonds break, and the material dissolves into a fluidic combination of positive and negative ions, into a gas of neutrally charged molecules, or into a gas of neutrally charged atoms. At very high temperatures, electrons break away from nuclei, and the combination of free electrons and stripped nuclei is called a *plasma*. In all circumstances, the vibrational or linear movements of particles represent thermal kinetic energy expressed as a temperature, whereas the bonds themselves represent thermal potential energy called *LATENT HEAT.*

## 17.3 SOLID MATERIAL

A body of solid material is constructed from a very large number of positively charged ions and negatively charged electrons. These particles join forces to form a body because this association takes less energy to maintain than does any other condition that can be imagined for them.

As might be expected, the energy content of a body of material is present in both its potential and kinetic forms. The potential form is represented by a particle occupying a specific, unchanging position within the lattice structure, whereas the kinetic is represented by particles in motion. Such movements can take on two different forms. In the first, particles continually change their position in space by moving through the lattice. In the second, particles vibrate within the constraints that bind them to their fixed location within the lattice. As a general statement, all of the ions and most of the electrons are bound to specific, fixed locations, whereas only a few of the latter are free to move through the lattice.

When there is no interchange of energy between a body and its surrounding environment, a state of equilibrium is ultimately reached in which all of the phenomonological characteristics are exactly the same for each unit-volume. These include temperature, electric potential, carrier population density, mean vibration amplitude, the mean velocity of the carriers, and the *MEAN FREE PATH* of the carriers. The latter term refers to the distance a carrier moves before it collides with a vibrating particle.

The term *MEAN VELOCITY* is derived from the statistical nature of carrier activities. Although each carrier travels in its own unique direction at its own unique velocity, the overall effect follows distribution theory in both time and space. Individual activities may fall anywhere within a wide range of possibilities, but the mean activity is a constant for a given set of conditions. The same logic applys to the vibratory activities of lattice particles.

The maintenance of equilibrium depends on various means of energy exchanges between all of the particles in a body of material. The lattice particles transmit energy through their bonds, whereas interchanges between lattice and carriers are accomplished as a continuing series of particle collisions in which either particle may gain or loose energy. A not unreasonable view of each unit-volume, then, is of a seething mass of widely swinging lattice particles busily banging away at a few carriers that dash madly about among the lattice interstices with random velocities and in random directions. All activities, including collisions, however, are statistical in nature and follow distribution theory. Although there is a continuing interchange of energy between individual particles, there is no net interchange between groups of particles once equilibrium has been established.

As a consequence, if the temperature of every unit-volume is not the same, heat will flow between the unit-volumes until it is the same. Furthermore, if the electric potential of every unit-volume is not the same, as a result of carrier populations that are not the same, carriers will travel between the unit-volumes until the potentials are the same. In short, gradients of any kind cannot be maintained within a body of homogeneous material except as a result of energy supplied from a source external to that body.

## 17.4 CARRIERS

By virtue of their negative electric charge, moving electrons carry electric currents through all solid materials. Electric conduction, then, consists of a general drift of a number of electrons in one particular direction that exceeds the number moving in any other direction. Such movements result in a net displacement of electric charge with a body of material.

Electrons can move by either of two quite different mechanisms, as illustrated in Fig. 17.1. In the first of these mechanisms, electrons free of all molecular bonds move throughout a volume of material without particular concern for any bonding constraints. The freely moving electrons carry their negative electric charges through the spaces between the nodal points of the lattice structure. The broad arrow in Fig. 17.1 represents a possible path of free electrons.

In the second mechanism, carriers move as a part of a lattice structure. Bonding electrons obtain their freedom from their bonds for only a short period of time. During this short interval, they move from one bond position to another in a series of discontinous steps. The bonding electron "hops" from a bond adjacent to one ion to a second nearby ion, which is missing one of its bonding electrons. As the bonding electron hops to fill the first missing bond, it leaves behind a similarly missing bond. This second missing bond is then filled by a seond hopping electron that leaves behind a third missing bond. The process continues in this manner in both time and space. Under conditions of equilibrium, the process is random, and there is no net movement of electric charge. On the other hand, electric conduction is accomplished as a total of all those particular individual electron "hops" that happen to result in a net movement of negative electric charge.

In the process of electric conduction, the electrons carrying their negative electric charges all move in the same direction, as illustrated in Fig. 17.1. The position of an unfilled bond, called a *HOLE*, however, moves in the opposite direction. Since an unfilled bond is represented by a position in space where a part of an ion's positive charge is not neutralized by an electron's negative charge, a positive charge is invariably associated with a hole. Since holes not only carry positive charges but move in directions opposite to those of hopping electrons, they assume the characteristics of positive carriers.

Electric conduction within any solid material is thus carried by electrons, by holes, or by a combination of both.

## 17.5 CARRIER AVAILABILITY

The processes of conduction are dependent on the availability of carrier electrons and/or carrier holes in a particular *POPULATION DENSITY*. This availability is a part of a particular material's unique chemical nature when it is exposed to a particular environment. Both the electrons and the holes exist in their own, individual characteristic population densities. Depending on the circumstances, these population densities can vary over a wide range of possibilities including zero density.

Carriers of either type are made available by two different mechanisms that may function either separately or together. In one of these, the energy of either the bonding electron itself or its associated ions becomes great enough to rupture the electron's bonds, thus releasing the electron from its bonding position. The rupturing energy can be supplied by electrical, mechanical, thermal, or electromagnetic means. This mechanism creates holes and free electrons as pairs. After these pairs loose energy, any electrons and holes that happen to meet as they both travel through the lattice on more or less independent paths combine again in a fixed lattice location and disappear as carriers.

All of these processes, however, are both simultaneous and statistical in nature. A certain number of bonding electrons are continually achieving enough energy to break loose, whereas another number of free electrons lose enough energy to return to bondage. In the process, a certain number of electrons are left free. With only temperature considered as a variable, the number free at any time is a very precise characteristic of a particular material at a particular temperature; the higher the temperature, the more carriers are available.

Semiconducting materials whose carrier population is measured in terms of electron–hole pairs are called *INTRINSIC* or *I-type* semiconductors.

The second mechanism is chemical in nature. Here impure ions are introduced into an otherwise uniform lattice structure at the time the lattice is originally formed—when it solidifies from the molten state. The bonding possibilities presented by an impure ion do not exactly match the bonding requirements of a particular lattice structure. Impure ions can be either surplus or deficient in their bonding requirements. If an impure ion is surplus, it provides a free electron; if deficient, it supplys a hole.

Semiconducting materials whose carrier population is measured in free electrons derived from impure ions with surplus bonding requirments are call *N-type*, whereas impure ions with a deficiency in bonding requirements creating holes are called *P-type*.

Metals offer very large numbers of free electrons; semimetals exhibit medium mixtures of electrons and holes; intrinsic semiconductors provide a few electrons and a few holes. N-type semiconductors have a few surplus electrons that depend on the number of impurity ions. P-type semiconductors have small numbers of holes that also depend on the number of impurity ions. Both N-type and P-type semiconductors become intrinsic at high temperatures.

It is thus seen that holes and electrons can exist in materials and environments in various quantities, either alone or together. Furthermore, the population densities of both holes and electrons and the ratios of holes to electrons vary under varying environmental conditions.

## 17.6 HOT/COLD CARRIERS

The temperature of a given volume of material is a measure of its energy content. This content consists of the cumulative effect of the individual energies of each molecular particle, carriers as well as the bonded ions. Although the temperature partly reflects a unit-volume constant, the energies of the constitutent particles are subject to distribution theory and individually represent a wide range of contantly changing possibilities.

The energy of a particle is measured as a product of one-half its mass times the square of its velocity. A particle with a high energy content manifests this fact by moving relatively fast. Conversely, a similar particle with less energy moves proportionally slower. The velocity can be either vibratory, as it is with bonded ions, or linear, as it may be with a carrier. In this sense, energy, temperature, and particle velocity can be represented by a common denominator. A fast moving particle, one with more energy, is said to be *HOT*, whereas a slow moving particle, one with less energy, is said to be *COLD*. The validity of this concept is demonstrated by the establishment of temperature gradients within bodies of material when fast moving carriers are separated from slow ones. Although "hot" and "cold" are relative terms, the anomalous difference between otherwise like particles in similar environments is represented solely by their different temperatures.

## 17.7 CARRIER MOVEMENTS

The general condition of both holes and electrons moving in response to electric forcing fields is illustrated by Fig. 17.2. Electrons move in one direction, whereas holes move in the opposite direction. Together they carry an electric current in the X direction in response to an electric forcing field applied in the same X direction. No forces are present that tend to separate one type of carrier from any other type. Both the mixture of hot and cold carriers and the ratio of electrons to holes are essentially the same for each unit-volume of material.

Figure 17.2 serves equally well to illustrate any ratio of holes to electrons, including those of either all holes or all electrons. It is only necessary to visualize any one of the carriers shown as being either a hole or an electron whether or not it is now so drawn. In response to other forcing fields, it is possible for either of these two types of carriers to travel alone or for both to travel together, as shown. If they travel together, the holes may travel in the opposite direction from the electrons, or they may move in the same direction.

There are, then, four different carrier-movement possibilities: electrons alone, holes alone, together opposite, and together same. In addition, it is possible for cold and hot carriers to move in the same direction or for them to move in opposite directions. This expands the carrier-movement possibilities to the following:

1. Electrons alone; hot and cold together—as in the case of electric current flowing in a metal or an N-type semiconductor
2. Electrons alone; hot and cold opposite—as in the case of heat flowing in a metal or an N-type semiconductor

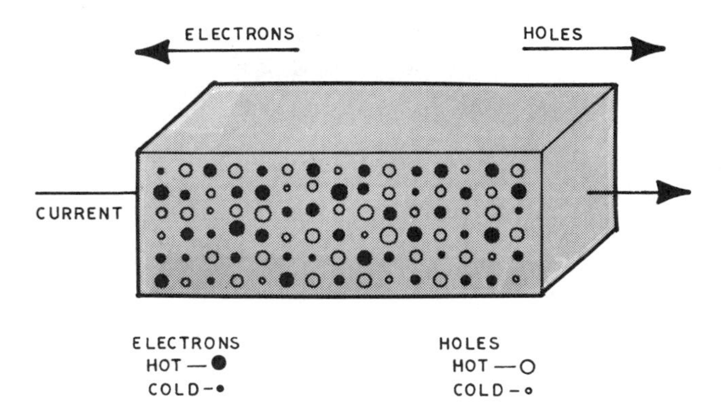


Fig. 17.2. Carriers in a solid: The negative electrons move in one direction; the positive holes, in the other.

3. Holes alone; hot and cold together—as in the case of electric current flowing in a P-type semiconductor
4. Holes alone; hot and cold opposite—as in the case of heat flowing in a P-type semiconductor
5. Together opposite; hot and cold together—as in the case of electric current flowing in either a semimetal or an intrinsic semiconductor
6. Together same; hot and cold opposite—as in the case of heat flowing in either a semimetal or an intrinsic semiconductor
7. Together same; hot and cold same—as in the case of redistribution in response to a pair's population density gradient, which can take place in I-type semiconductors.

## 17.8 FIELD EFFECT

In the *FIELD EFFECT*, carriers are crowded into a smaller effective cross section of a conductor with a resulting increase in conductor resistance. Under some conditions of carrier concentration, it is even possible to stop the flow of carriers in the X direction completely through the effects of voltages applied in the Y direction.

Recalling the discussion of carrier movements in response to longitudinal electric forcing fields, consider the application of a second electric field

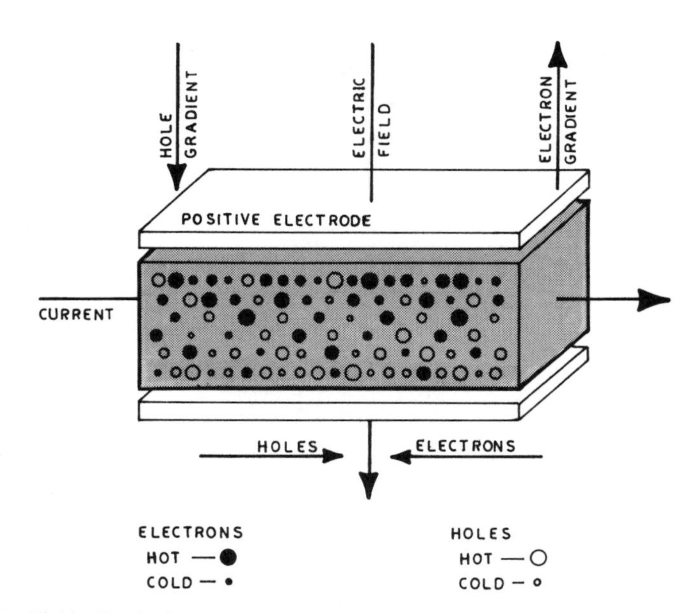


Fig. 17.3. Field effect: If an electric field is applied at right angles to current flow, both electrons and holes will be forced into a voltage gradient opposing that field.

in the Y direction at right angles to the original longitudinal X field. This second field is applied through an insulating barrier, and although it exercises some influence on the carriers, no current can flow in the Y direction in response to a Y field.

Under these circumstances, the carriers redistribute themselves within the material, a response of both holes and electrons that establishes a voltage gradient that is in direct opposition to the imposed Y field. Electrons concentrate near the positive electrodes and thin out near the negative electrode, as shown in Fig. 17.3. If holes are present, they move in the opposite direction from the electrons concentrating near the negative electrode and dissipating near the positive electrode. In all circumstances, the result if the same—the establishment of a voltage gradient within the material in opposition to the imposed Y electric field.

There is, then, a separation of electric charges within a material in which the ratio of holes to electrons is no longer the same in every unit-volume. The mixture of cold and hot carriers, however, remains constant throughout.

## 17.9 LORENTZ FORCES

Reconsider, now, the concepts of Sec. 10.1 (Wire in a Magnetic Field). There a wire cut across lines of magnetic flux as it moved through space, thus imposing a mechanical "forcing field" on the electrons in the wire. These electrons were forced to move through the magnetic field because they were a part of the wire and moved when it moved. They were restricted in their forward motion by whatever forces impeded the motion of the wire, but they were not exposed to electric resistance in the direction of wire motion.

Now, a Lorentz voltage is generated between the two ends of such a wire (Y direction) that is proportional to the velocity of motion (X direction) and to the strength of the magnetic field (Z direction). The Lorentz forces push electrons along the wire in one direction, concentrating more of them at one end than at the other. The voltage that results from this enforced charge separation tries to pull the electrons back into a homogeneous distribution, and an equilibrium is achieved with a particular voltage that is proportional to the velocity of motion.

If current is allowed to flow through the wire, an IR voltage is created in opposition to the originating Lorentz voltage.

## 17.10 FORCING FIELDS

Carriers contained with a volume of solid material are moved by several different kinds of forcing fields.

First, a body of solid material can be moved physically through a magnetic field in response to a mechanical forcing field. In these circumstances, the electrons and the holes move in the same direction along with the bulk of the material. Lorentz forces then deflect the holes in a direction opposite to that of the electron deflections and perpendicular to the direction of motion. Lorentz voltages occur in directions that are perpendicular to mechanical forcing fields as these are shown in Fig. 10.1.

Second, carriers move in response to electric field gradients. Holes move in one direction, whereas electrons move in the opposite direction. Lorentz forces then deflect both holes and electrons in the same direction. Hall voltages occur in directions that are perpendicular to electric forcing fields, as shown in Figs. 17.8, 17.9, and 17.13.

Third, carriers move in conjunction with a flow of heat. Phonons are one mechanism used to transmit heat through solid materials. As they move through a material, they drag or push both holes and electrons in the same direction. Lorentz forces then deflect the holes in a direction opposite to that of the electron deflections. Nernst voltages occur in directions that are perpendicular to heat flux movements, and these, in turn, are a response to thermal forcing fields, as shown in Figs. 17.11 and 17.12.

Fourth, carriers have a second response to thermal gradients, and there is a tendency to equalize the energy content of all carriers throughout a given volume of material. If anything happens to increase the energy content at any one point in space, carriers respond in a process of equalization. Cold carriers, either holes or electrons, move toward a source of heat, whereas hot carriers move away from such a source. Lorentz forces then deflect the hot holes in a direction opposite that of the hot electrons, whereas the cold holes are deflected with the hot electrons and the cold electrons with the hot holes. Since the hot carriers move faster than the cold ones, voltage gradients result. This, along with the voltage gradients that result from phonon drag, is the Thomson part of a thermoelectric effect. Nernst voltages occur in directions perpendicular to thermal forcing fields, as shown in Figs. 17.11 and 17.12.

Fifth, carriers move in response to population density gradients. If anything happens to increase the intrinsic carrier population density for any given material at any absolute energy level, carriers move away from a point of concentration toward a point of lesser density. This motion is analogous to a response to a local change in the pressure of a gas, in which the population density of carriers is somewhat similar to the molecular density of the gas. Since holes and electrons are commonly created as pairs in these circumstances, their local pressures increase together. If so, they move off together in the same general direction. Lorentz forces then deflect holes in a direction opposite to that of electron deflections.

PEM voltages occur when electron–hole pairs are created by photon–material interactions, as shown in Fig. 17.21.

## 17.11 DYNAMIC MECHANISMS OF RESISTANCE

Electric currents are carried through solid materials by free electrons moving through the interstices of lattice structures. This concept is depicted by the electron beam of Fig. 17.1. In a manner analogous to fluid flow, the electric current may be envisioned as the passage of some finite number of electrons in a unit of time, either a few electrons moving rapidly or many electrons moving slowly. As used here, the term "free" is only relative. Although free electrons are not restricted to one point in space as ions and bonding electrons are, they are involved with other processes that influence their movements. In fact, free electrons are only "free" because they are a vital part of these other processes, and their freedom is constrained by their participation in these processes.

The flow of an electric current is opposed by these other processes. Energy is abstracted from an electron stream and transmitted to the structure through which this stream flows by means of the mechanisms provided by these other processes. In this case, an electric resistance is somewhat analogous to a fluid's viscosity. Since energy is consumed in overcoming electric resistance, an electric current can only be sustained when this energy is continually supplied by an external forcing field. Electric resistance is a measure of this energy transfer.

Three of the above mentioned processes are relevant to this discussion:

In the first, each separate electron is accompanied by an identical electric charge. These electric charges exercise repelling forces, one against every other, that tend to distribute the free electrons uniformly throughout a body of material. Or, to look at this same phenomenon in another way, free electrons exposed to a voltage gradient move from a unit-volume of relatively negative potential to a unit-volume of relatively positive potential. Since each electron carries a negative electric charge, this movement lowers the positive potential and raises the negative potential. Free electrons continue to move in response to any voltage difference to which they may be exposed until all unit-volumes are at the same potential. In these circumstances, a free electron population density gradient is directly equivalent to a voltage gradient because a population density gradient creates the latter. As a consequence of such activity, a nonuniform distribution of free electrons can be maintained only if an external force balances these internal redistribution forces.

In the second, the temperature of a unit-volume of a solid material is a manifestation of an integration of various kinetic energies represented by the vibratory velocities of individual lattice particles and by individual free electron velocities. Free electrons are thus in constant motion as a result of a material's particular temperature. This motion, of course, is constrained by the objective of uniform density. Motions imposed by an externally applied forcing field must be added to these basic thermal movements.

In the third, the continuing interchange of energy between vibrating ions transmitted along structural bonds and the interchange of energy accomplished in collisions between lattice ions and free electrons are basic mechanisms for achieving and maintaining a common temperature throughout a body of solid material. These mechanisms keep the energy content of any particular particle, ion, or electron in a constant state of energy interchange with adjacent particles. As a part of this process, a free electron travels in a certain direction, at a certain velocity, for a certain distance until it collides with a lattice ion. The collision alters its energy content so that the electron travels at a different velocity, in a different direction, for a different distance until it collides with a second ion. This process of travel-and-collision for thermal reasons continues ad infinitum. A single free electron is thus constantly moving through the interstices of a lattice in a disjointed, random series of small trajectories spaced by ionic collisions.

The actual directions, velocities, and distances between collisions of all free electrons conform to distribution theory in both a temporal and a spatial sense. The mean direction is zero in the absence of a forcing field, whereas both the mean velocity and the mean distance between collisions are characteristic of a particular material at a particular temperature. (Mean distance is here used synonymously with mean free path.) In short, by their thermally imposed collisions, free electrons are constantly interchanging energy with vibrating lattice ions. Because of this mechanism, the energy content of any group of particles, ions, or electrons undergoes a continuous redistribution to all other particles.

A forcing field causes a "seething mass" of electrons to drift slowly in one direction without diminishing its random activities. In fact, the random activities are increased slightly by a drift velocity since the energy necessary to maintain the drift is added to the otherwise random velocities. This added velocity, representing added energy, is transmitted to the lattice through increases in both the mean collision rate and the mean energy transmitted in each collision. As a result of the electric current flow, latttice vibratory activities increase. The increase is measured as an increase in temperature.

Free electrons respond to forcing fields by adding unidirectional vectors to each random trajectory. These vectors represent accelerating forces as long as electrons are free to move. Each electron responds to such forces by increasing its velocity in the vector direction. Acceleration is effective only over the distance between collisions however, since the velocity in the vector direction is significantly reduced after each collision. Free electrons then move in the vector direction in a series of accelerating-decelerating steps. The net result is a movement in the vector direction at some effective velocity, which is a function of both the strength of the forcing field and the mean distance between collisions. Under these circumstances, the vector previously mentioned represents effective velocity rather than acceleration.

An analogy can be found in the motion of a free-falling-body in the earth's atmosphere. The body is exposed to gravity—a constant accelerating force—and opposed by the viscosity of the air—a force that is a function of velocity. The free-falling body reaches a terminal velocity when the accelerating force equals the viscosity force. The free-falling effective velocity, then, is a function of both the accelerating force of gravity and the decelerating force of viscosity.

In solid materials, the following things occur as the temperature increases: The amplitude of ionic oscillation increases, the collision rate increases, and the mean distance between collisions of free electrons and the lattice ions decreases. As a result, electric resistance increases.

## 17.12 STATIC MECHANISMS OF RESISTANCE

In Fig. 17.1, a beam of free electrons passes through a crystal lattice without interacting with the ionic particles that make up the lattice structure. The beam may be envisioned as proceeding unimpeded down an empty "corridor." As discussed previously, this would be impossible if the ions were vibrating, since vibrating ions would periodically intrude in the corridor and collide with the beam.

Near absolute zero, ions do not vibrate. In the absence of ionic vibration, collisions may cease, and the dynamic mechanims for energy transfer from the free electrons to the lattice structure would be lacking. As a matter of fact, certain unblemished crystalline materials do not have an electric resistance when their temperatures are near absolute zero. Crystals in general, however, *are* blemished. The spatial periodicity of the lattice is not completely uniform throughout the body of material, and the imperfections impose barriers in the path of an electron beam. Such barriers scatter electrons from the beam with results similar to those achieved in collisions between vibrating ions and the free electrons.

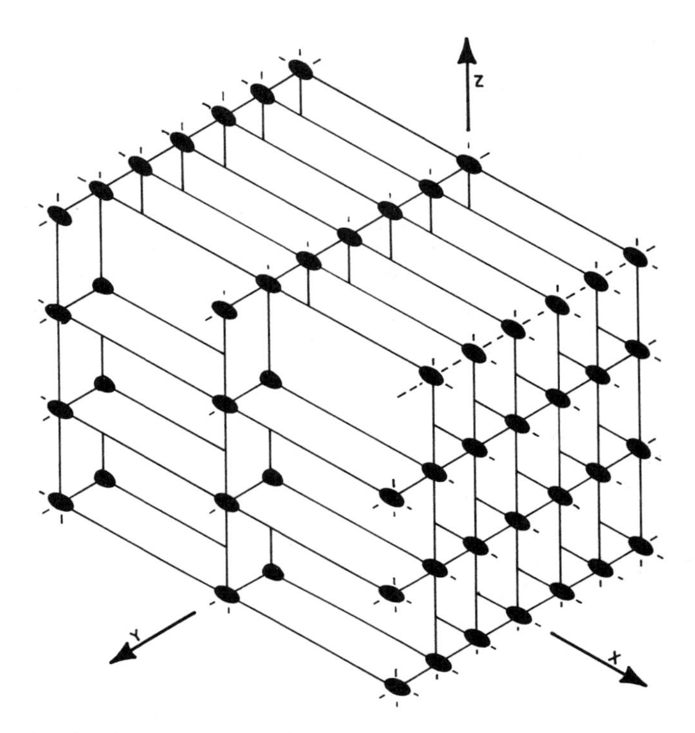


Fig. 17.4. Directional characteristics of lattice structures: Spacings between crystal nodes may vary significantly in each of the crystallographic-axis directions.

The possible interruptions to spatial periodicity are numerous: An ion can be missing from the location where it is supposed to be—a vacancy; an ion can be dislodged from its assigned location—a dislocation; the ions at various nodal points can be sufficiently different from one another to block the corridor—as in an alloy or an interruption of the spatial periodicity by a stacking fault.

Figure 17.4 is a simplified expansion of Fig. 17.1. It purports to illustrate a crystalline lattice in which the spatial periodicity significantly differs in the X, Y, and Z directions and allows one to envision the circumstances in which the amplitude of a given ionic vibration will have completely different effects on an electron beam depending on the axis along which the beam travels. This, of course, results from the fact that amplitude represents a different percentage of the lattice period in each of the three different directions. The same can be said for the various static mechanisms. In fact, materials can be found in which the X, Y, and Z resistivities differ by orders of magnitude.

## 17.13 RESISTANCE

If an electric forcing field or "voltage gradient" is established within a body of solid material, certain phenomenologic consequences may be observed.

First, an electric current flows through that body if the necessary carriers are present. If the material is an insulator, which by definition is devoid of carriers, no current can flow. If the material is a metal, which by definition has an abundance of carriers, the flow of current, according to Ohm's Law, is proportional to the voltage and the inverse of the constant of proportionality, which is called *RESISTANCE.* Each particular type of material has its own characteristic resistance. *OHM'S LAW* is stated mathematically as follows:

OHM'S LAW $$I = E/R \quad \text{(Eq. 17.1)}$$

where $E$ is the voltage gradient that causes a current to flow; $I$, the current that flows; and $R$, the resistance experienced by the current flow.

If the material is a semiconductor, which by definition has a limited number of carriers (and only a few at that), the flow of current does not necessarily follow Ohm's Law because the number of available carriers is subject to change under varying environmental conditions as well as restricted by particular environmental conditions.

Second, the resistance of a body of material to the flow of an electric current is a function of both the dimensions of that body and a material characteristic called *RESISTIVITY.* Resistance may be stated mathematically as follows:

RESISTANCE $$R = rl/A \quad \text{(Eq. 17.2)}$$

where $R$ is the resistance to the flow of an electric current when $l$ is the length of the body, $A$ is the cross-sectional area of the body, and $r$ is the material's resistivity.

Resistivity may be regarded as a per-unit-volume resistance when a body is constructed as a 1-meter cube so that the length of path traversed by the current is 1 meter and the cross-sectional area traversed is equal to 1 square meter.

As shown by Eq. 17.2, the resistance experienced by an electric current increases either if the path traversed by the current increases in length or if the cross-sectional area through which the current flows decreases.

Third, heat is added to a conductor when an electric current experiences resistance in flowing through it. According to *JOULE'S LAW*, the amount of heat transmitted is as follows:

JOULE'S LAW $$H = kRI^2t$$ (Eq. 17.3)

where $H$ is the amount of heat transmitted to the material; $I$, the current; $R$, the resistance; $t$, the time of current flow; and $k$, a constant of proportionality depending on the units used to describe the other factors.

## 17.14 MOBILITY

*MOBILITY* is an expression used to describe the ease with which carriers are moved through solid materials by forcing fields. The significance of the "mobility" concept to this discussion lies in the fundamental difference between an "ohmic voltage" and a "galvanomagnetic voltage."

As described by Eq. 17.1, an ohmic voltage is proportional to electric current, which, in turn, depends both on the number of carriers present and on the mean velocity of these carriers in the forcing-field direction. A galvanomagnetic voltage, on the other hand, depends on the number of carriers moving in the primary forcing-field direction, on their mean velocity in this direction, and on the mobility of these same carriers in response to secondary Lorentz forces that are directed at right angles to the primary forcing-field direction. The number of carriers present does not affect a galvanomagnetic voltage, only the number actually participating in primary current flow.

From a mathematical standpoint, mobility is the magnitude of a carrier's drift velocity achieved in any forcing field's direction divided by the strength of that forcing field, or

$$Mobility = Velocity/Voltage$$

Here the velocity achieved is the effective velocity previously discussed in Secs. 17.11 and 17.12.

Under any circumstances, molecular or otherwise, the velocity achieved by a body on experiencing acceleration is the product of that acceleration and the distance traveled, or

$$Velocity = Distance \times Acceleration$$

Moreover, the acceleration experienced by any body is the magnitude of the forcing field to which it is exposed divided by its mass, or

$$Acceleration = Voltage/Mass$$

By combining the above quantities and choosing appropriate units, we obtain

$$Mobility = Velocity/Voltage = Distance/Mass$$

The mobility of a carrier is thus proportional to the mean distance between collisions and inversely proportional to its mass.

A broad range of carrier mobilities is encountered in various materials because both the mean distance and the effective mass are characteristics of a particular material exposed to a particular environment. The variation of mean distance with lattice structure, lattice defects, impurities, crystal axis directions, and temperature is discussed in Secs. 17.11 and 17.12. The possibility of mass variations may not be immediately obvious since the mass of any particular object that humans might encounter in their daily experience does not change with time or other variables and all electrons can be shown to have exactly the same characteristics in every respect. Our previous equations reveal, however, that

$$Mass = Voltage/Acceleration$$

Thus, if the acceleration experienced by a carrier, while exposed to the same voltage differs under different circumstances, that carrier's mass will be different.

When an electron is part of a structure, either in bonding or relatively free, it does not respond to an accelerating voltage to the same degree it would if it were completely free. The molecular fields to which structural electrons are exposed, and to which they contribute, affect their view of an externally applied accelerating force. The accelerating field must be shared with the local fields if the acceleration experienced depends on the strength of the local fields as well as of the accelerating field. As a consequence, the mass of a particular electron appears to be a function of the local fields, and, as such, is labeled an *EFFECTIVE MASS*.

## 17.15 CONDUCTIVITY

Conductivity is a per-unit characteristic of materials that is reciprocal to their resistivity (see Sec. 17.13) and quantifies the ability of the carriers present to carry a current. It may be expressed as follows:

$$Conductivity = Per\text{-}unit\ current/Per\text{-}unit\ voltage$$

Since

$$Per\text{-}unit\ current = Carrier\ density \times Carrier\ velocity$$

combining the above, we obtain

$$Conductivity = (Density \times Velocity)/Voltage$$

with the result that

CONDUCTIVITY $$c = mn$$ (Eq. 17.4)

where $c$ describes the conductivity of a material when $m$ is the mobility of the carriers present and $n$ is the carrier density.

## 17.16 GALVANOMAGNETORESISTANCE

The term *magnetoresistance* describes a change in electric resistance experienced when a conductor is placed in a magnetic field.

*GALVANOMAGNETORESISTANCE* is only one of several possible resistive responses to magnetic fields. In the Gauss Effect (see Sec. 18.2), a magnetic field alters the configuration of a crystal lattice, thereby af-

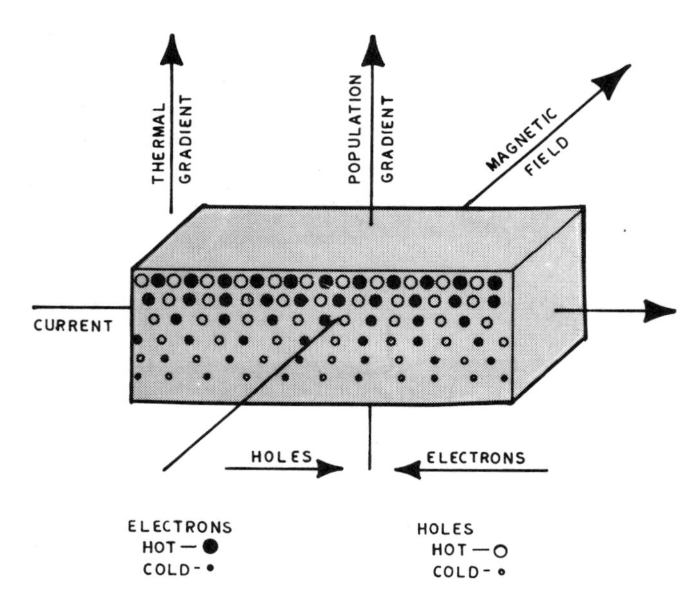


Fig. 17.5. Adiabatic galvanomagnetoresistance: Since Lorentz forces push the fast carriers of a current-carrying body harder than the slow ones, a thermal gradient is established when such a body is immersed in a magnetic field.

fecting carrier mobility, whereas in a Corbino device (see Sec. 17.29), the forward current generates an opposing back electromotive force.

The overall effect of galvanomagnetoresistance is divided into two component parts. These are a reduction in carrier forward velocity as a result of the carriers being forced to move sideways as well as forward and a reduction in the effective cross-sectional area of the conductor as a result of the carriers being crowded to one side. If a carrier moves at an angle to the direction of current flow and the mean free path in the angular direction remains the same as it would have been in the current direction, the effective forward mean free path decreases in the direction of current flow. To the degree that conductivity depends on the mean free path, a reduction in the latter also entails a reduction in conductivity. The greater the carrier mobility, the greater the tendency toward deflection, and hence the greater the effect of a magnetic field on conductivity.

Galvanomagnetoresistance can be detected in almost any type of material, but it appears to have fairly large values in semimetals because of their relatively large carrier population densities and high carrier mobilities. The semimetal bismuth has a particularly large coefficient of galvanomagnetoresistance.

Galvanomagnetoresistance is associated with a thermal separation of carriers in which hot carriers are deflected more than cold carriers. The term *adiabatic* refers to those circumstances in which this separation is maintained, in equilibrium, as a thermal gradient at right angles to the direction of the primary current flow. Under such conditions, there is no heat flowing through the material. Depending on the type of carriers present, the *ADIABATIC GALVANOMAGNETORESISTANCE* illustrated in Fig. 17.5 might represent that exhibited by either a semimetal or an intrinsic semiconductor.

In *ISOTHERMAL GALVANOMAGNETORESISTANCE*, heat is allowed to flow in response to the thermal gradient just described. A back emf associated with this heat flow increases the value of magnetoresistance over that which would otherwise be detected in an adiabatic state.

## 17.17 MAGNETORESISTIVE DEVICE

Figure 17.6 illustrates a possible configuration for a magnetoresistive device. By rotating the magnetic structure as shown, any part of the resistive element may be either exposed to, or not exposed to, a magnetic field. The resistance of such a device thus becomes a function of the angle of rotation.

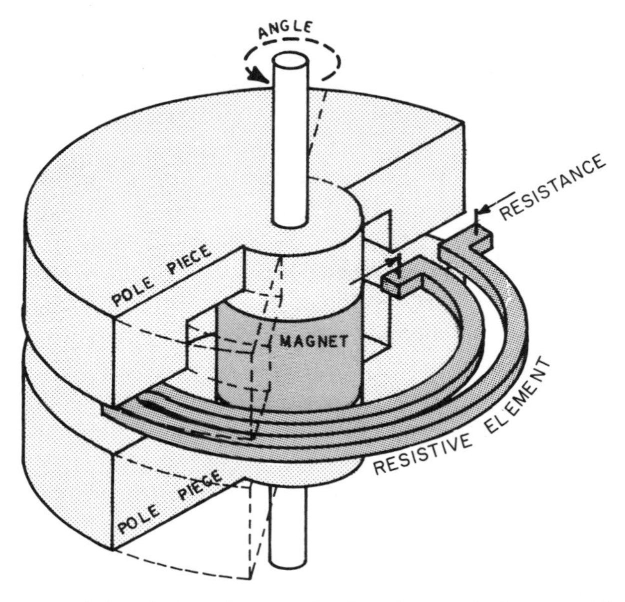


Fig. 17.6. Magnetoresistive device: As a result of carriers pushed to one side by Lorentz forces, the resistance of a conductor is increased when it is placed in a magnetic field.

## 17.18 TWO-MATERIAL MAGNETORESISTANCE

The phenomenon of magnetoresistance can be enhanced by combining two intrinsic semiconductor materials, each representing a different carrier lifetime, into one element, as illustrated by Fig. 17.7. In the absence of a magnetic field, the resistance of such a device is determined by the material with the longer carrier lifetime. Resistance, in this case, is a function of a characteristic carrier population density, and this, in turn, is determined by carrier lifetime. In the presence of an appropriately directed magnetic field, both electrons and holes are swept from the material of longer lifetime into the material of shorter lifetime, where they combine and are lost as carriers.

In a device consisting of only one intrinsic material, magnetoresistance is primarily a consequence of crowding carriers to one side of a conductor. Although the increase in population density on that one side does increase the recombination rate and thus reduces the number of carriers present, this is a secondary function. The loss of carriers caused by recombination can be made a more significant factor and the degree of magnetoresistance increased by combining two materials of the type described.

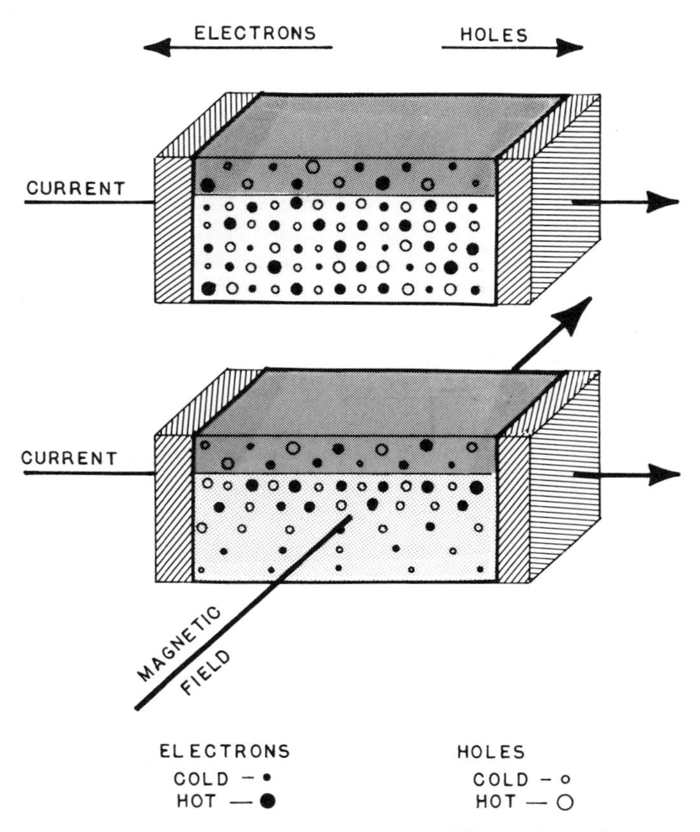


Fig. 17.7. Two-material galvanomagnetoresistance: An ability to change the magnetoresistance of a body may be enhanced when the conducting path is fabricated from materials of markedly different magnetoresistances.

## 17.19 ADIABATIC NEGATIVE HALL EFFECT

A voltage may be induced in a current-carrying conductor by placing that conductor in a magnetic field. This voltage will appear in a direction that is perpendicular to the directions of both the electric current and the magnetic field. Its generation is called the *HALL EFFECT,* whereas the voltage itself is called a *HALL VOLTAGE.*

As shown in Fig. 17.2, the carriers of an electric current are of two types, the positive carriers moving in one direction and the negative carriers in the opposite direction. As shown in Fig. 2.2, negative carriers traveling in one direction through a magnetic field experience a force that tends to deflect them from a linear path. At the same time, positive carriers traveling in the opposite direction through the same magnetic field tend to be deflected in the same direction. This deflection by Lorentz forces of all current carriers toward the same side of a conductor establishes a carrier population gradient in which more carriers are found per unit volume on one side of a conductor than on the other.

Figure 17.5 shows the general result when both types of carriers are present in equal quantities. Here the potential gradients established by each type of carrier are in opposition, and their effects cannot be detected externally. On the other hand, if one type of carrier is more numerous than the other type, the carrier population gradient results in a Hall voltage gradient, and a Hall voltage can be detected across the width of the conductor. Figure 17.8 pictorializes a condition in which all carriers are electrons carrying a negative charge, and Fig. 17.9 shows one with only positive holes.

If no transverse current is possible in response to a Hall voltage, the Lorentz forces and the Hall voltage establish equilibrium. In this case, the Lorentz forces seek to form the gradient across the conductor, whereas a Hall voltage tries to re-establish homogeneiety in the carrier density throughout the volume of the conductor. The *HALL ELECTRIC FIELD*, which is transverse to both the current and magnetic field directions, is determined as follows:

HALL ELECTRIC FIELD $$F_H = K_H BJ \qquad \text{(Eq. 17.5)}$$

where $F_H$ is the Hall electric field (voltage per unit conductor thickness), $K_H$ is the *HALL COEFFICIENT* (whose sign and value can vary widely under varying circumstances), $B$ is the strength of the magnetic field, and $J$ is the current density flowing through the conductor (current per unit cross-sectional area).

The word "adiabatic" is used to describe those circumstances in which no heat flows either into, or out of, a system. In Figs. 17.8 and 17.9, a layer of insulation on both sides of a conductor keeps both heat and electric current from flowing in the transverse direction.

Since a Hall voltage depends on a nonhomogeneous distribution of carriers, it can be sustained within a body of material only as a result of energy supplied from a source external to that body. This energy is supplied by the electric field that drives the primary current through the material.

Two voltage gradients are established in a galvanomagnetic material. The primary voltage gradient is a result of the primary current density times the resistivity of the material, whereas the Hall voltage gradient is a result of the primary current density times the Hall coefficient. As these two gradients are at right angles to each other, they may be considered to have a vector sum whose effect is directed at some angle to the direction

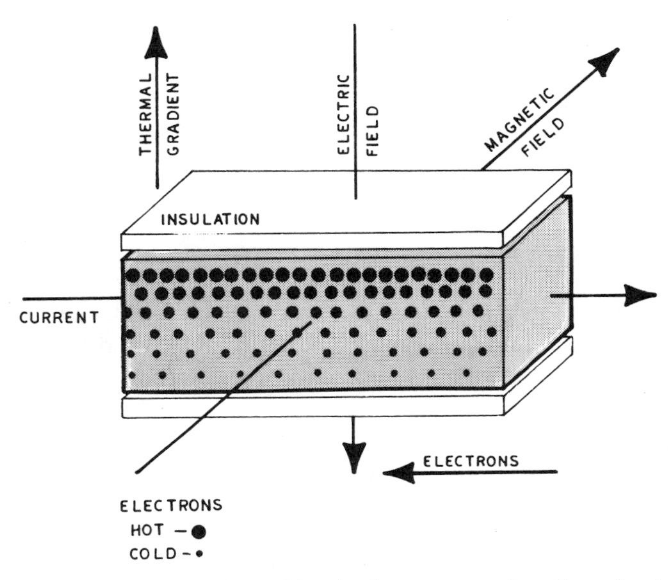


Fig. 17.8. Adiabatic negative Hall Effect: With only electrons as carriers, thermal and electrical gradients are in opposite directions.

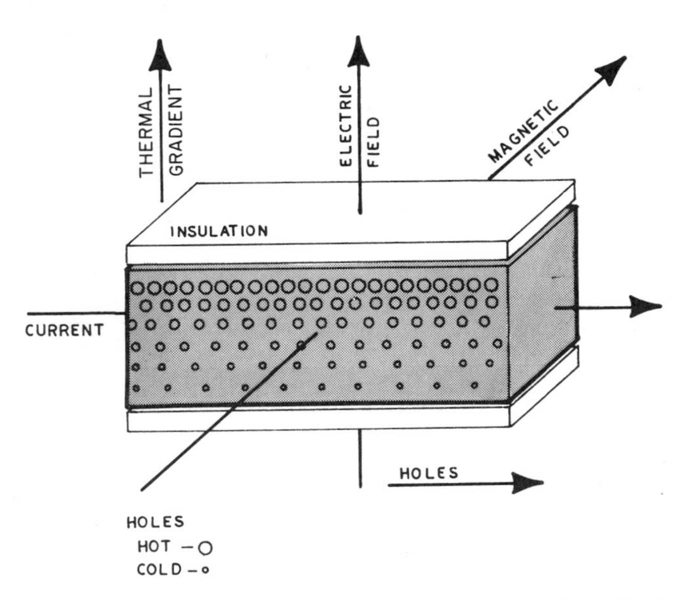


Fig. 17.9. Adiabatic positive Hall Effect: With only holes as carriers, thermal and electrical gradients are in the same direction.

of the primary current flow. This angle, measured as the ratio of the electric field generated *across* the current to the electric field generated *by* the current, is called the *HALL ANGLE*. The Hall angle may be either positive or negative in relation to the direction of current flow depending on whether positive or negative carriers dominate.

## 17.20 ADIABATIC POSITIVE HALL EFFECT

The Hall Effect is a majority carrier mechanism that depends on the bulk properties of the conducting material. With metals and N-type semiconductors, the carriers are electrons, and, with p-type semiconductors, the carriers are holes.

As discussed in Sec. 17.19 and illustrated in Fig. 17.5, the holes of a carried current are forced toward the same side of a conductor as the electrons are. If the holes and electrons occur in equal quantities, they creat two Hall voltages in opposition to each other. If the quantities differ, then one or the other Hall voltage dominates and may be measured in a Hall device.

For positive carriers, the direction of Hall voltage needed to oppose Lorentz carrier deflections is opposite that needed for negative ones. With metals and N-type semiconductors, it is considered to be negative; with P-type semiconductors, positive. If both types of carriers are present, as shown in Fig. 17.5, but their effects are not equal, the result might be either a positive or negative Hall voltage or even a change from one to the other under varying field or temperature conditions.

## 17.21 SUHL EFFECT

The orientation of magnetons within a semiconducting material affects the rate at which holes can be injected into that material. The manipulation of magneton directions for this purpose is called the *SUHL EFFECT* (see Fig. 17.10).

## 17.22 ETTINGHAUSEN EFFECT

The ETTINGHAUSEN EFFECT is the thermal gradient established in conjunction with both a Hall voltage and galvanomagnetoresistance. It is indicated in Figs. 17.5, 17.8, and 17.9 by the different energy and different velocity of the carriers. This thermal gradient is always oriented in the same direction as dictated by the current and the magnetic field. In a material where electrons are the majority carrier, it is in the opposite di-

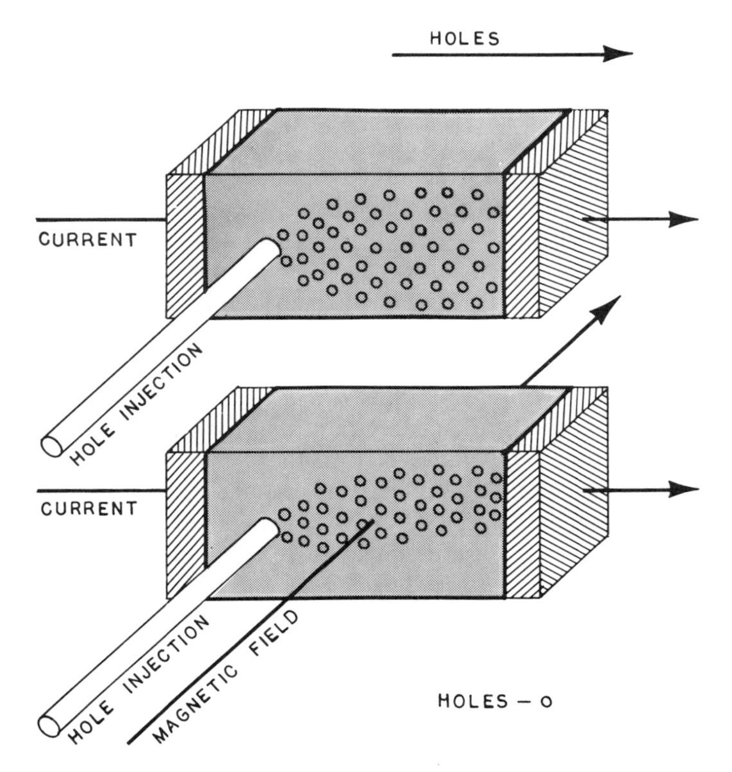


Fig. 17.10. Suhl Effect: The efficiency of hole injection in a positive semiconductor is a function of magnetic field strength.

rection to the electric field, and in a material where holes are the majority carrier, it is in the same direction as the electric field.

In systems where this thermal gradient is not allowed to support a flow of heat, the processes are called *adiabatic*. If heat is allowed to flow to the maximum degree possible, the processes are called *isothermal*. Most devices operate somewhere in between these two extremes.

The leads attached to a Hall device are often made of a material different from the Hall sensitive material. As the Ettinghausen Effect heats one side of the device and cools the other, a thermocouple is created that generates a Seebeck or *FALSE HALL VOLTAGE*. This voltage functions as a zero offset voltage to the true Hall voltage.

## 17.23 HEAT FLOW

When a body of homogeneous material is in a state of equilibrium and no changes of any kind are in process, the energy content of each unit-volume within that body is the same. Under this configuration of uniformity, potential energy is represented by one constant and kinetic energy by another. Moreover, the phenomenologic characteristics of each unit-volume—including temperature, carrier population density, mean activity of lattice vibration, and carrier mean velocity—are also the same.

Not all of these variables are independent, however, since temperature is a manifestation of kinetic energy as this is embodied in both lattice vibration and carrier velocity. Although a particular ion within the lattice structure of a material may vibrate with more violence than some other ion, or a particular carrier may move with more velocity than some other carrier, the temperature of a unit-volume of material is an indication of the mean conditions within that unit-volume. Each material has its own, unique, characteristic set of mean conditions for every particular temperature. If anything happens to disturb these mean conditions so that the kinetic energy content of one unit-volume becomes more than that of some other unit-volume, an energy transfer occurs spontaneously between the two unit-volumes in an attempt to restore equilibrium. This energy transfer is accomplished as an exchange of heat between the unit-volumes.

The term "heat" is commonly used to describe two different, but closely related, phenomena. In the first, heat is the total energy content of a unit-volume of material represented by the sum of the kinetic and the potential energies. In the second, heat is used to describe energy in the process of transfer between one unit-volume and surrounding unit-volumes as a result of temperature differences between unit-volumes.

In any event, if a temperature difference exists within a body of material, there will be a flow of heat from a point of high temperature to a point of low temperature until the temperature difference no longer exists. This transfer is accomplished because heat leaving a point of high temperature reduces that temperature and heat ariving at a point of low temperature raises that temperature. A temperature difference thus represents a *THERMAL FORCING FIELD* directed to accomplish a transfer of heat.

Heat is transported through solid materials by several different mechanisms. In the first, the lattice vibrations that occur at any point within a body of material are transmitted throughout that body by wave propagation through the system of elastic bonds. As a means of visualizing this concept, consider the model illustrated in Fig. 17.1. Here one ion is shown to be vibrating while all others remain stationary. This, of course, is an impossible condition since all of the ions in a body of material are held together by one system of elastic bonds, and if one ion within these bonds vibrates, then all of the others will vibrate in sympathy.

Imagine an initial static situation, however, in which all ions are at rest in their normal positions, with all bonds in balance. Now suppose that someone reaches into this system and manually displaces one ion from its normal position. Doing so will create a second static situation. Since all bonds are elastic, the deflection of one primary ion stresses every bond in the system and displaces every other ion to some degree. Of course, the more remote the bond from the point of imposed stress, the less its strain, and the more remote the ion, the less its displacement. Nevertheless, all bonds and ions are affected by the arbitrary displacement of one ion.

Now, if the ion that was "manually" displaced is suddenly released, it will respond dynamically by vibrating around its normal position with an amplitude approximately twice as great as its static displacement. Some time after this primary ion is released, every other ion experiences an equivalent release and starts to vibrate around its normal position. Since the displacements of these ions are less than the primary displacement, the amplitude of their vibrations will also be less.

This, however, is not the end of the story by any means, for the amplitude of maximum vibration, which is determined by the displacement of the primary ion, is propagated throughout the body of material as an oscillatory wave, and every ion in the system experiences the maximum amplitude of vibration at one time or another. The energy of molecules with more violent vibrations at one point in a lattice is transmitted by this means to molecules of less violence at all other points in that lattice.

Although this movement of vibratory disturbances has the nature of an oscillatory wave, it conforms to Plank's Quantum Theory, in which the quantized unit is called a *PHONON*. In this case, a phonon might be visualized as a unit-package of heat that moves through solid materials by way of their lattice structure with the objective of eliminating temperature differences. Following the logic of heat transfer, phonons move away from points of high temperature toward points of low temperature. Since phonons are derived from the oscillatory motion, in contrast to the linearly progressive displacement, of particles, their movements through materials have no direct electric consequences, and their paths are not influenced by Lorentz forces.

The second mechanism of heat transfer occurs only if carriers are present. In this case, the carriers act as "floating particles" swept along by a stream of phonons. Since they are dragged along by the phonons, they move in the same direction, that is, from positions of high temperature to positions of low temperature. Moreover, since carriers from positions of high temperature continue to maintain the high velocities they achieved in contributing to that high temperature, their arrival in unit-volumes of low temperature raises that low temperature. This principle applies to both hot/cold holes and hot/cold electrons. As a result, carriers of both types tend to concentrate in cold unit-volumes. In such circumstances, carrier movements are subject to Lorentz forces, and their concentration in cold unit-volumes have electric consequences, as will be discussed.

The third mechanism of heat transfer also depends on the presence of carriers but in this case is associated with the random movements of carriers. All carriers within a body of material are in constant motion, a motion maintained by a continuing interchange of energy between the carriers moving through the lattice and the ions vibrating around their fixed nodal positions in the structure of the lattice. In the absence of forcing fields, this motion is random both in direction and in velocity. Although the velocity of any particular carrier may be random, the mean velocity of all carriers within a unit-volume is constant. At the same time, although the direction in which a particular carrier moves is random, the mean direction of all carriers within a unit-volume is zero! Under all conditions of equilibrium there is, then, no net displacement of carriers. Although the mean carrier movement is zero, however, carriers in any one unit-volume exchange places with carriers in all other unit-volumes over a period of time. The whole lattice structure, and the carriers moving within that structure, are thus kept in a continuing state of energy interchange.

The continuous movements of both phonons and carriers throughout the lattice tend to maintain all of these interchanges in a state of equilibrium. If this equilibrium is disturbed in any way, the carriers in one unit-volume move with a different mean velocity than carriers in some other unit-volume. Since the carriers are constantly exchanging places, however, the exchanges tend to restore equilibrium. The result is a movement of hot carriers away from, and a movement of cold carriers toward, sources of heat.

Since hot carriers move with more velocity than cold carriers, carriers tend to be depleted in hot unit-volumes and concentrated in cold unit-volumes. Although the mechanism is different, the result is the same as that achieved by phonon drag—a carrier population density gradient set up in opposition to the thermal gradient. If the characteristic carrier population densities are the same for both temperatures and only one type of carrier is present, a voltage gradient will be created between hot and cold unit-volumes that is directed at re-establishing the uniformity of population density. This voltage is called the *THOMSON EFFECT*.

In both the second and third mechanisms, heat is transported by carriers. This fact explains why metals, which have a plethora of carriers, are good conductors of heat, whereas insulators, with no carriers at all, are poor conductors of heat.

The fourth mechanism of heat transfer involves the formation of electron-hole pairs in hot unit-volumes and their recombination in cold unit-volumes. Electrons and holes occur together when a bonding electron breaks loose from its bond to move out freely through the interstices of the lattice. When this happens, the bond becomes a hole at the point at which the electron broke loose. The population density of electron/hole pairs is a phenomenologic characteristic of a particular material at a particular temperature. At higher temperatures, more electrons break loose to create more pairs. Once created, both the holes and the electrons move, as do all carriers, from the high-temperature unit-volumes toward the low-temperature unit-volumes.

In these circumstances, however, the arrival of pairs in the low-temperature unit-volumes increases the population density of the pairs beyond that which is characteristic of the low temperature. The pairs then recombine and disappear as carriers. Just as the formation of pairs in the high-temperature unit-volumes absorbs heat, their recombination in the low-temperature unit-volumes releases heat. This absorption or release of heat (a transition between kinetic and potential energy) is a fundamental result of the formation or rupture of any type of chemical bond. The net result is a flow of heat from points of high temperature to points of low temperature.

Although the flow of heat provided by this mechanism is somewhat different, the movement of both types of carriers is the same as mentioned previously—from points of high temperature toward points of low temperature—and once again it is subject to Lorentz forces. In short, heat is transported through solid materials both by phonons propagating along the lattice bonds and by carriers moving through the lattice interstices.

## 17.24 ADIABATIC NEGATIVE NERNST EFFECT

Heat flows in a conductor in response to a thermal gradient in much the same fashion as current flows in response to a voltage gradient. Within the body of a solid, the heat may be carried by a number of different mechanisms, as described in the previous section. Since all types of "hot" carriers move away from heat sources, electrons and holes move in the same direction, in contrast with the "opposite" carrier movements characteristic of both the Hall Effect and galvanomagnetoresistance. Moving in the same direction allows them to be separated by magnetic fields, that is, the positive holes and negative electrons are deflected in opposite directions. This separation, if both types of carrier are present, or simple deflection, if only one type is present, gives rise to the *NERNST EFFECT* and a resulting *NERNST VOLTAGE*. (See Fig. 17.11.)

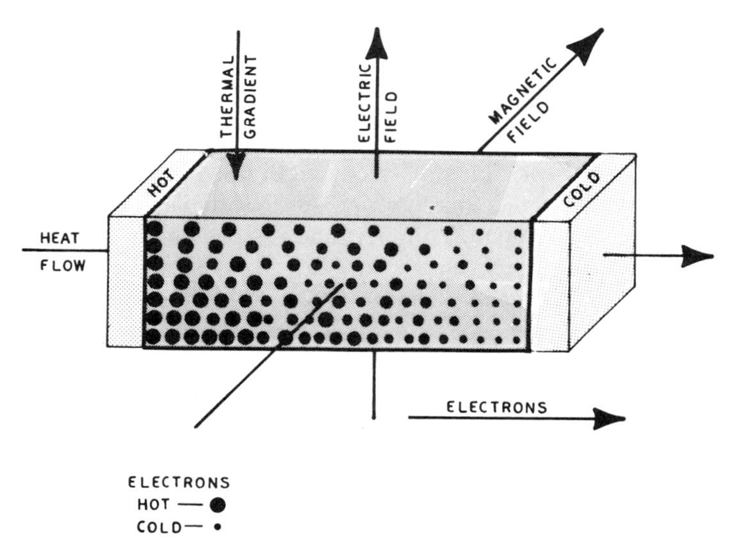


Fig. 17.11. Adiabatic negative Nernst Effect: With only electrons as heat carriers, the electrical and thermal fields are in opposite directions.

The *NERNST ELECTRIC FIELD*, which is transverse both to the heat flow direction and to the magnetic field direction, is mathematically stated as follows:

NERNST ELECTRIC FIELD $\qquad F_n = K_n WB \qquad$ (Eq. 17.6)

where $F_n$ is the electric field in volts per unit thickness; $K_n$, the NERNST COEFFICIENT of the material; $W$, a term descriptive of heat flow (in calories per unit area); and $B$, the strength of the magnetic field.

## 17.25 ADIABATIC POSITIVE NERNST EFFECT

When heat flows through a material immersed in a magnetic field, a Nernst voltage is established if any carriers are present to give it support. Since this voltage is always oriented in the same direction regardless of carrier type, the Nernst Effect is not divisible into positive and negative classes. The Righi-Leduc Effect (see Sec. 17.26) is associated with the Nernst Effect, however, and its direction does depend on the type of carriers present.

The use of the word "adiabatic" to describe a heat flow condition may be a bit confusing at first glance. However, this usage refers to the nonflow

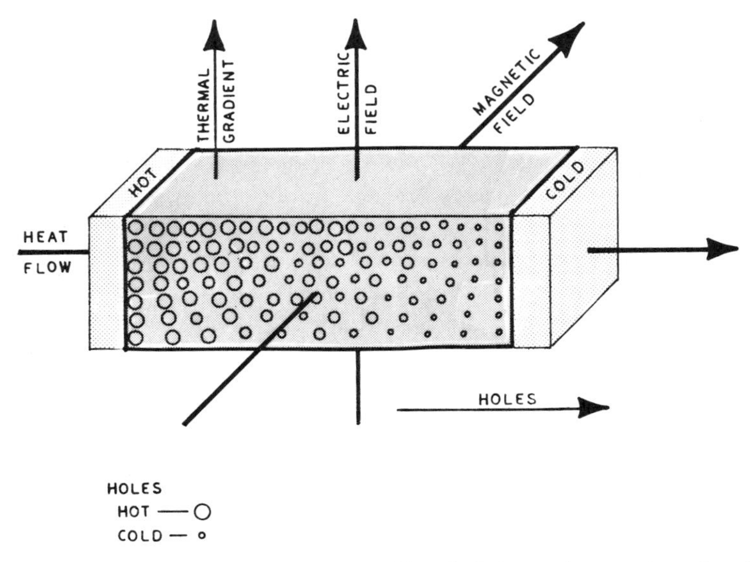


Fig. 17.12. Adiabatic positive Nernst Effect: With only holes as heat carriers, the electric and thermal fields are in the same direction.

of heat in response to the Righi-Leduc Effect which is established at right angles to the primary flow of heat.

There is an interesting analogy between the Hall Effect and the Nernst Effect. In the Hall Effect, the direction of a Hall voltage depends on the type of carrier present, whereas the direction of the Ettinghausen thermal gradient is always the same. In the Nernst Effect, the Nernst voltage is always oriented in the same direction, but the orientation of the Righi-Leduc thermal gradient depends on the type of carriers present.

Since a Nernst voltage depends on a nonhomogenous distribution of carriers, it can be sustained within a body of material only as a result of energy supplied from a source external to that body. The energy comes from a temperature difference that drives a flow of heat through the material. (See Fig. 17.12.)

## 17.26 RIGHI-LEDUC EFFECT

The *RIGHI-LEDUC EFFECT* is the thermal gradient established in conjunction with a Nernst voltage under the circumstances shown in Figs. 17.11 and 17.12. These circumstances depend on the type of carriers present and on the direction of their motion in a magnetic field. Basically, however, "hot" carriers move away from sources of heat, whereas "cold" carriers move toward such sources. The "hot" positive carriers are deflected in one direction and the "hot" negative carriers in the other, whereas the "cold" positive carriers move with the "hot" negative carriers and the "cold" negatives with the "hot" positives. Since the "hot" carriers move faster than the cold, the effect of their movements dominate.

If only one type of carrier is present, its movements establish a transverse temperature gradient that results from a longitudinal flow of heat. If both types of carriers are present, the "hot" holes are deflected in a direction opposite to that of the "hot" electrons, and, if their effects are equivalent, there is no Righi-Leduc thermal gradient.

As the temperature of a P-type material increases a transition is made to an I-type material at some point. Since the mobility of electrons is generally greater than that of holes, their effects dominate at high temperatures. In such materials, the Righi-Leduc thermal gradient is oriented in one direction at low temperatures and reverses to the other direction at high temperatures.

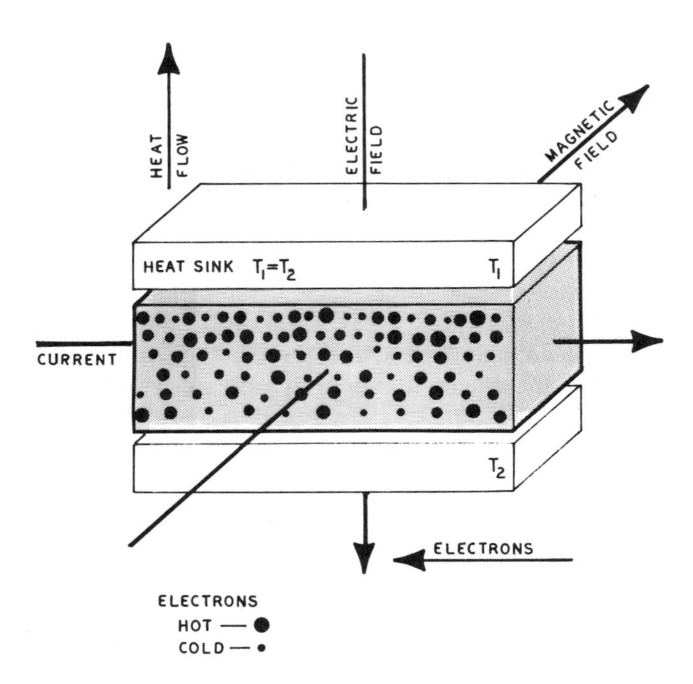


Fig. 17.13. Isothermal negative Hall Effect: If a temperature gradient maintained by electron displacement is dissipated by a heat sink, the voltage gradient caused by Lorentz forces differs from what it would otherwise be.

## 17.27 ISOTHERMAL HALL EFFECTS

*ISOTHERMAL HALL EFFECTS*, either positive (Fig. 17.14) or negative (Fig. 17.13), combine an adiabatic Hall Effect with an adiabatic Nernst Effect. As illustrated in the vector diagram of Fig. 17.15, the heat flowing in response to the Ettinghausen Effect creates a *LONGITUDINAL NERNST VOLTAGE* in opposition to the originating forcing field. The longitudinal Nernst voltage increases the resistance of a conductor, decreases the current flow, and reduces the magnitude of a Hall voltage.

Figure 17.13 shows the insulating plates of Figs. 17.8 and 17.9 replaced by heat sinks to allow a flow of heat in response to the Ettinghausen Effect but not to allow a flow of current in response to the Hall Effect.

## 17.28 HALL CURRENT

If a conducting path is available, a Hall voltage will function as a conventional electromotive force (emf) and drive a current at right angles to

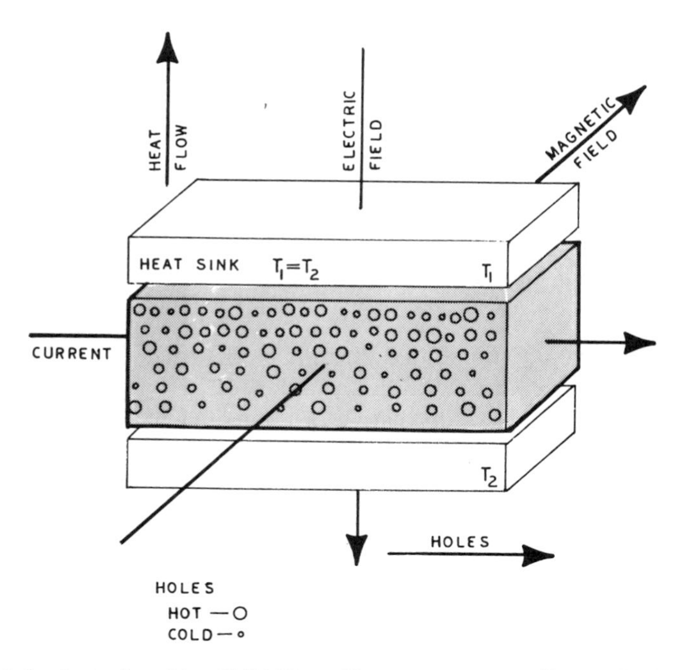


Fig. 17.14. Isothermal positive Hall Effect: If a temperature gradient maintained by hole displacement is dissipated in a heat sink, the voltage gradient caused by Lorentz forces differs from what it would otherwise be.

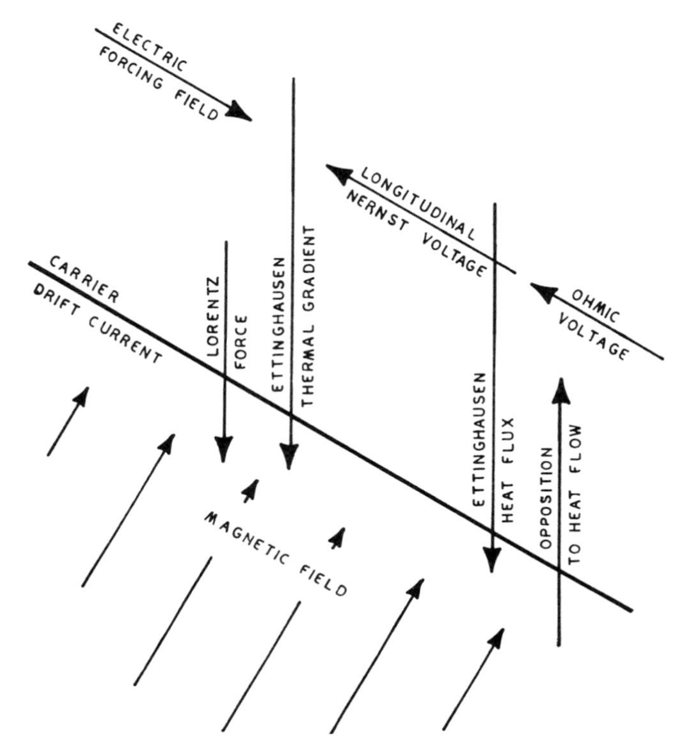


Fig. 17.15. Longitudinal Nernst voltage: Vectors may be used to illustrate the various forces to which an electronic carrier is exposed when it moves through a conducting material immersed in a magnetic field.

the primary current. This added current is identified as a *HALL CURRENT*.

As the Hall current is a derivative of the primary current flowing in a magnetic field, it too flows in the same magnetic field and is subject to Lorentz forces in accordance with the same laws. As shown in Fig. 17.16, the Hall current generates a *LONGITUDINAL HALL VOLTAGE* that directly opposes the originating forcing field. The effect of this Hall back emf is to increase the resistance seen by the primary emf, which in turn reduces the primary current flow and the derived Hall voltage. This change in resistance is called the *HALL RESISTANCE*. It is separate from magnetoresistance and is carried to its logical maximum in a Corbino Device, as described in Sec. 17.29.

The Hall current is also exposed to an ohmic impedance that generates an IR voltage in opposition to the Hall voltage. If the Hall current is allowed

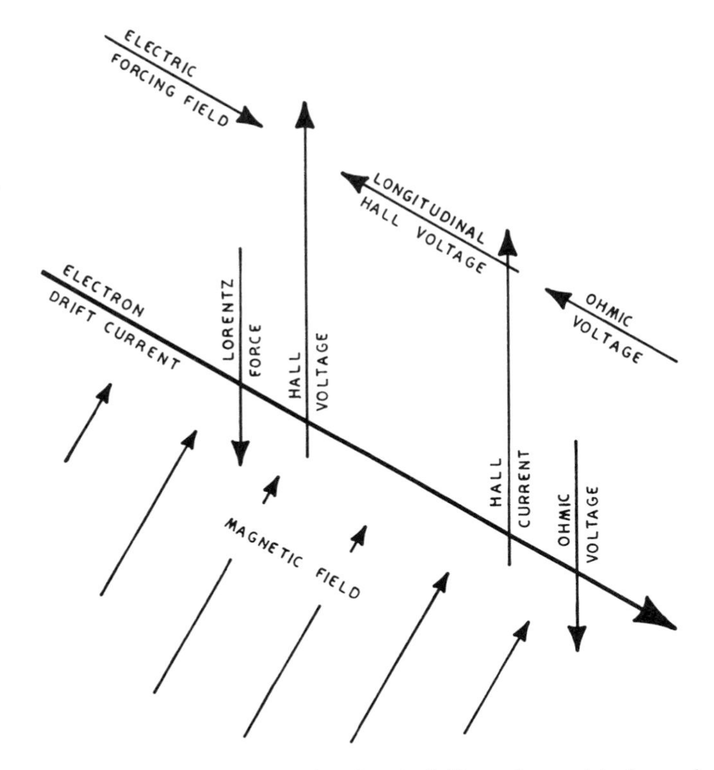


Fig. 17.16. Longitudinal Hall voltage: An electric field can be used to force electrons to drift as an electric current through a conductor. A Hall voltage is generated in such a conductor when it is placed in a magnetic field. If a Hall current is allowed to flow in response to the Hall voltage, it, too, is affected by the magnetic field and creates a second, longitudinal Hall voltage that opposes the original forcing field.

to flow, a new balance is established in the galvanomagnetic material that alters all factors—voltages, currents, thermal gradients, heat flows, etc.

## 17.29 CORBINO DEVICE

If the original Hall voltage is shorted out and replaced by a secondary current at right angles to the primary current, a new longitudinal Hall voltage is created that opposes the primary current. This Hall back emf is directly opposed to the driving voltage. The net effect is an increase in the effective resistance of the original circuit.

To short out the Hall voltage completely, it is necessary to go to a geometrical structure such as the CORBINO DISK (see Fig. 17.17). Here the secondary current circulates along a circumferential path as a result

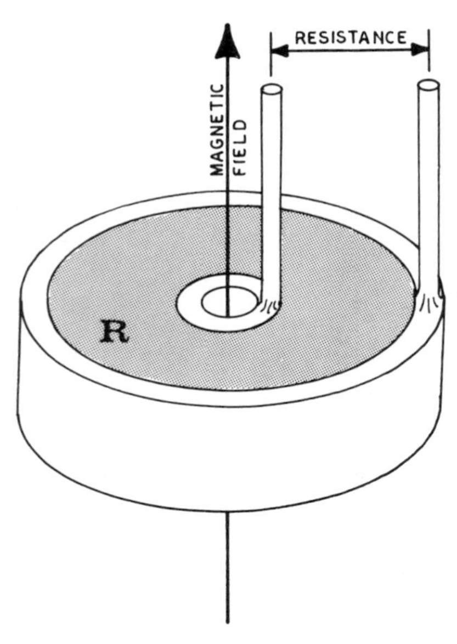


Fig. 17.17. Corbino device: The resistance of a body fabricated from a Hall material is significantly increased by the configuration shown here.

of the primary current's following a radial path. The change of resistance is proportional to the square of the magnetic field for small fields but deviates from this pattern at very high field intensities. The Corbino resistance is dependent on the Hall voltage, whereas magnetoresistance is not.

Following the same general logic, heat will flow from the outer to the inner ring of a Corbino device, and the rate of flow will vary with magnetic field intensity.

## 17.30 LIQUID CORBINO DEVICE

Magnetoelectric and galvanomagnetic effects are all based on transverse electric fields established by Lorentz forces. Whether carriers are positive or negative, whether they move in solids or liquids, and regardless of why they move, all experience the same force and respond in the same general way. Thus, in a *LIQUID CORBINO DEVICE,* a secondary current of positive and negative ions circulates in a circumferential path as a result of a primary current flowing in a radial path just as it does in Fig. 17.17.

## 17.31 NEGATIVE HALL DEVICE

A discussion of general galvanomagnetic principles deals with per-unit quantities—voltage gradients, current densities, etc. On the other hand, a Hall device deals with absolute quantities. A specific material with exact dimensions and carrying a finite current develops an individual Hall voltage when exposed to a singular set of environmental conditions—including a precise magnetic field. Such practical considerations profoundly affect the performance of Hall devices, as follows:

1. *Choice of Hall material.* The Hall Effect is a majority carrier mechanism that depends on bulk material properties. It is independent of surface effects, junction leakage currents, and junction threshold voltages. To obtain a maximum output voltage, the active element must have a large Hall coefficient, which requires high carrier mobilities. In addition, since thermal noise is a limit to device sensitivity, the resistivity should be kept as low as possible to prevent excessive heating of the driving current. These requirements are optimized in an N-type semiconductor.

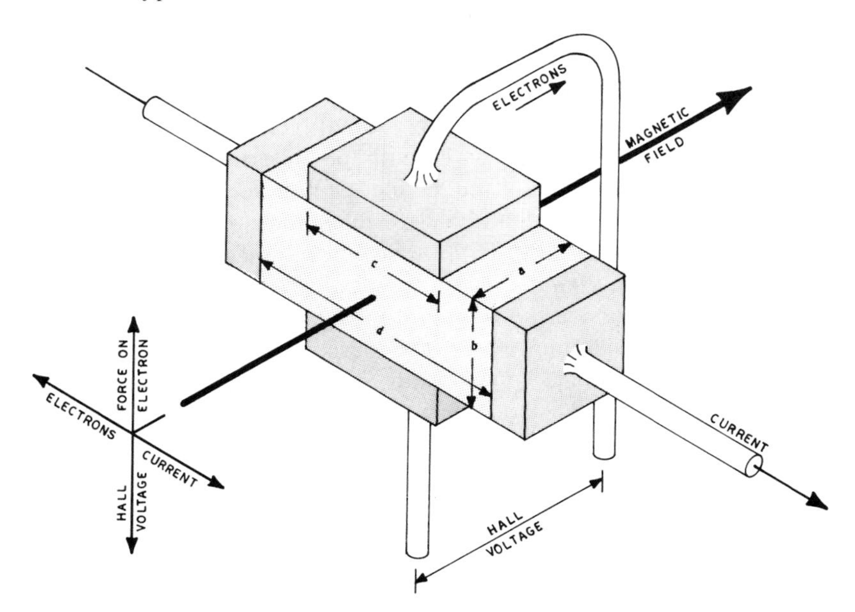


Fig. 17.18. Negative Hall device: The Hall voltage generated by a practical device fabricated from a Hall material differs significantly from the theoretical because of thermocouple effects, heat flow consequences, the problems of attaching leads, and the like.

2. *Dimensions*. A block of Hall sensitive material has a thickness $a$, a width $b$, and a length $d$. If the device illustrated in Fig. 17.18 is designed for a particular resistance, the Hall voltage will tend to have larger values with an increase in the $b$ dimension. Keeping the resistance constant would require a decrease in the $d$ or $a$ dimension. In either circumstance, the metering current would flow from an increased source impedance. This indicates a series of trade-offs offering various optimum configurations for devices designed for various purposes.
3. *Configuration*. To be useful, a Hall device must be equipped with electrodes to inject the driving current and to extract a metering current. The dimensions of these electrodes and their placement on the Hall-sensitive material significantly influences device performance. The longitudinal $c$ dimension of the hall electrode pad shorts out a part of the longitudinal voltage gradient and redistributes both the direction and the intensity of the current flowing through the device. Then, too, although the hall electrodes are intended to detect only a Hall voltage they will measure a portion of the longitudinal voltage gradient as well as the Hall voltage if they are placed longitudinally in an out-of-balance position.
4. *Polygalvanomagnetics*. To some degree, Hall electrodes act as heat sinks that allow a flow of heat in response to the Ettinghausen Effect. The addition of the resulting longitudinal Nernst voltage to the longitudinal Hall voltage associated with the metering current increases the apparent resistance of the device. Of course, magnetoresistance is added to both the Hall and Nernst resistances. As these quantities are subject to change, so is the resistance subject to change.
5. *Temperature coefficients*. Initially, the resistance of an N-type Hall material increases as the temperature increases because of the decrease in carrier mobility with the increase in temperature. Above a certain temperature, however, an N-type semiconductor becomes an I-type, and the resistance decreases as the carrier population density increases. These effects of resistance change may be avoided if the driving current is derived from a constant current source and if the Hall voltage is balanced against an opposing voltage. At the same time, the Hall voltage itself decreases continually with increasing temperature since its magnitude is dependent on both carrier mobility and current density but not upon population density.
6. *Thermal effects*. A zero field residual voltage is a result of a Hall electrode misalignment. This voltage will change significantly with slight changes in the driving current. It is almost impossible to predict consistently and accurately the rate and direction of drift in this offset

voltage. The effect can be minimized only by perfecting the alignment. A zero field residual voltage is also a result of thermoelectric effects arising from the differences of materials used in the Hall circuit. The effects of these may be minimized by keeping the Hall circuit at a constant and uniform temperature.

7. *Magnetic field.* The Hall voltage is proportional to the product of the magnitude of the driving current and the strength of the magnetic field at right angles to the direction of the driving current. If the driving current and the magnetic field are not at right angles, the Hall voltage will be reduced by an amount proportional to the sine of the angle of misalignment.
8. *Positive Hall device.* Since the mobility of holes is usually less than the mobility of electrons, there is little incentive to construct positive Hall devices even though they have some characteristics that are quite different from those of negative Hall devices. The first of these, of course, is the completely opposite orientation of the Hall voltages in the two. In fact, the existence of a positive Hall voltage is one dramatic proof of the existence of holes. As a second difference, the Hall voltage in a positive Hall device decreases in value with an increase in temperature as it does in a negative hall device but now extends all the way to zero. Above a certain temperature, the Hall voltage reverses in sign, and the device becomes negative. This temperature, called the *HALL REVERSAL POINT,* lies somewhere in the zone where P-type material becomes I-type. Even when holes and electrons occur together as pairs, the greater mobility of the electrons is the dominant force in the creation of Hall voltages.

## 17.32 HALL MULTIPLYING DEVICE

In a *HALL MULTIPLYING DEVICE,* the magnetic field required by the Hall Effect is derived from the field surrounding a current flowing through a conductor. In these circumstances, the strength of the magnetic field is proportional to the magnitude of this current. Furthermore, a Hall voltage is proportional to both the strength of the magnetic field and the magnitude of the driving current. As a direct consequence, the Hall voltage ($E_h$) in a Hall multiplying device is proportional to the product of the magnetic field generating current ($I_1$) and the Hall driving current ($I_2$) The Hall voltage will respond to this product with either current operating at any frequency, including direct currents at zero frequency.

As shown in Fig. 17.19, a magnetic field concentrator in the form of a ferromagnetic core may be added around the conductor carrying $I_1$. Although this concentrator will significantly increase the effect of $I_1$ on the

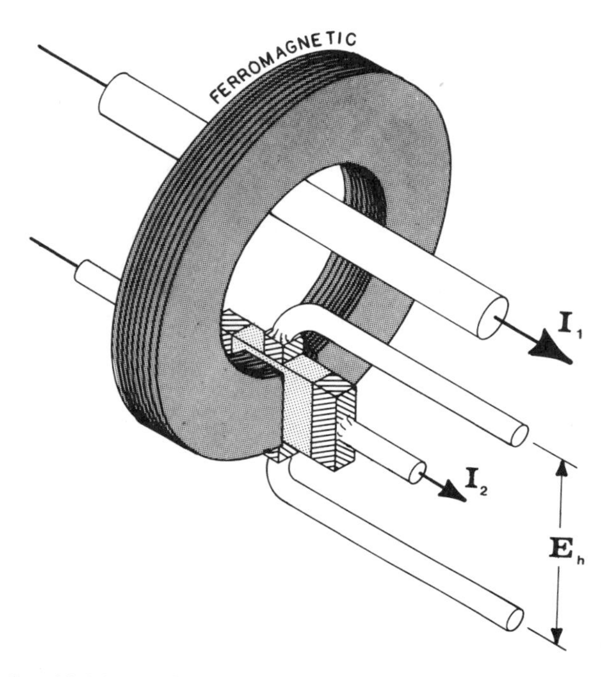


Fig. 17.19. Hall multiplying device: When a Biot-Savart field surrounding a current-carrying wire is applied to a body fabricated from Hall material, the Hall voltage is a function of the product of two currents, that of the primary conductor and that through the Hall device.

product $E_h$, it adds an inductive reactance to the load seen by $I_1$ and introduces hysteresis into $E_h$. In other circumstances, both $I_1$ and $I_2$ drive pure resistive loads.

## 17.33 CRYSTALLINE ANISOTROPY

If a forcing field is in the X direction and a magnetic field in the Y direction, a galvanomagnetic effect might reasonably be expected for the Z direction. If these conditions are established within a crystal, however, the orientation of the crystal axes can alter the results profoundly since the carrier mobility may be many times greater in one axis direction than in another. Figure 17.4 illustrates one lattice configuration where this could be possible. If the "easy" axis does not conform with the Z direction, the maximum galvanomagnetic voltage will be found in a direction somewhere between the Z direction and the "easy" axis direction. In a Hall Effect, this is called a *QUADRATIC HALL VOLTAGE*, and the component extending in the magnetic field direction is called a *PLANER HALL VOLTAGE*.

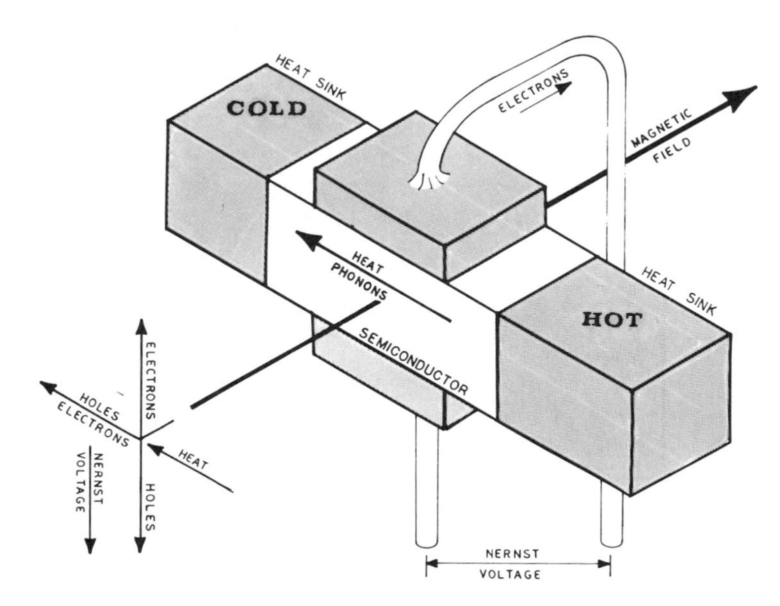


Fig. 17.20. Negative Nernst device: The Nernst voltage generated by a device fabricated from a Nernst material differs significantly from the theoretical because of thermocouple effects, heat flow consequences, the problems of attaching leads, and the like.

## 17.34 NEGATIVE NERNST DEVICE

Since a negative Nernst device is similar to a negative Hall device both in concept and in construction (see Fig. 17.20), many of the previous comments apply here as well. In a Nernst device, however, the holes and the electrons travel together instead of in opposition as they do in a Hall device. This unidirectional flow is the basis for some significant differences.

For instance, since a Nernst voltage is always oriented in the same direction, there is no positive Nernst voltage. Then, too, there is no equivalent of a Hall reversal point in respect to voltage, although there is an analogous *NERNST REVERSAL POINT* where the heat flow from the Righi-Leduc Effect changes direction.

## 17.35 PHOTOELECTROMAGNETIC EFFECT (PEM EFFECT)

In I-type semiconducting materials, electron–hole pairs are a consequence of phonon-bonding energy interactions. The *PHOTOELECTROMAGNETIC (PEM) EFFECT* is comprised of the voltage gradients established when pair components—holes and electrons—are separated by Lorentz forces, as shown by Fig. 17.21.

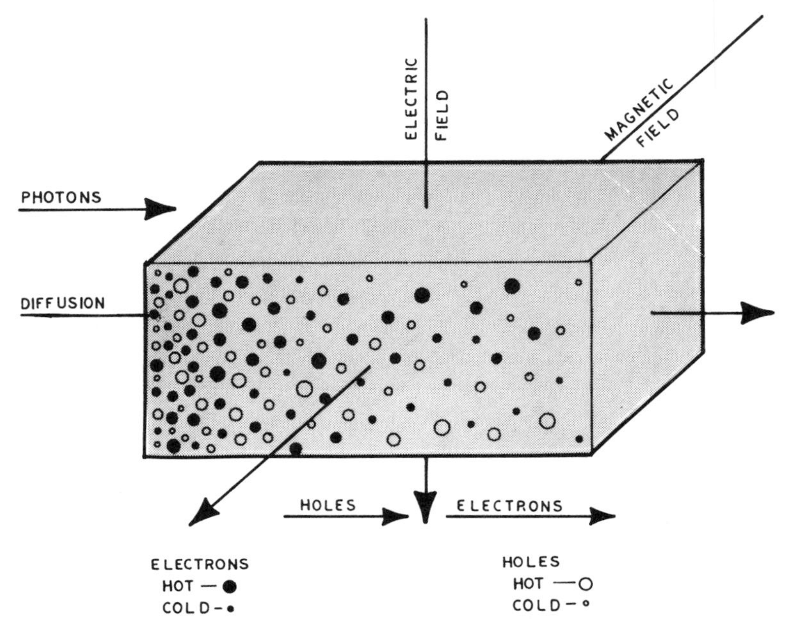


Fig. 17.21. Photoelectromagnetic effect: New carriers released in pairs by photon action move away from their point of origin to try to maintain a characteristic population density. Lorentz forces on them establish voltage gradients at right angles to their movement.

Electromagnetic radiation incident on the surfaces of semiconducting materials is strongly absorbed by these materials if the radiant frequency is equivalent to the bonding energy of their structural electrons.

When radiant energy interacts with electron bonds, the latter are broken by the energy equivalence of radiant energy frequencies, and electron–hole pairs are created. The creation of these pairs near the surface of a body raises the carrier population density near the surface beyond what would otherwise be thermally characteristic. The carriers thus produced diffuse into the crystal in response to an electron-hole population gradient. The movements involved in this diffusion occur at some average velocity, in a direction normal to the surface, and for some characteristic lifetime.

When magnetic fields are applied transversely to the diffusion direction, electrons are deflected in one direction and holes in the other. A potential difference corresponding to a Nernst voltage then occurs across the material at right angles both to the magnetic field direction and the direction of diffusion. The strength of this PEM voltage is determined by the product of carrier mobility and the intensity of the magnetic field. As the "hot" electrons are deflected in the opposite direction to the "hot" holes, a significant thermal gradient does not accompany the voltage gradient.

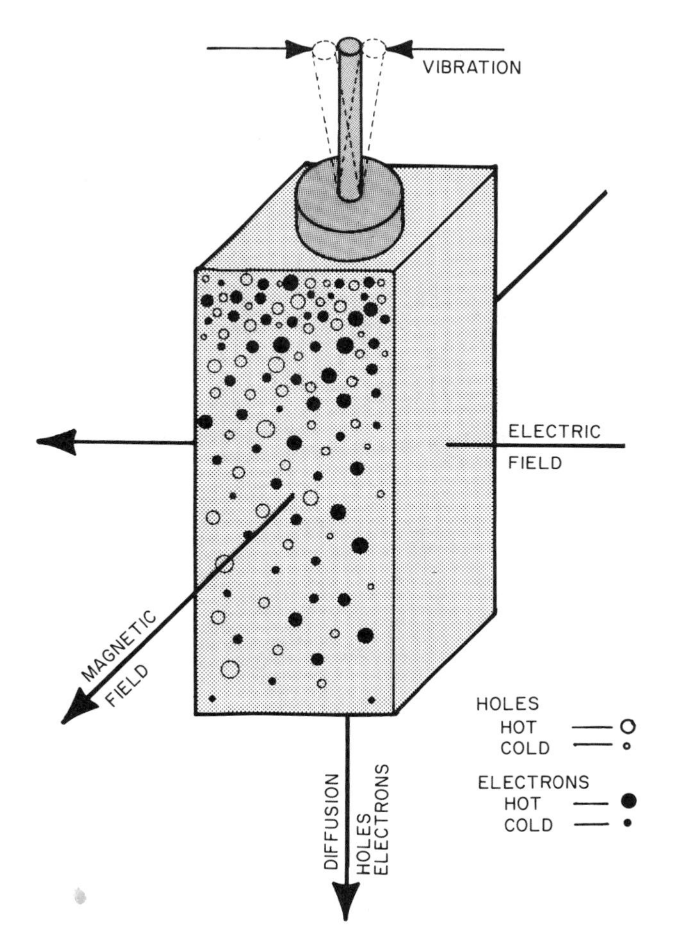


Fig. 17.22. Mechano-electromagnetic effect: New carriers released in pairs by mechanical strain move away from their point of origin to try to maintain a characteristic population density. Lorentz forces on them establish voltage gradients at right angles to their movement.

A PEM photon detector is a relatively noisefree device because it can be operated at cryogenic temperatures, thus minimizing thermal noise.

## 17.36 MECHANOELECTROMAGNETIC EFFECT

In the *MECHANOELECTROMAGNETIC EFFECT* (see Fig. 17.22), electron–hole pairs are created by bond ruptures resulting from dynamic mechanical deformations of lattice structures. If mechanical strain is concentrated near one locality, carrier population gradients result that can be manipulated in a manner similar to that utilized in the PEM Effect.

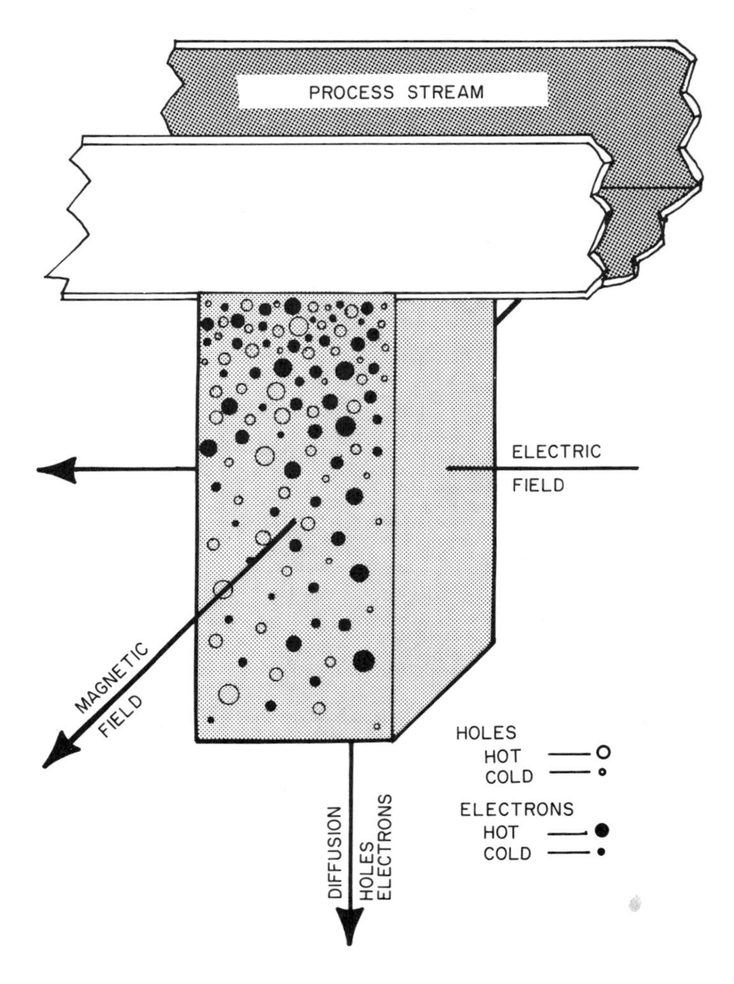


Fig. 17.23. Chemoelectromagnetic effect: New carriers released in pairs by various chemical processes move away from their point of origin to try to maintain a characteristic population density. Lorentz forces on them establish voltage gradients at right angles to their movement.

## 17.37 CHEMOELECTROMAGNETIC EFFECT

Carriers are created by some chemical processes. If these processes interact with certain semiconductor surfaces, a *CHEMOELECTROMAGNETIC EFFECT* can result which is analogous to the PEM Effect (see Fig. 17.23).

# 18
# MAGNETON ORDER EFFECTS

Changes in the electric characteristics of materials that are derived from changes in patterns of magneton alignment are here classified as MAGNETON ORDER EFFECTS. As an example, the electrical resistance of a body is influenced by the structural order of the material from which that body is fabricated. This order includes both the positional order of the various constitutent nuclei and electrons and the directional order of these same particles functioning as magnetons.

The location of elements in lattice structures, lattice spacings, the regularity of lattice structures, the thermal movements of lattice elements within their bonds, and so forth, contribute to a material's resistivity. Any changes in these factors are reflected as changes in resistivity. In addition, as the directions assumed by magneton axes are a part of lattice organizations, these directions also affect resistivity phenomena. Forces that influence directivity patterns have resistivity implications.

The organization of magneton axes has two effects. In the first, magneton directions affect lattice spacings and other lattice characteristics seen by conduction electrons. That is, magneton directions affect the Coulombic forces of attraction, or repulsion, between individual lattice components. These forces impose strains on lattice spacings. If they are changed in any way, changing strains result in changed lattice spacings.

In the second, magnetons themselves act as scatter centers for conduction electrons. Here, if the magneton directions are disordered, their scattering effect can be more pronounced than if the pattern is ordered.

Changes in magneton alignments will, then, change the positional nature of lattice structures as well as the scatter mechanism based on magneton alignments. Other electric characteristics of materials are also affected by magneton alignments. As any factor that influences magneton directions also affects lattice structures to some degree, variations in all electric characteristics are sure to follow.

Magnetic fields can be used to impose order on what would otherwise be thermally disordered patterns of magneton alignment. As magneton directions respond to magnetic fields of any source, an imposed field influences the electric characteristics of a material to the degree of this response.

Magneton order effects are in contrast to *GALVANOMAGNETIC EFFECTS*. The latter are limited to the consequences of Lorentz forces, whereas the former result from both Coulomb forces and directional order.

As both Magneton order and galvanomagnetics are responses to imposed magnetic fields, both can occur simultaneously.

## 18.1 NEGATIVE MAGNETORESISTANCE

The decrease in resistance with increasing magnetic field strength that is experienced by a paramagnetic conductor when the current through that conductor and a magnetic field lie parallel to each other is called *NEGATIVE MAGNETORESISTANCE*. This effect is maximized in paramagnetic semiconductors.

In magnetic conductors, bonded spinning electrons provide the magnetic properties, whereas the environment seen by itinerant electrons provides the electric properties. These two systems interact as conduction electrons are scattered by localized magnetons. In this case, scattering is a function

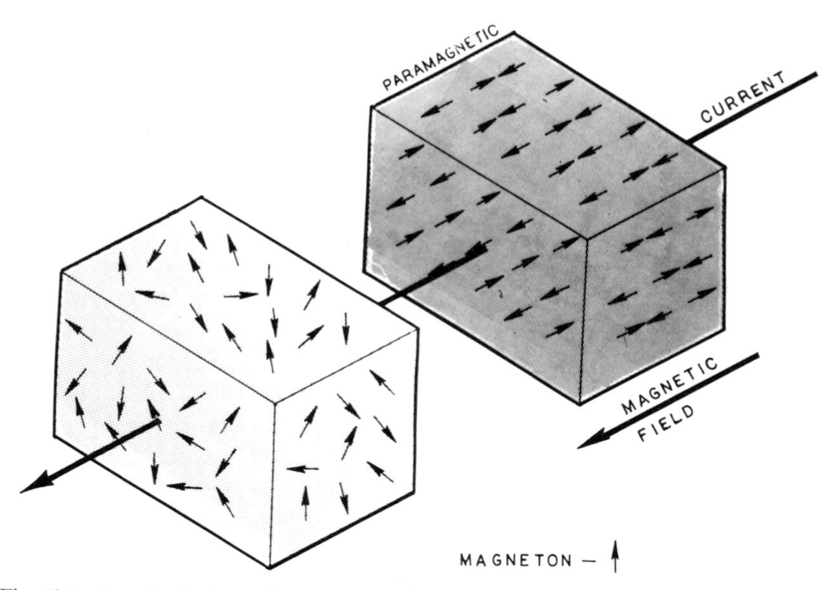


Fig. 18.1 Longitudinal negative magnetoresistance: The resistivity of paramagnetic materials is reduced when the constituent magnetons are aligned in an orderly array, particularly in the direction of carrier movement.

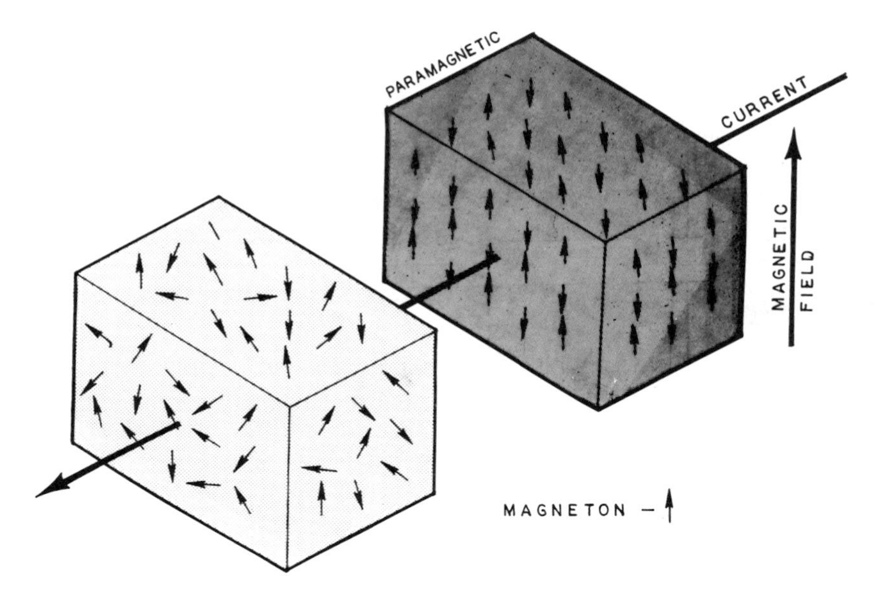


Fig. 18.2 Transverse negative magnetoresistance: The resistivity of paramagnetic materials is reduced when the constituent magnetons are aligned in an orderly array.

of magneton directional-order (or "spin-order"). The greater the disorder, the greater the scatter, and hence the greater a material's resistivity.

Figure 18.1 shows one condition of order and a second condition of disorder, illustrating the general nature of negative magnetoresistance. Since the imposed magnetic field lies parallel to the direction of current flow, the Lorentz forces are not functional. Under these circumstances and at low field intensities, resistivity decreases with increasing field strength. When this parallel field is very strong, the effects described in Chapter 21 dominate, and negative magnetoresistance is overpowered. If the conducting body is very small, negative magnetoresistance can be dominated by the size effects described in Chapter 20.

Figure 18.2 indicates a different pattern of magneton order/disorder that yields a significantly different characteristic magnetoresistance. With the magnetic field and electric current at right angles to each other, magneton directional-order and galvanomagnetic effects are combined. In this case, at low magnetic field intensities, resistivity increases with increasing field strength because of the galvanomagnetic resistance of Sec. 17.16.

As in all ordered systems, there is some temperature below which order is established spontaneously and above which thermal activities destroy order. In magnetic conductors, directional order is a function of temperature, localized exchange forces, and imposed magnetic fields.

The Gauss Effect (see Sec. 18.2) describes the magnetoresistance of a ferromagnetic conductor. The Cabrera-Torroja Effect (see Sec. 5.5) describes the magnetoresistance of a conductor during a Curie transition between ferromagnetism and paramagnetism.

The term "magnetoresistance" is thus applied to a number of different phenomena: Gauss, Cabrera-Torroja, Negative Hall, Corbino, Size, Field Strength, etc.

## 18.2 GAUSS EFFECT

The resistivity of a ferromagnetic conductor changes in the presence of a magnetic field. When the current through such a conductor and the magnetic field direction lie parallel to each other, the result is called the *GAUSS EFFECT*. In this effect, an external magnetic field imposes a pattern of order on the magneton alignments of a ferromagnetic material that differs from the order pattern established by the self-aligning exchange forces acting alone. (See Fig. 18.3.)

Although negative magnetoresistance and the Gauss Effect are related, the latter is a phenomenon based on domain alignments, whereas the former is based on individual magneton alignments. Since the phenomenon

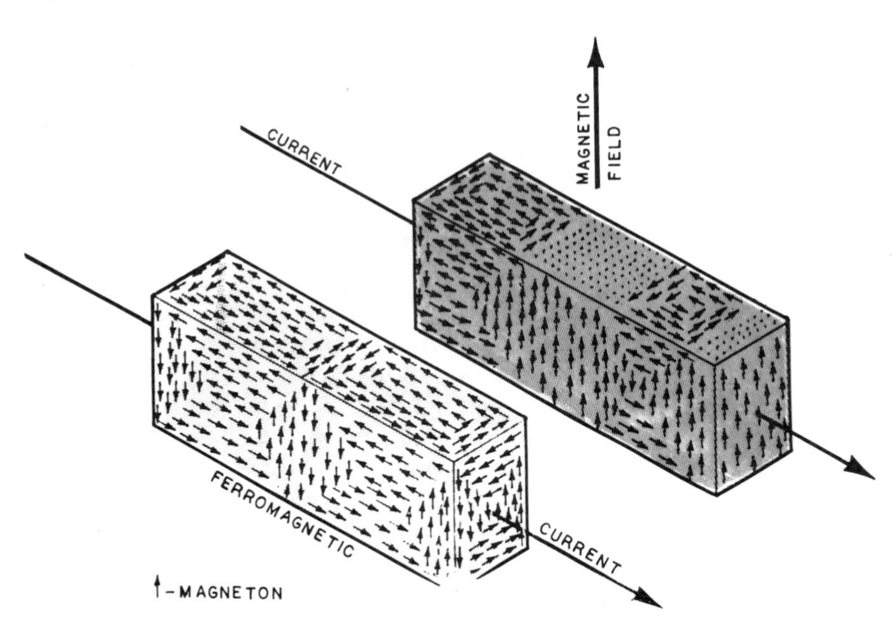


Fig. 18.3 Gauss Effect: The resistance of a ferromagnetic conductor is changed when the magneton array pattern is changed by an imposed magnetic field.

of magnetostriction also occurs when ferromagnetic materials are exposed to magnetic fields, the lattice spacing changes that result from magnetostrictive dimensional changes are an integral part of the Gauss Effect.

Although each domain of a ferromagnetic material has a unidirectional magneton alignment pattern in the absence of an imposed field, the alignments of adjacent domains form patterns of field opposition. The net effect of all magnetion alignments in a ferromagnetic material is, then, very like the net effect of completely random individual alignments. As is true of any pattern of random alignments, an imposed field increases the number of magnetons that are aligned in one direction.

The resulting magneton alignment pattern and the associated magnetostrictive dimensional changes are significantly different under different field strengths and different field directions. As a result, the Gauss Effect in different materials exposed to different magnetic fields exhibits differences in both sign and magnitude. A resistance can increase with an increasingly weak magnetic field and then decrease with an increasingly strong magnetic field (or vice versa). The *GAUSS REVERSAL* occurs at a particular field strength where changes in field strength do not effect resistance.

In the *TRANSVERSE GAUSS EFFECT*, the imposed magnetic field is transverse to the current flowing through a ferromagnetic conductor, and the response is usually opposite to that of the Gauss Effect. That is, the resistance decreases with an increasingly weak field and increases with an increasingly strong field if the reverse is true for the Gauss Effect. The Transverse Gauss Effect combines the effects of magneton order, galvanomagnetics, and magnetostriction. The resistivity changes experienced are a result of all three mechanisms acting simultaneously.

## 18.3 ELASTORESISTANCE

Changes in resistance with changing stress are experienced in ferromagnetic conductors. When the current through such a conductor and the stress are parallel to each other, the resulting change in resistance is called *ELASTORESISTANCE.* Elastoresistance is analogous to the Gauss Effect, but, in this case, domain directivity is influenced by mechanical strain rather than by imposed magnetic fields.

In *TRANSVERSE ELASTORESISTANCE*, a change in resistance accompanies a stress imposed at right angles to current flow.

## 18.4 FERROMAGNETIC HALL EFFECT

There really is no such thing as a "Ferromagnetic Hall Effect" as such. Rather the Hall Effect occurs in ferromagnetic materials just as it does in any conducting material. However, the *FERROMAGNETIC HALL EFFECT* combines the Hall and Gauss Effects in such a way that the Hall mobility is a function of the Gauss Effect. In fact, the Hall Effect can also combine with the Kondo Effect or any other phenomenon that influences Hall mobility.

## 18.5 ELECTROFERRIMAGNETIC EFFECT

The permittivity changes with changing magnetic field strength that are experienced by insulating ferrimagnetic materials exposed to both electric and magnetic fields are here called the *ELECTROFERRIMAGNETIC EFFECT*. (See Fig. 18.4.)

The lattice structures of ferrite crystals are influenced by domain formations just as the lattices of any ferromagnetic material are. Just as lattice structures also affect the dielectric characteristics of insulating materials, so do domain configurations affect the permittivities of ferrites.

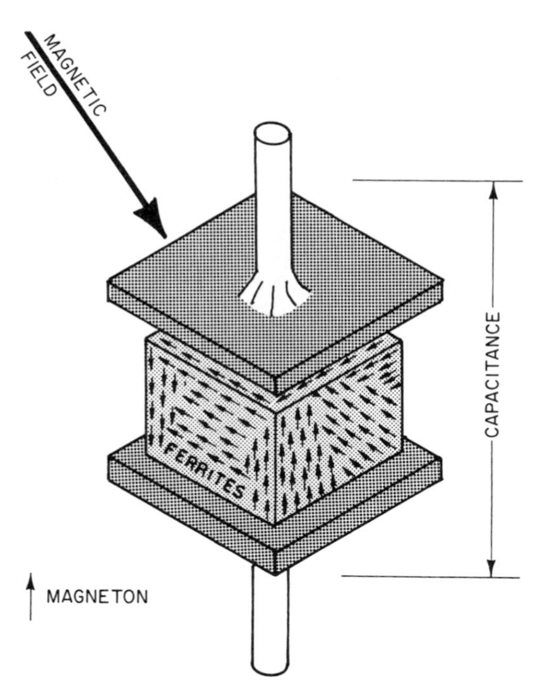


Fig. 18.4 Electroferrimagnetic effect: The permittivity of a ferrite is a function of the magnetic domain pattern.

Depending on the nature of a dielectric constant, the Electroferrimagnetic Effect can be either a scalar or a tensor function.

Electroferrimagnetism is related to the thermoferrimagnetism of Sec. 5.6. Thermoferrimagnetism is a phenomenon based on a Curie transformation, whereas electroferrimagnetism occurs at temperatures below Curie transformations.

## 18.6 HEAT TRANSPORT EFFECTS

Section 17.23 on "Heat Flow" describes the various mechanisms involved in moving heat through a crystalline structure. The efficiency of heat passage from lattice component to lattice component is affected by their relative order or disorder.

Just as magnetic fields affect component order, so will the transportation of heat through solid materials be affected by magnetic fields. The various circumstances that influence this transfer are not too different from those that affect current flow. As a result, there is a whole family of magnetic heat transport phenomena that are analogous to those of the electronic family discussed here.

# 19 HYSTERETIC EFFECTS

*HYSTERETIC EFFECTS* are derived from the characteristics of hysteresis loops. Although all hysteresis loops have the same general shape, the various parameters that determine this shape have different magnitudes in different materials. The most significant of these parameters and their descriptive terminology are outlined in Fig. 4.2. All of these parameters are symmetrical, with the mirror image of a negative value having exactly the same magnitude as its positive equivalent. Both positive and negative values are precisely reproducible under identical environmental conditions.

Because of the symmetry of hysteresis loops, a specimen of material can be magnetized in diametrically opposite directions with exactly the same magnetic results. On the other hand, because of crystal line energy (the tendency for magnetons to align with crystal axes), magnetic materials are not isotropic, and magnetization in some other direction will not conform to the same hysteresis loop.

Different materials display a wide variety of hysteretic factors—coercivity, permeability, saturation, and the like.

The concept of retentivity—the ability of a material to maintain a flow of magnetic flux after the removal of an externally imposed magnetizing force—is used to classify magnetic materials. In *hard magnetic materials* retentivity is relatively large, whereas in *soft magnetic materials* retentivity is minimal. Soft materials are commonly used to direct, or to channel, magnetic flux, whereas hard materials are used as sources of magnetomotive force to drive flux through magnetic circuits.

The area enclosed by a driven hysterisis loop is an indication of the energy consumed from the driving force. In this respect, hard materials consume more energy than do soft materials. If energy consumption is a significant consideration, soft materials are chosen because of the very little area enclosed by their minor loops.

## 19.1 PERMEABILITY

PERMEABILITY is an indication of the ease with which magnetic flux can be driven through magnetic materials. It is defined by Eq. 2.7 as the ratio of flux density flowing through a material divided by the magnetizing force that is required to achieve that flux density. Symbolically, this is commonly indicated as $B/H$.

$B/H$ is an indication of absolute, or static, permeability. When magnetic materials are used in applications that depend on dynamic conditions, the changes that occur in flux density in response to changes in the magnetizing force are of interest.

Because of a hysteresis loop's nonlinear shape, the permeability of a given material has different values for different flux densities. In fact, values for the same flux density will differ depending on the material's magnetic history.

Terms relevant to permeability variations include the following: *DIFFERENTIAL PERMEABILITY* is the slope of a major magnetizing curve at a particular point. *RECOIL PERMEABILITY* is the slope of a minor loop in the vicinity of an operating point. *INCREMENTAL PERMEABILITY* is the ratio of a small cyclical change in flux density divided by the corresponding cyclical change in magnetizing force. Finally, *MAXIMUM PERMEABILITY* exists at that point on a hysteresis loop where a maximum change in flux density is experienced in response to a unit change in magnetizing force.

Equation 13.1 indicates the relationship between coil inductance and core permeability. If the permeability of such a core is changed by any means, the coil inductance will change as a consequence. As described in Chapter 12, many circuit techniques utilized to detect changes in coil inductance, and hence changes in core reluctance, may be used to detect the permeability changes associated with hysteresis.

## 19.2 OPERATING POINT

Each element in a magnetic circuit has what is known as an *OPERATING POINT*. This is defined by the hysteretic coordinates of a specific flux density ($B$) and a specific magnetizing force ($H$). Operating points can lie anywhere either on or inside the confines of a major loop. In permanent magnetic circuits, they are more or less fixed locations, whereas in cyclical circumstances they are usually defined as the centroids of minor loops.

Magnets consisting of appropriately shaped bodies of hard magnetic material are commonly used to drive magnetic flux through magnetic circuits against demagnetizing forces. A vertical line drawn on a conventional $B/H$ diagram to the left of the zero magnetizing force axis represents a constant demagnetizing force. An air gap in a magnetic circuit is one example of such a demagnetizing force. A particular air gap is represented by a particular vertical line. The operating point of a magnet used to drive flux through an air gap can thus lie anywhere on the vertical line repre-

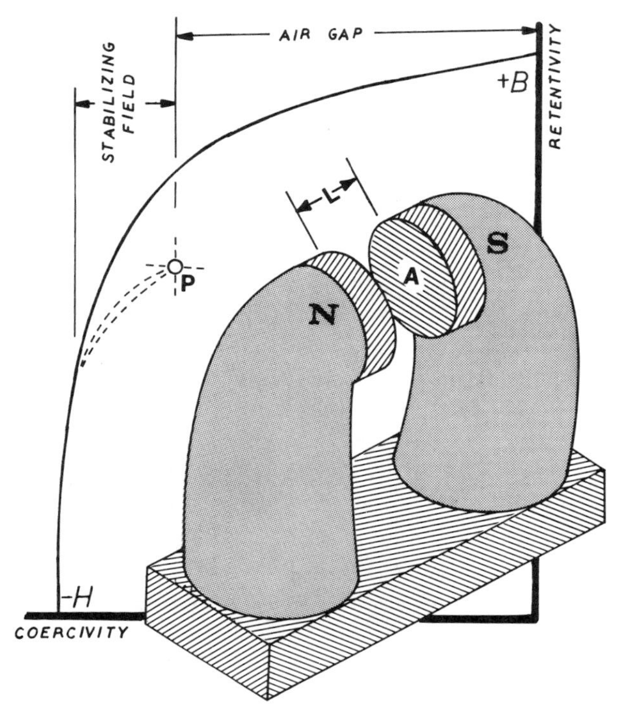


Fig. 19.1 Permanent magnet: The magnetic characteristics of a magnetic circuit containing an air gap depend on the magnetic history of the path it takes to reach its operating point.

senting that air gap. The particular point on this vertical line depends on the magnetic history of the driving magnet's magnetic material.

An operating point is chosen by first saturating a magnet assembly, including a magnet, pole pieces, and air gap as shown in Fig. 19.1. If the maximum flux density through the air gap is to be achieved, saturation must take place while the magnet is operating in place as part of its final assembly. Once the saturating force is removed, the flux density within the magnetic material will slide back down along its major loop until it encounters the vertical line representing the air gap.

In some assemblies, saturation cannot be reached in the presence of an air gap. When this is the case, a temporary magnetic shunt must be placed around the air gap (the shunt is removed after saturation has been achieved). A stabilizing field is then added to demagnetize the magnet further. This continues the flux density slide, along the confines of the major loop, to a new vertical line that represents the sum of the air gap and the stabilizing field. When the stabilizing field is removed, the flux density rises along a minor loop to a second intersection, but at a lower level, on the original air-gap line.

As long as a stabilized magnet is never exposed to an extraneous field that is greater than the stabilizing field, the device in which the magnet functions remains on a specific minor loop. If the flux density in the magnetic material is driven off the operating point for any reason, it will follow the outline of this minor loop and always return to the same operating point after the extraneous demagnetizing force has been removed.

A particular minor loop will be destroyed if the material experiences a demagnetizing force that is greater than the stabilizing field. Removing a magnet from its assembled magnetic circuit will impose just such a destabilizing force. Once a minor loop has been destroyed, the device in which it functions operates on another minor loop and at a different stable point, where the magnetizing force of the new point drives less flux through the air gap. The desired minor loop and its associated operating point can be recovered only through resaturation and restabilization.

An alternating current flowing through a coil surrounding a ferromagnetic material represents an alternating magnetizing–demagnetizing force. Depending on the magnitude of this current, such a force drives the material through some kind of hysteresis loop, either a major or a minor one. If the drive response follows a minor loop, this loop can be located within the major loop by a bias force.

## 19.3 PERMANENT MAGNETISM

Permanent magnetism is a consequence of retentivity. In a *PERMANENT MAGNET*, magnetic flux continues to pass through a magnetic circuit after all extrinsic magnetizing forces have been removed.

Permanent magnets are commonly used to drive magnetic flux through air gaps. An air gap functions in this case as a demagnetizing force. Figure 19.1 shows that part of a hysteresis loop in which flux density is decreased by an increasing demagnetizing force. The effect of a particular demagnetizing force (a particular air gap) in a particular device is represented by a vertical line, or "ordinate."

Figure 19.1 also illustrates the general components of permanent magnetic circuits. Here a body of "hard" magnetic material (shown in two parts) is used to drive an air gap of area $A$ and length $L$ through pole pieces of "soft" magnetic material. The "hard" material is delineated by shading; the "soft" material, by hatching.

Soft magnetic materials are used in permanent magnetic circuits for two reasons. First, many *magnetically* hard materials are also physically hard and brittle and are therefore difficult to machine or otherwise fabricate.

In addition, if their coercivity is to be maximized, they must be cast in a magnetic field that duplicates the direction that is to be established in the final product. A soft material is then used in conjunction with the hard to achieve an intricacy of shape that requires only flat surfaces ground on hard materials. The second reason is that magnetically hard materials tend to lack homogeneity. If they are used to interface air gaps, the flux density through the gaps will not be uniform. On the other hand, soft materials tend toward homogeneity. They can be used to "smooth out" the flux in an air gap so that uniformity of its density is maximized.

A permanent magnet has an operating point. This point, shown in Fig. 19.1 as the point $P$, represents the driving force, $H$, required to drive flux at a particular density, $B$, against demagnetizing forces. When a permanent magnet is used to drive an air gap, the length of that gap is the main demagnetizing force present in the magnetic circuit. As $H$ represents a per-unit-length of magnet, the actual length of magnet required in a circuit is proportional to the length of the driven air gap. In addition, all of the flux that passes through the gap and all of the "leakage" flux that passes around it must be supplied by the flux density $B$ represented by the operating point $P$. This B is also a per-unit value and represents a per-unit-area of magnetic material. The actual area of magnetic material required is proportional to the area of the air gap plus an equivalent area for the leakage flux.

A permanent magnet circuit is designed for an optimum operating point taking two things into consideration. First of all, $P$ is chosen to minimize the volume of hard material used. At all other operating points the hard material is either longer than it needs to be or has a larger area than necessary. Length and area together establish an *energy product*, $B \times H$. In this case, the larger the multiple, $B \times H$, the smaller the amount of hard material required. The hysteresis loops of different materials exhibit different maximum energy products. ALNICO V can reach 5 MGOe, samarium extends this to 15 MGOe, and experimental materials have reached as high as 30 MGOe.

Second, $P$ is chosen to maximize stability. As mentioned previously, $P$ represents a point on a minor hysteresis loop. If $P$ is chosen to be near the edge of the major loop, extraneous demagnetizing forces will move it down the major loop where it becomes a part of a second minor loop that is not capable of supplying the intended gap flux density. The point $P$ must then be chosen to lie at a distance that is farther from the edge of the major loop than is represented by the magnitude of any extraneous demagnetizing force that might be experienced by the magnet after its original adjustment.

The body of hard material is often tapered to have a decrease in area

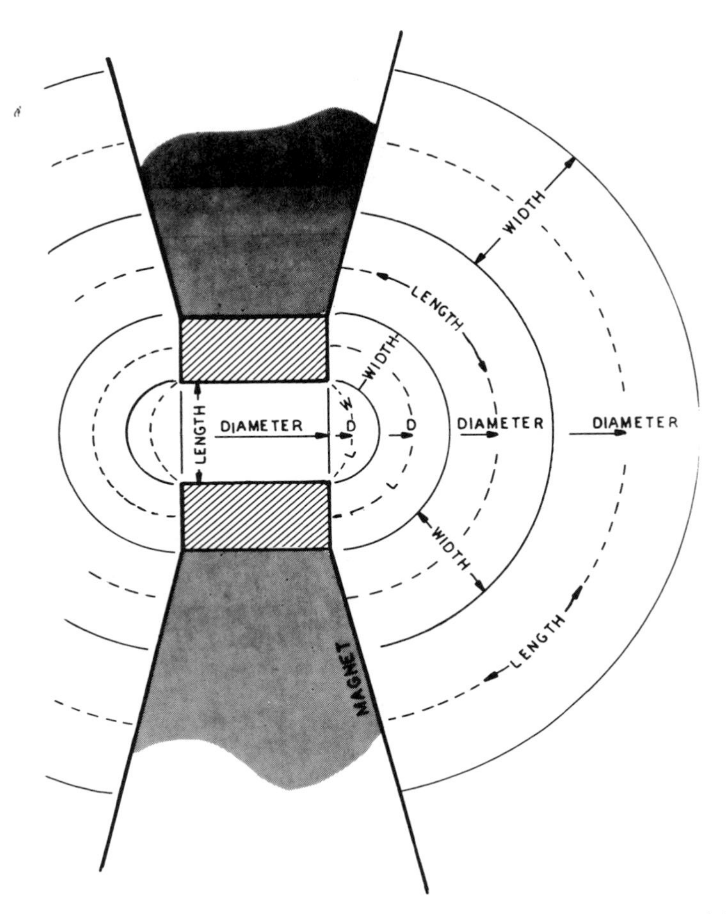


Fig. 19.2 Air-gap leakage paths: All useful air gaps are surrounded by unwanted, but characteristic, leakage paths that must be supplied with magnetic flux if the design flux density is to be maintained in the gap.

as an air gap is approached. As shown by Fig. 19.2, the greater the distance from the air gap, the greater the leakage flux. Since it takes magnetic material to support leakage flux as well as gap flux, the amount of magnetic material increases the further away from the gap it extends.

If a magnetic field is concentrated near the surface of a body of hard magnetic material, the magnetons in a localized unit volume are reoriented in the direction of the local field so that they no longer contribute to the general effort. As the operating point is achieved by summing the effects of all unit volumes, that sum is changed if the contribution of one component unit volume is changed.

Just touching the surface of a hard material with any type of ferromagnetic material, either hard or soft, may be enough to upset the intended operating point. What is more, this type of demagnetization does not contribute to a stabilizing process. To prevent local concentrations of demagnetizing fields, the hard materials in a magnetic circuit are often covered with a nonmagnetic coating of sufficient thickness to prevent ferromagnetic bodies from approaching close enough to their surface to cause this type of damage.

## 19.4 AIR GAPS

Air gaps in permanent magnetic circuits represent demagnetizing forces. From a magnetic point of view, they are more complex than they appear to be on first appraisal. The reason is that parallel flux lines oppose each other and, whenever they have a chance, will bulge out the sides of air gaps into patterns that are characteristic for each gap configuration.

In Fig. 19.2, the leakage pattern is divided into a number of simplified paths. Although these are not the actual paths followed, their total effect, if enough are chosen, gives a reasonable measure of leakage flux in terms of air-gap flux. The problem, of course, is one of integration. The larger the number of paths of smaller area chosen for the calculation, the more accurate the conclusions are likely to be. As a general statement, however, extreme accuracy is not required in such determinations because minor inaccuracies can be easily compensated for by selecting the proper operating point.

The total flux flowing in a magnetic circuit (as supplied by the $B$ part of operating point $P$) is the sum of the flux in the air gap plus the flux flowing in each leakage path. As the flux flowing in any path is proportional to the area, and inversely proportional to the length, of that path, leakage can be calculated by measuring the length of each path (the $L$s) from a representative sketch, as shown by Fig. 19.2, and calculating the areas from the widths and knowledge of the air-gap plan view. If the air-gap plan view is circular, the area of each leakage path is one of a series of annular rings.

## 19.5 FLUX GUIDE

In a flux guide configuration in which an air gap is surrounded by a shield constructed from a ferromagnetic material and in which magnetization is the same for every unit-volume, the magnetic field within the shield will be uniform.

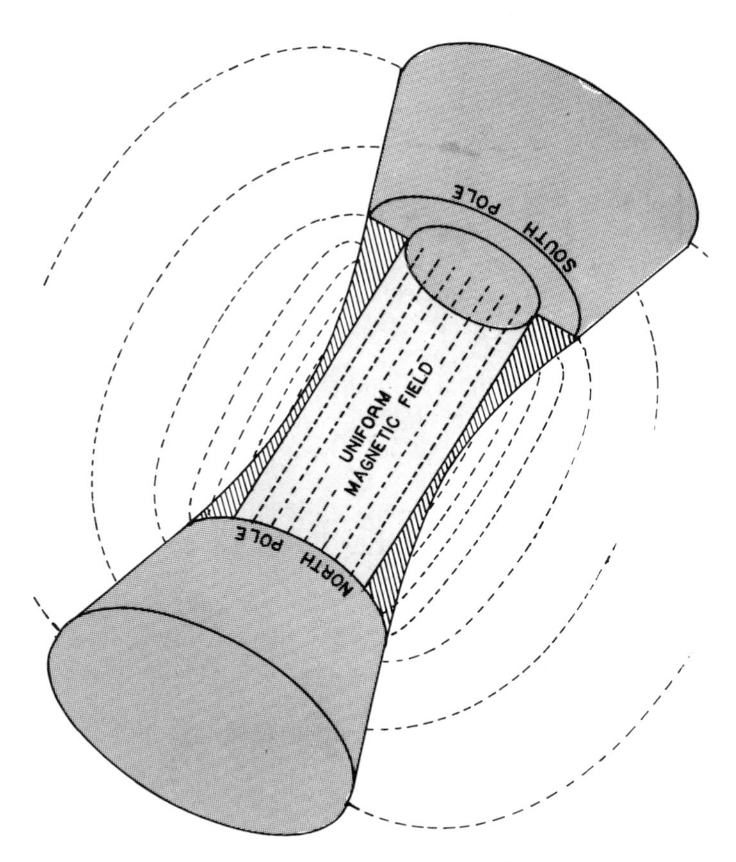


Fig. 19.3 Flux guide: The only way of maintaining a uniform magnetic field over an air gap whose length is large in relation to its diameter is to shield the effects of leakage paths with a flux guide.

Because of leakage flux, which tends to concentrate near pole-piece edges, the cross section of a flux guide must be variable, as shown in Fig. 19.3.

## 19.6 SQUARE HYSTERESIS LOOPS

In *SQUARE HYSTERESIS LOOPS,* the slope change of a major hysteresis loop, from low to high flux densities, takes place over a relatively small increment of magnetizing force. By concentrating slope changes (hence permeability changes) at the "knee" of a square loop, a very small change in magnetizing force can be detected as a relatively large change in permeability.

*SQUARENESS* is defined as the ratio of the difference in flux density between two retentivities, divided by the difference in flux densities experienced between the two associated saturation points. Figures 4.2, 19.1, and 19.5 show loops of varying degrees of squareness.

## 19.7 MEMORY CORES

Because of the symmetry of hysteresis loops, a magnetic circuit can be operated in either of two diametrically opposed directions with exactly the same results. The principle involved here can be used to store numbers in digital computer memories.

The mechanics of all digital computers are based on the binary number system, which involves the manipulation of only two numbers—"ones" and "zeros". These two symbols can be represented by any two-state condition—physical, electric, or magnetic. Small, toroidal (doughnut-shaped) magnetic cores with square hysteresis loops and enough coercivity to withstand extraneous demagnetizing forces can be used in computer memories to store either a "one" or a "zero."

Figure 19.4 illustrates a simple configuration wherein three wires are passed through one of these cores. One of these wires is used to set the core either to a "zero" or to a "one" (by magnetizing it in one or the other direction, clockwise or counterclockwise). This is accomplished by a pulse of driving current on the "write" bus that is of a direction required

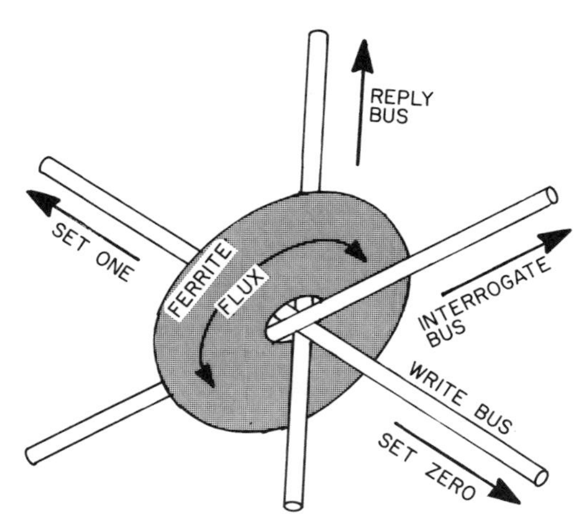


Fig. 19.4 Memory core: The flux in a magnetic circuit can be set in either one of two directions that can be assigned the value of zero or one in accordance with binary logic.

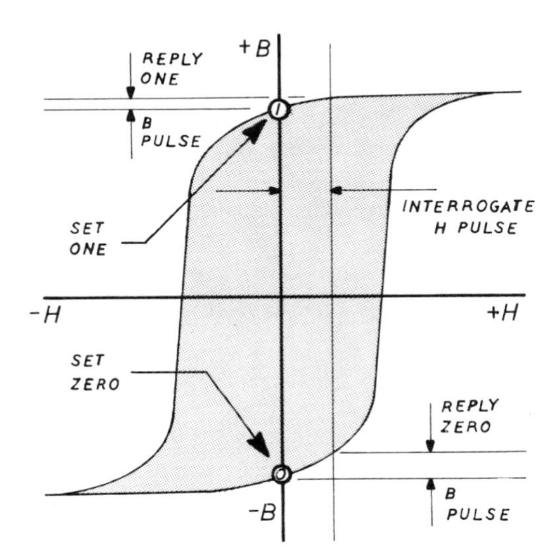


Fig. 19.5 Square loop in a memory core: The direction of core magnetization can be sensed by a pulse of constant magnitude that tends to slightly magnetize or demagnetize the resident magnetic condition.

for either clockwise or counterclockwise magnetization and of sufficient magnitude to drive the core to saturation. The second wire is used to interrogate the condition of the core, whereas the third wire provides a signal that is the reply to this interrogation.

Once a toroid core has been driven to saturation and the magnetizing force is removed, the flux through this core assumes that value on the major hysteresis loop for which the magnetizing force is zero. Figure 4.2 labels this point as the *retentivity* of the material from which that core is fabricated. The material condition represented by retentivity can be used as an operating point in a toroidal specimen because a toroid is not subject to intrinsic demagnetizing forces, that is, it does not have a demagnetizing air gap. When applied to a specimen, this condition is called the *MAJOR REMANENT STATE.* When a specimen is driven off the major hysteresis loop and is operating on a minor loop, the magnetizing-force axis intercept is called a *MINOR REMANENT STATE.*

In a toroid core with a square hysteresis loop such as that shown in Fig. 19.5, the operating point falls on a minor hysteresis loop in which the major and minor remanent states are very close together. A stored number is then represented by any remanent state that is very near the major remanent state. In Fig. 19.5, this is labeled as either the "set one" or the "set zero" point.

A pulse of electric current on the interrogation bus (the interrogate H pulse of Fig. 19.5) is sensed by the core as either a magnetizing or a demagnetizing force. If the force is magnetizing, the core is driven toward saturation and returns to the major remanent state, via the major loop, after the magnetizing pulse has passed. If the force is demagnetizing, the core is driven down the major hysteresis loop and returns on a minor loop to a minor remanent condition slightly less than the major remanent condition. In either circumstance, the flux through the core experiences a change in the form of a pulse that follows the interrogation current pulse. A pulse of voltage that is proportional to the flux pulse then appears on the "reply" bus.

If the material from which the toroid is fabricated has a square loop, a magnetizing pulse causes a much smaller change in flux density than does a demagnetizing pulse. The voltage pulse that results from the smaller of these two flux pulses can be discriminated against and eliminated from consideration by the system. The direction of core magnetization can then be sensed by the presence or absence of a reply pulse in response to an interrogation pulse. The maximum difference between the two possible responses to an interrogation pulse and the proximity of the major and minor remanent states are persuasive reasons for using square loops in core memories.

In core memory application, an interrogation pulse has the same general characteristics as a switching, or writing, pulse. Since the interrogation pulse should not switch the device, however, its magnitude must be kept below that which causes switching. On the other hand, the greater the flux change with an interrogation pulse, the more pronounced the reply signal. A switching pulse should then create the maximum possible flux change without switching the device.

*SWITCHING FLUX* is the maximum change in flux density that a device can experience without its state being switched.

A memory core is an element in pulse-oriented circuits. As such, its response to pulse-type signals is a significant factor in the performance of such circuits. The speed with which a core can respond to pulse-type signals is determined by several factors. For one thing, eddy currents are obvious sources of response delays, but these currents can be eliminated by using ferrite materials. There are two other delaying factors, however, that are intrinsic to the magnetic nature of ferromagnetic materials.

At low and medium flux densities, magnetization changes are primarily accomplished through wall movements. At high flux densities, a complete reorientation of magneton vectors is the inhibiting factor. The leading edge of a switching pulse first initiates wall movement and then attempts to turn the magnetons completely over. After the domain magnetons do flip over, the walls move back to their original position but with their own magnetons also flipped over. In this double action, the walls move in much less time than is required by the magneton-flipping process.

Wall movements involve only a few degrees of change in magneton orientation, whereas flipping requires a 180°-degree change. A voltage pulse on the "reply" bus in response to a square pulse on the "write" bus is characterized by a sharp leading spike that is created by the fast moving walls. This is followed by a more general rise, which is the slower response to the domain flips. This type of device cannot be switched faster than the magneton flipping times allow. *DYNAMIC REMAGNETIZATION* is the phenomen in which the switching time is limited by domain rotation.

The magnitude of a pulse of flux is the factor that ultimately determines the magnitude of the voltage pulse used in core memories. *FLYBACK FLUX* is the change in flux density that is experienced when a magnetizing force is reduced to zero.

## 19.8 VARIABLE MHU MAGNETOMETER

In ferromagnetic materials with square hysteresis loops, applying very slight changes in magnetizing force at the "knees" of hysteresis loops will yield relatively large changes in permeability. A magnetomotive bias can be used to locate an operating point at which small increments of magnetizing force will yield maximum changes in permeability. In *VARIABLE MHU MAGNETOMETERS*, the increment of magnetizing force involved is the ambient magnetic field in which the device is immersed.

Consider a rod fabricated of a square-loop material exposed to an alternating magnetic field. In Fig. 19.6, this is shown as a lightly shaded rod driven through a heavily shaded yoke of different magnetic material. In this case, the characteristics of the magnetic circuit are determined almost exclusively by the rod. Figure 19.7 shows the hysteretic results of this configuration as a minor loop driven over a "range" by an alternating magnetic field around an operating point $P$ that is determined by a magnetomotive-bias field. The minor loop is shown shaded in Fig. 19.7. The choice of minor loop is determined by the saturation-demagnetization history of the rod's material.

The inductance of the driving coil of Fig. 19.6 is determined by the recoil permeability centered on operating point $P$. If small increments of magnetizing force are added to the bias field, the operating point shifts slightly, the recoil permeability changes, and the inductance of the driving coil follows the permeability change. In the vicinity of the knee of a square loop, an increment added to the bias field is detected by a maximum shift in the value of the coil inductance.

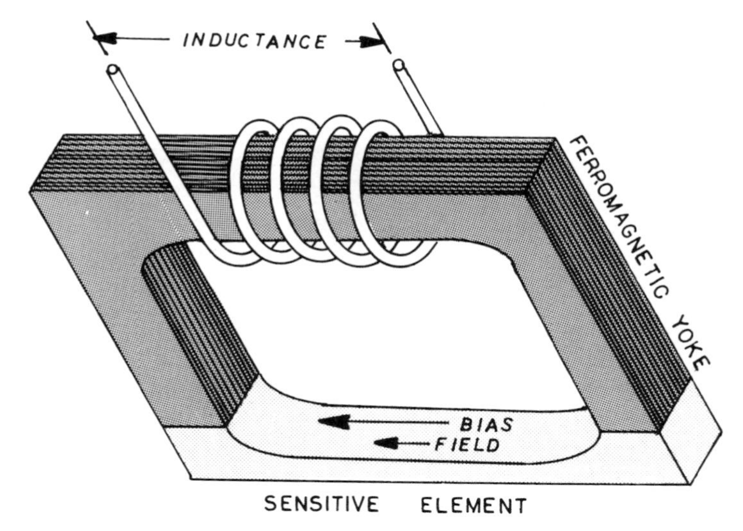


Fig. 19.6 Variable mhu magnetometer: Slight changes in the magnitude of an immersion field can cause significant changes in the permittivity of a "square loop" material biased to operate near the knee of its hysteresis loop. The inductance of a coil sharing the magnetic circuit with the sensitive element experiences changes in inductance that track those changes in permittivity.

As discussed in Chap. 12, a change in coil inductance can be detected by a shift in the frequency of an L/C tuned oscillator. The coil inductance of Fig. 19.6 can be used as the L of such a tuned oscillator. A variable mhu magnetometer can be used to measure both the magnitude and the direction of any magnetic field into which it is immersed. In this case, the magnitude of a shift is an indication of the strength of the magnetic field, whereas the direction of the shift indicates the direction of the magnetic field.

The immersion field is represented by a vector that has both magnitude and direction. Since the bias field is directed along the axis of the ferromagnetic rod, it, too, represents a vector whose direction lies along the rod axis. When the magnetometer assembly is rotated in an immersion field, the oscillator shift indicates two maximum excursions—one when the two vectors add and the other when the immersion field is subtracted from the bias field. One of these excursions maximizes the oscillator frequency whereas the other minimizes this frequency.

The required magnetic bias, which both places the minor loop at the desired location within the major loop and provides the vector direction, can be supplied by adding a permanent magnet, by having a direct current

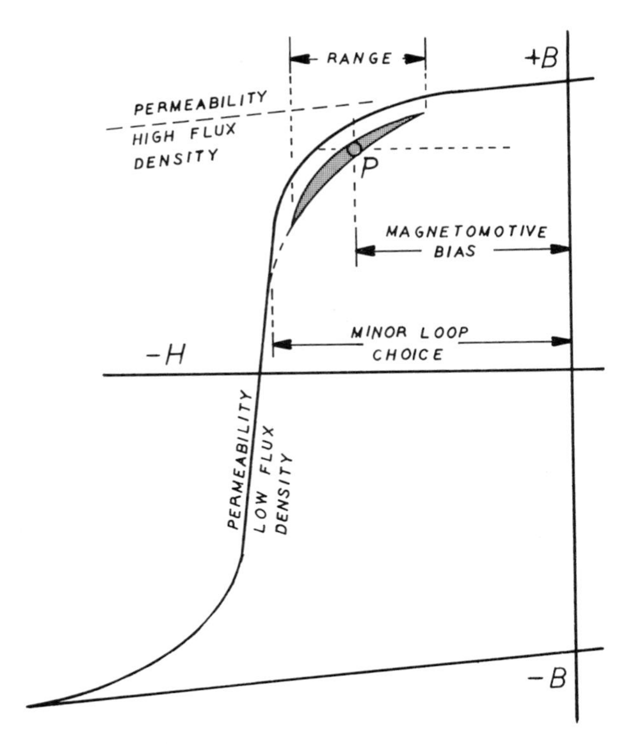


Fig. 19.7 Over the operating range of a square-loop material, significant permittivity changes result from slight changes in magnetizing force.

flow through the inductance coil, or by the retentivity of the ferromagnetic rod itself.

## 19.9 BOUCKE CONFIGURATION

Permeability is a function of the strength of magnetic fields into which permeable bodies are immersed. A measure of permeability can then be used as an indication of field strength.

The inductance of a coil is a function of the permeability of that coil's core material. As coil inductance varies with permeability and permeability varies with field strength, coil inductance can be used to indicate field strength. Although the transfer function of field strength to inductance is a nonlinear function, it is reproducible and subject to calibration.

In a *BOUCKE CONFIGURATION*, a flat washer formed from compressed, finely powdered ferromagnetic material is used as the core of a

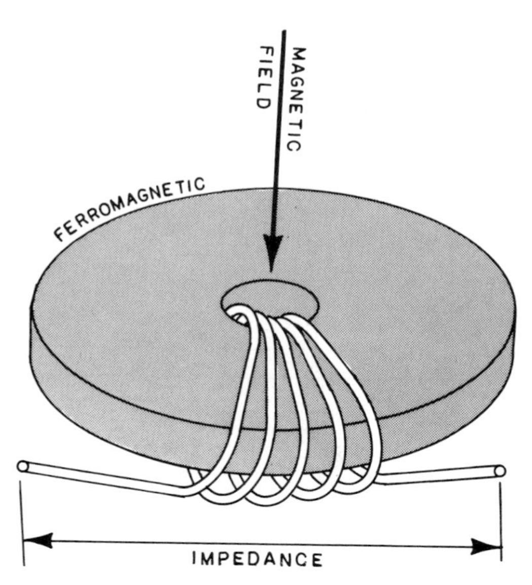


Fig. 19.8 Boucke configuration: Once calibrated, the magnetization of any material in any direction can be measured by noting the impedance of a coil surrounding a portion of that material.

torroid coil (see Fig. 19.8). It is possible to construct an electrostatically shielded assembly 6 mm in diameter and 1 mm thick with an inductance of several microhenrys. With such a device, a magnetic field variation of 0.01 percent has been observed in the range of 300 to 2,000 oersteds.

## 19.10 PEAKING STRIP

When the major hysteresis loop of a ferromagnetic material is exercised by a source of sinusoidally varying energy, the time of magnetization-direction reversal can be specified in terms of the phase of oscillation. If a body constructed from a ferromagnetic material is exposed simultaneously to both oscillating and constant immersion fields, the time of magnetization reversal is displaced in phase. If the phase displacement is nulled by a third constant imposed field, the imposed field is equal and opposite to the immersion field.

In Fig. 19.9, a ferromagnetic wire (or PEAKING STRIP) is placed in the center of three coils. One coil supplies the oscillating drive current, the second detects the instant of magnetization reversal, and the third, which is driven by a direct current, provides the nulling field. The current flowing in the third coil can be calibrated in terms of immersion field strength.

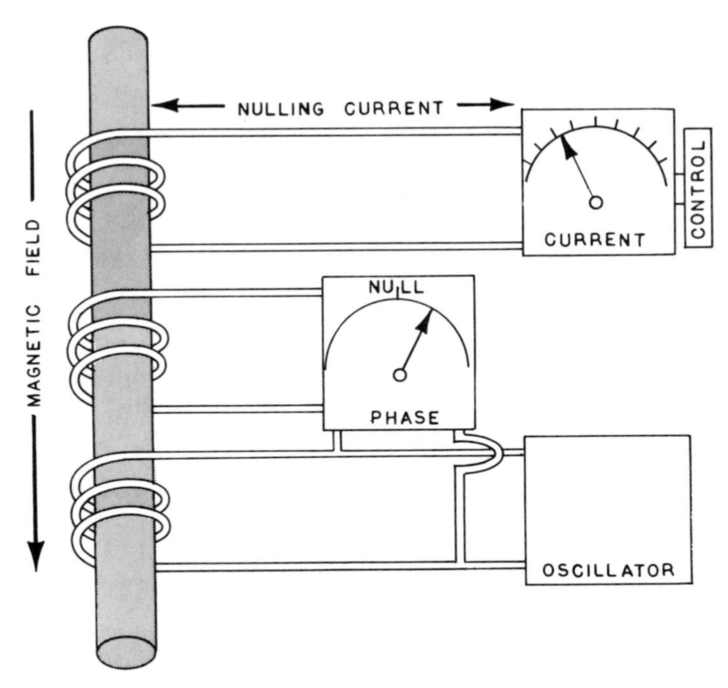


Fig. 19.9 Peaking strip: A nulling current can be equated to magnetic field strength.

## 19.11 MAGNETIC AMPLIFIER

In *MAGNETIC AMPLIFIERS*, small changes in the current flowing through control windings (acting as increments of magnetizing force) are used to establish relatively large changes in the self-inductance of load windings.

In the simple configuration of Fig. 19.10, two coils are wound on a single ferromagnetic core. One of these two windings functions as the inductance of an electric circuit consisting of an inductance in series with a resistance. If this circuit is driven by an alternating voltage, an alternating current flows through both the resistance and the inductance as shown by the vector diagram of Fig. 11.6. This winding is called the "load" winding. If the inductance of this coil is changed, the current through the resistance changes in response. In this case, the inductance of the load winding is changed by changing the permeability of the magnetic material from which the device is fabricated.

The second coil is called the "control" winding. A direct current flowing through this winding can be used to position the minor hysteresis loop

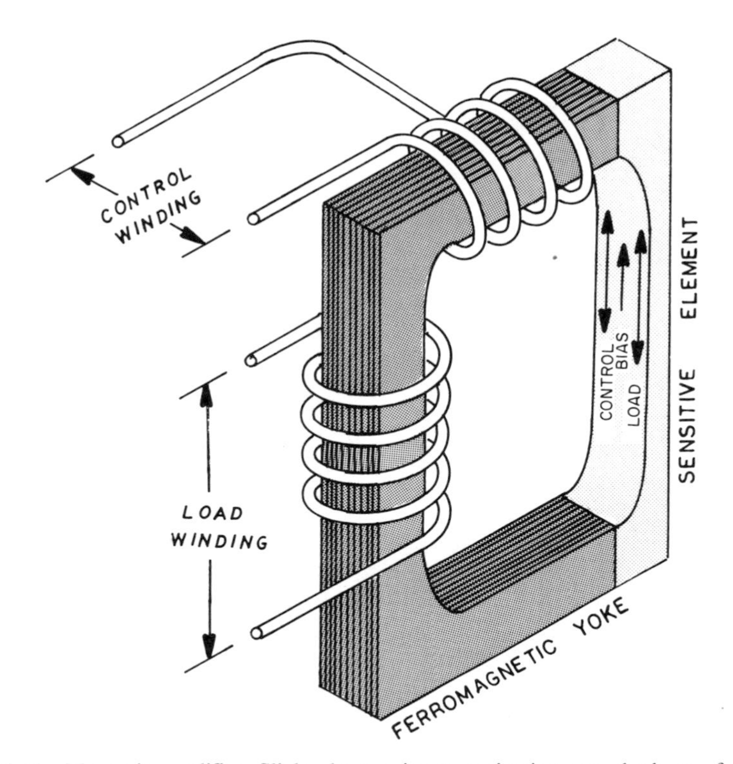


Fig. 19.10 Magnetic amplifier: Slight changes in magnetization near the knee of a square hysteresis loop yield large changes in permittivity; small changes in current flowing through a control winding induce large changes in the inductance of a load winding.

created by the alternating load current. Changes in control current are detected as changes in the inductance of the load coil, which, in turn, cause changes in the load current flowing through the resistive load. This is an amplification process because relatively small changes in control energy can yield much larger changes in load energy.

These same principles are used by *MAGNETIC MODULATORS*. If a magnetic core is biased appropriately and an alternating control current at low frequency is added to a bias current, the control current will modulate the alternating load current driven at a higher frequency. A magnetic modulator is sometimes called a *MAGNETTOR*.

To maximize the performance of a magnetic amplifier, the ferromagnetic core should have the following characteristics:

1. The hysteresis loop should be square.
2. The incremental permeability should be as large as possible and should remain more or less constant in value with increasing field

strength until saturation is approached. Near saturation, the incremental permeability should decrease rapidly to zero as the field strength is further increased.
3. The resistivity should be high and the coercibility low in order to minimize both eddy current and hysteretic losses.
4. The saturation flux density should be as large as possible in order to minimize the size of the device.
5. Magnetic characteristics should be stable under varying conditions of temperature and strain.

Figure 19.10 is organized to show the affinity of magnetic amplifiers with both flux-gate and variable-mhu magnetometers. Magnetometers, however, are concerned with unidirectional extraneous fields. Whereas amplifiers create their own fields internally. As a result, magnetometers must limit their response to that of one linear element driven by a nonresponsive yoke, whereas a magnetic amplifier makes use of all the elements of its yoke. In fact, the configuration of Fig. 19.11 is undesirable for a magnetic amplifier because of its sensitivity to extrinsic fields.

## 19.12 FLUX-GATE MAGNETOMETER

In *FLUX-GATE MAGNETOMETERS*, the square hysteresis loops of two ferromagnetic rods operating in a magnetic bridge circuit are used to evaluate immersion fields without drawing on magnetic bias. By eliminating the need for magnetic bias, these magnetometers improve both the accuracy and stability of those measurements obtained with a single ferromagnetic rod as described in Sec. 19.8.

The two identical magnetic circuits of flux-gate magnetometers are driven by identical, alternately demagnetizing–magnetizing driving forces. Although identical conditions can never be perfectly achieved, they can be approached in these circumstances. The effects of these two driving forces can be cancelled by subtracting one from the other.

As shown by Fig. 19.11, subtraction is accomplished electromagnetically by winding a single coil around both magnetic paths. If this coil is assembled so that the alternating fluxes from each of the two circuits flow in opposite directions through the single coil, no voltage is generated in that coil.

In Fig. 19.12, magnetizing currents (forces) $A$ and $B$ are converted by the transfer characteristics of the two identical hysteresis loops into the flux variations $A$ and $B$. These two hysteresis loops are shown as one (which is highly stylized) for the purpose of accentuating basic principles. The symmetrical alternations of the two driving forces created by the $A$ and $B$ currents are automatically centered on their symmetrical hysteresis

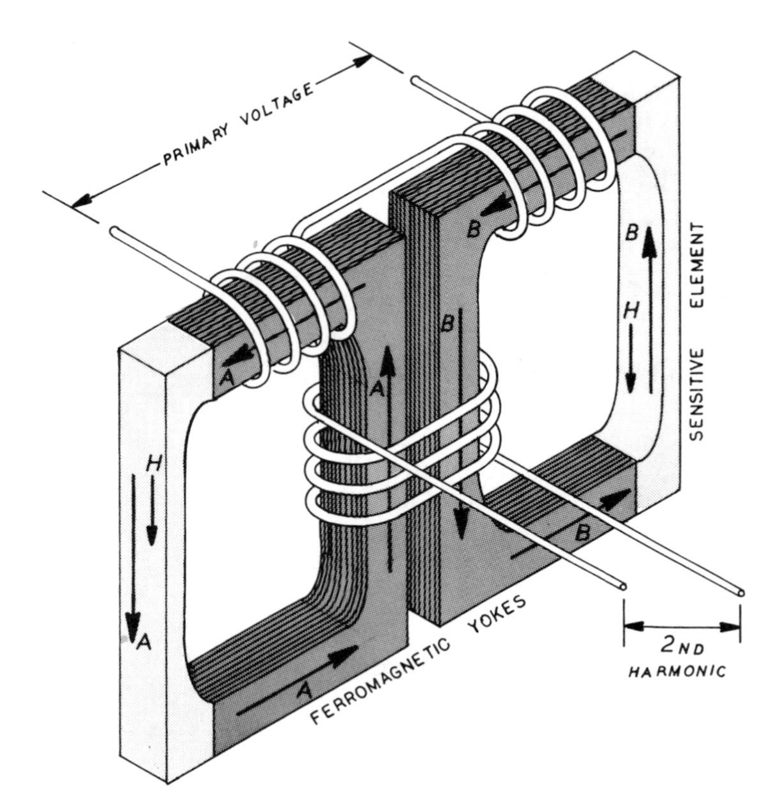


Fig. 19.11 Flux gate magnetometer: The symmetrical "square loops" of two ferromagnetic rods operating in a bridge circuit are used to evaluate small immersion fields.

loops. When the $B$ flux is subtracted from the $A$ flux in the common coil, the net result is a zero flux variation through this coil and hence a zero induced voltage.

Since the transfer characteristics of hysteresis loops are not linear, the shapes of the $A$ and $B$ flux variations are not the same as those of the $A$ and $B$ driving currents. As long as all factors in the system are symmetrical, however, the flux variations of the $A$ and $B$ paths will still cancel in the shared coil.

As shown by Fig. 19.11, an immersion field can be added to the driving field in one circuit and subtracted from the driving field in the other circuit. In Fig. 19.12, a magnetizing field $H$ is shown added to sinusoidal magnetizing force $B$. As shown by Fig. 19.11, a single field $H$ can be added to $A$ and subtracted from $B$ because of the phase difference between $A$ and $B$.

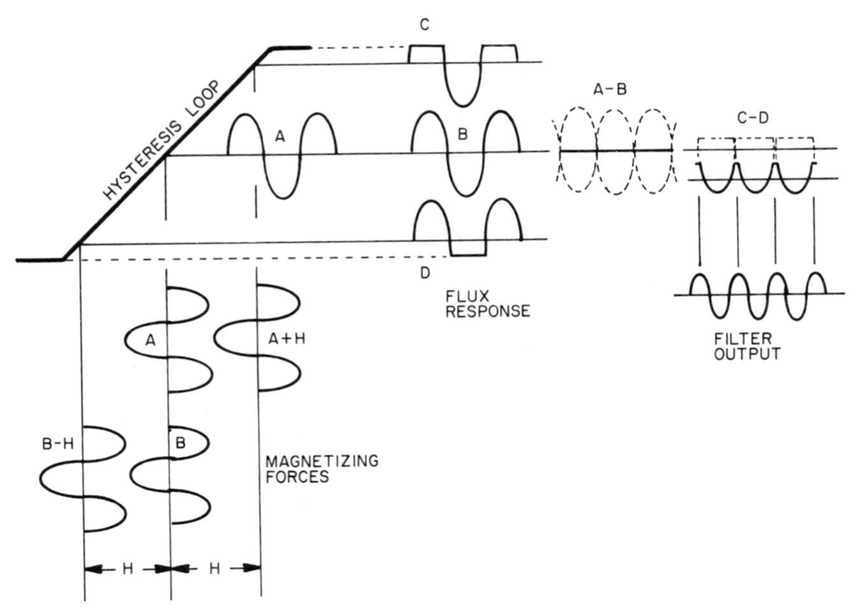


Fig. 19.12 Flux gate wave forms: Movement in either direction from a magnetizing force centered on the knee of a square hysteresis loop converts an ac driving current into a series of dc pulses. The latter can be converted into a signal current of twice the driving-current frequency.

When an extrinsic field of strength and direction *H* is present, the positive-going cycles of *A* are flattened into *C*, as shown in Fig. 19.12, whereas the negative-going cycles of *B* are similarly flattened into *D*. When the flux variations of *D* are then subtracted from those of *C*, a time-varying flux remains in the common path that induces a voltage of similar wave form in the common coil.

This time-varying flux is constructed from the even harmonics of the fundamental frequency. If the output of the common coil is filtered to reject harmonics above the second harmonic, the magnitude of the second harmonic (see Fig. 19.11) can be calibrated in terms of the strength of the immersion field *H*. The sensitivity of such measurements can be maximized by using a carefully regulated driving current to drive a magnetic material with a square loop so that the driving flux maximums come very close to the knee of the square loop. Under these circumstances, the addition of the *H* field to the *A* and *B* fields has the maximum flattening effect and generates the maximum magnitude of the second harmonic detected by the secondary coil. The magnitude of the second harmonic is a measure of the strength of the vector field $H$, and the direction of this vector is indicated by the phase of the second harmonic.

Because the two magnetic circuits cannot be made absolutely identical, there will always be some second harmonic voltage generated in the shared coil. This acts as a zero offset to the signal output and may distort the linearity of the response.

## 19.13 MAGNETIC BUBBLES

Over a restricted range of imposed magnetic field intensity, cylindrical domains are formed in very thin sheets of some crystalline materials. In these cylindrical domains, the constituent magnetons are oriented in a direction normal to the plane of the sheet and opposite to that of the surrounding medium. Viewed on end, these cylinders have the appearance of small circles. Such circular domains are called *MAGNETIC BUBBLES*.

Multiple domains form in ferromagnetic crystals in a minimum-energy configuration. Constituent components of this energy minimum include the crystal line energy (the tendency for magnetons to align with a crystal axis), the domain wall energy, and the energy required to support an external magnetic field. In some very small crystals, the external field energy is relatively slight, whereas the energy required to support a domain wall would be relatively large. Under such circumstances, a single crystal can support itself as a single domain.

Under similar-size constraints, certain thin films can support domain walls oriented at right angles to the plane of the film but not walls that are parallel to that film. The result is a family of domains, some oriented in one direction and some in the other direction.

Under no-field conditions, as shown at the right in Fig. 19.13, a characteristic "serpentine" pattern is formed in which the area covered by the domains oriented in one direction is more or less the same as the area covered by the domains oriented in the other direction. As always, when a field is imposed on a domain configuration, the domains oriented with the imposed field grow, whereas those oriented against shrink. The serpentine pattern exhibited in the absence of an imposed field becomes thinner in one domain direction and thicker in the other direction. At some particular field strength, the serpentine pattern breaks up, with the remnant forming bubbles. If the imposed field becomes strong enough, the bubbles disappear, and the entire film becomes one domain. This bubble disappearance is called *ANNIHILATION*.

Bubbles are of different diameters when exposed to different field strengths. As the strength of an imposed field decreases, bubbles become

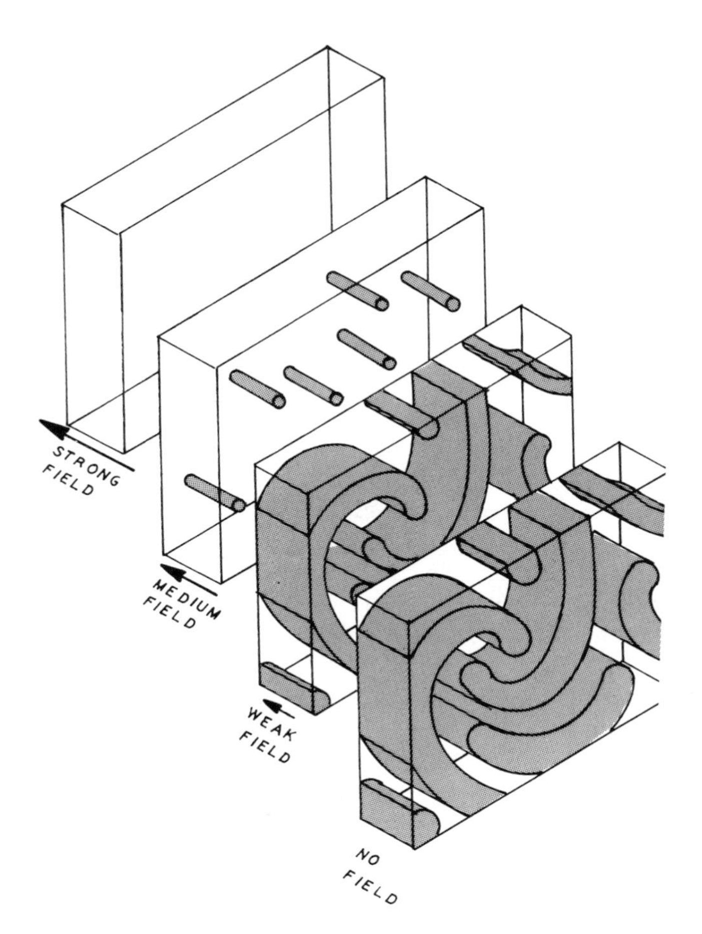


Fig. 19.13 Magnetic bubbles: In some ferrite films, plus/minus domains form where magneton orientations are directed at right angles to the plane of the film. In an unbiased condition, the areas of plus orientation equal those of minus orientation but their relative sizes vary when subjected to a bias field. Over a very short range of such bias, the domains in one direction take on a rodlike form, which in plane view is said to resemble a bubble.

larger in diameter. Bubbles are maintained in such films by a bias field whose intensity is set somewhere in between the annihilation field and the merging field. Below this bubble-supporting range, a number of bubbles merge back into some version of the typical serpentine pattern.

Once bubbles have been established in a film, they can be easily moved because their movements involve wall movements only. In fact, if a bubble is exposed to a field gradient, it moves toward an area where it sees minimum energy. The velocity of this movement is proportional to the ratio of the field difference across the bubble diameter to the coercivity of the magnetic medium.

## 19.14 MAGMETER

A *MAGMETER* is the trade name for a device that functions as a frequency detector that delivers d-c voltages proportional to frequency. In a magmeter, the primary coil of a transformer is driven by an alternating voltage of variable frequency and of such magnitude that the core is completely saturated at a very early part of each alternating cycle.

In the first half of an input cycle, as shown by Fig. 19.14, the core is saturated in one direction, whereas in the second half it is driven to its maximum in the opposite direction.

A secondary voltage is generated only while the flux through the core is changing. No secondary voltage can be detected when the core is saturated in either direction. The secondary voltage is then a series of pulses whose widths and heights, and hence areas, are more or less fixed by the saturation cycle and whose repetition rate is a function of frequency.

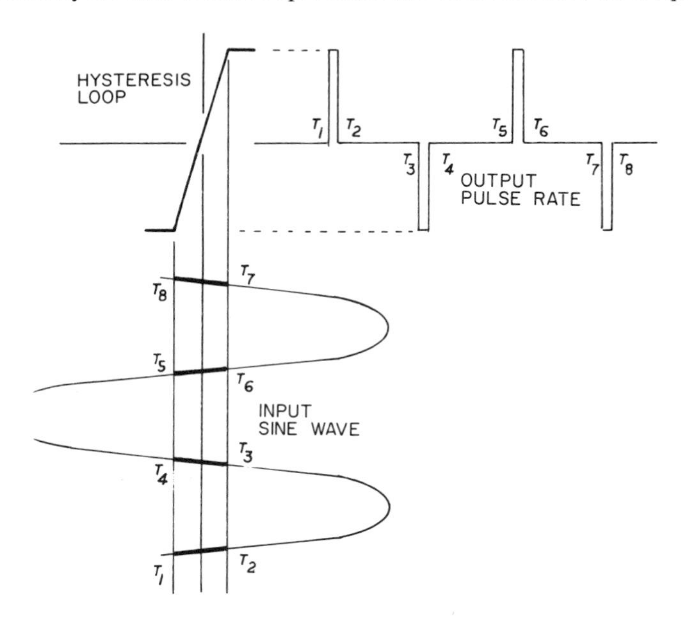


Fig. 19.14 Magmeter wave form: If the core of a transformer is constructed from a square-loop material that is driven into saturation at an early part of the energizing cycle, the ac current of any frequency is transformed into a series of equal area-pulses at that same frequency. Rectification converts these pulses into a dc current proportional to frequency.

As the frequency of driving voltage is increased, the number of fixed-area pulses developed per unit time in the secondary increases proportionally. If diodes are employed to rectify the secondary voltage pulses, the output is a d-c voltage whose magnitude is proportional to the frequency of the driving voltage.

## 19.15 BOBECK EFFECT

In the *BOBECK EFFECT*, magneton alignments are wound in a helix around a twisted ferromagnetic rod (see Fig. 19.15). A voltage is generated between the two ends of this rod whenever these magnetons are exposed to an increment of magnetizing force. The Bobeck Effect responds to the method of hysteresis loop stimulation illustrated in Fig. 19.5. In this case, the magnitude and direction of the voltage pulse that appears between the rod ends—for a standard increment of magnetizing–demagnetizing force applied by a surrounding coil—depend on the direction in which the helix is magnetized.

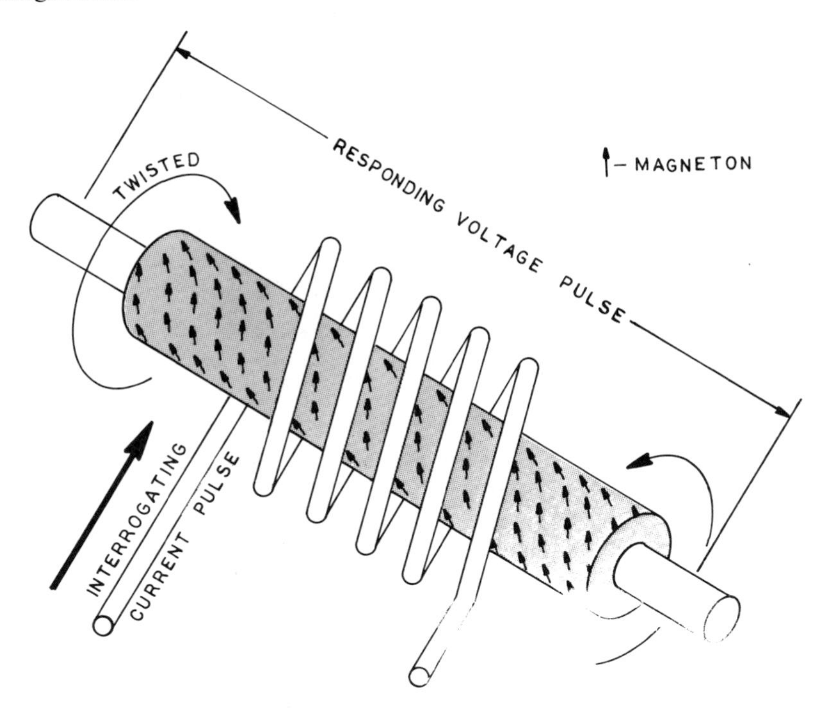


Fig. 19.15 Bobeck effect: When magnetons are oriented in a helical pattern around a ferromagnetic rod by mechanically twisting that rod, a voltage is generated in response to a change in magnetizing force.

A ferromagnetic rod can be used to carry electric current in either direction and can also be magnetized in either direction. As long as the axis of current flow coincides with the axis of magnetization, there is no particular interaction between the magnetic field and electric current; that is, a voltage is not generated between the two ends of such a rod when its magnetization is changed.

Magneton alignments, however, are responsive to mechanical strain. If the rod in question is exposed to torsional stress, the magneton axes follow the torsional strain, and the magnetic axis of the rod is deformed into a helix. As a helix, the magnetic axis no longer conforms to the axis of current flow. Under these circumstances, a voltage is generated between the two ends of the rod when the magnetization is changed. This arrangement is the inverse of a coil of wire wound around a magnetic rod. In a Bobeck configuration, magnetic material is wrapped around a current-carrying conductor. The interaction between the magnetized material and the current-carrying conductor is the same in both circumstances.

Although the Wiedemann, Wertheim, and Bobeck Effects are related, they are different. In the Inverse Wiedemann Effect, a current-carrying ferromagnetic rod is axially magnetized when it is twisted. In the Wertheim Effect, a voltage is generated between the two ends of a magnetized rod while it is being twisted. In the Bobeck Effect, a voltage is generated between two ends of a twisted, magnetized rod when it is exposed to an increment of magnetizing force.

The Bobeck Effect has been used in computer memories. There is no single material, however, that maximizes the magnetic and electric benefits of the Bobeck Effect. As a result, composite materials are fabricated by plating a highly coercive material on the surface of a wire of low resistance.

## 19.16 WIEGAND EFFECT

The *WIEGAND EFFECT* is a sudden flux change that occurs within a Wiegand wire under certain circumstances of imposed magnetomotive force. It represents a discontinuity in a hysteresis loop not found in any other material.

The Wiegand process involves fabricating short lengths of otherwise homogenous wire whose outer shells have much higher coercivities than their inner cores because these vicalloy wires have been cold worked by twisting them back and forth under tension, followed by appropriate heat treatment. This procedure sets up permanent helically oriented strains in the wires' surfaces. A typical Wiegand wire is 3m long and 0.25mm in diameter.

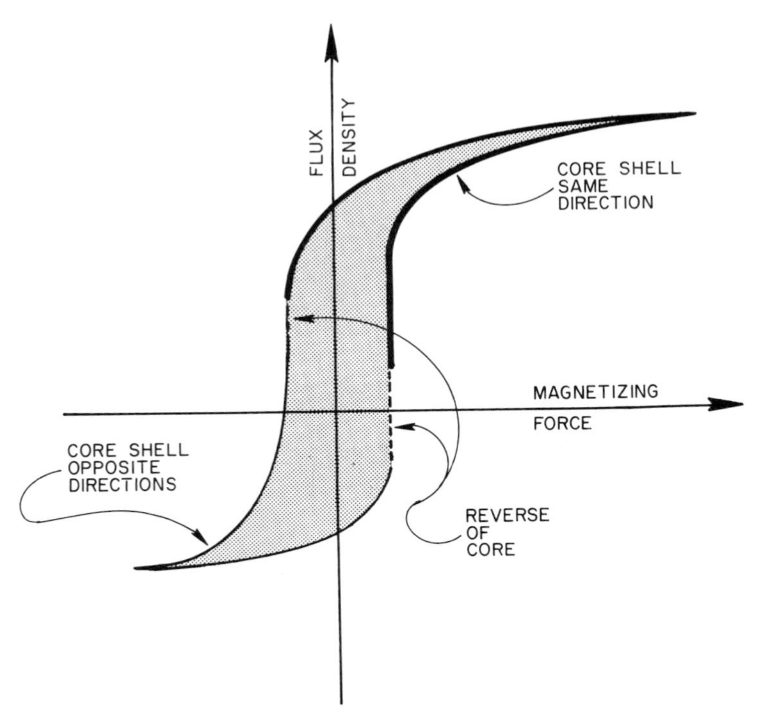


Fig. 19.16 Wiegand Effect: A twisted wire whose core and shell have different coercivities develops a discontinuity in its hysteresis loop when the higher coercivity of the shell suddenly imposes a reversal of magnetization in the core.

Because of the different coercivities (and the shell-to-core configuration), the inner core of such a device can support a different polarity from that of the outer shell. As a result, four magnetic configurations are possible: both inner core and outer shell in the first direction; inner core in the first direction and outer shell in the second; both core and shell in the second direction; and inner core in the second and outer shell in the first.

As shown by Fig. 19.16, it is possible to switch the inner core's direction back and forth without affecting the outer shell's direction. In what amounts to a carefully chosen minor hysteresis loop, a small jump in flux is experienced when the inner core's polarity is reversed in relation to that of the outer shell, whereas a return to the same polarity gives a much larger jump. Because of the helical structure of the cold-worked pattern, a voltage is generated across the two ends of such a wire during each polarity change. A Wiegand wire can also be driven over a range that

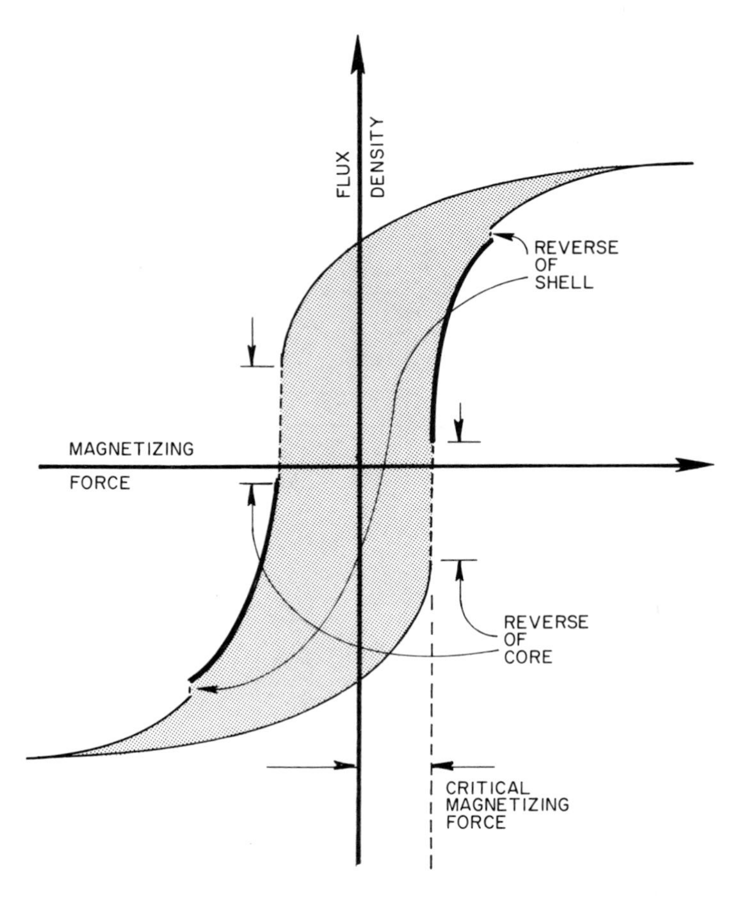


Fig. 19.17 Wiegand hysteresis loop: A Wiegand element consists of a twisted wire whose core and shell have different coercivities. The hysteresis loop of such an element experiences discontinuities whenever the higher coercivity of the shell imposes a reversal of magnetization on the core.

exercises all hysteretic possibilities for both the outer shell and inner core. This is represented by the major loop of Fig. 19.17.

In either circumstance, the sudden change in flux through a Wiegand wire experienced during a Wiegand jump can be used to induce an electric pulse in a surrounding coil. In fact, the flux change is so sudden that the characteristics of the resulting electric pulse are unique, that is, impossible to reproduce by any other means. Figure 19.18 illustrates a Wiegand configuration in which an electric pulse is generated each time a permanent magnet of appropriate strength passes in the vicinity.

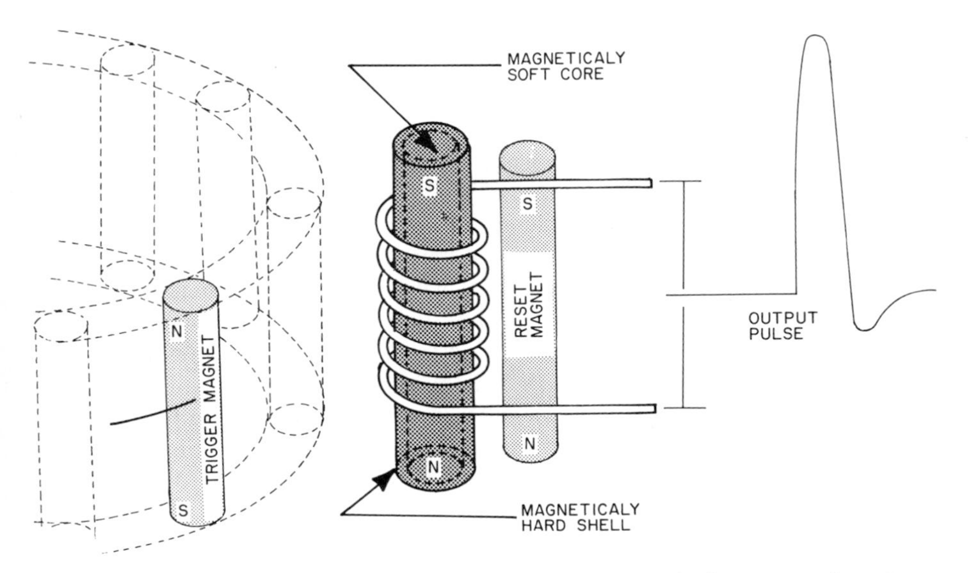


Fig. 19.18 Wiegand assembly: The sudden change in core magnetization representing a discontinuity in a hysteresis loop can generate a uniquely shaped pulse in a surrounding coil.

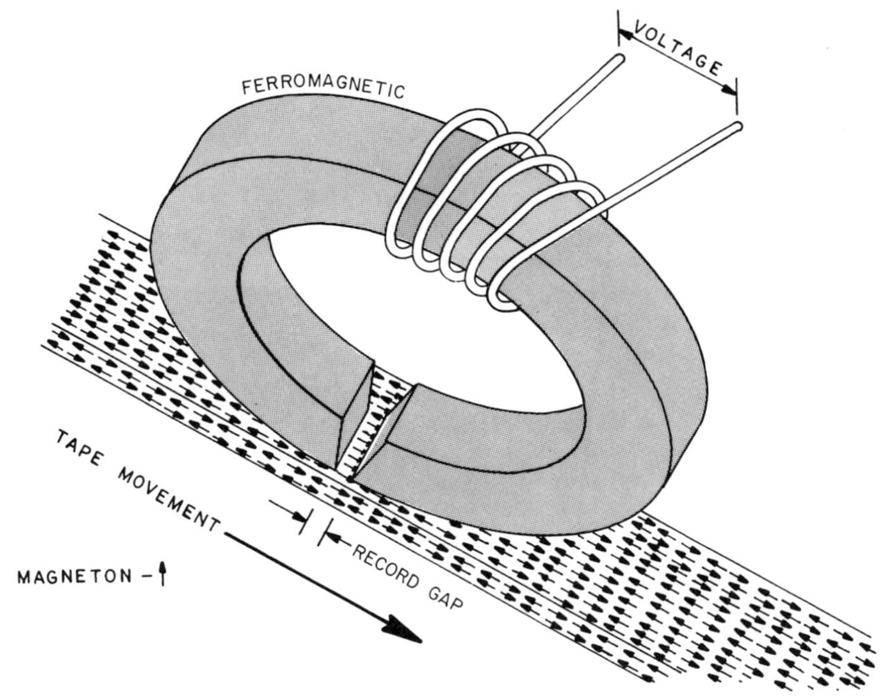


Fig. 19.19 Magnetic recording: Magnetons can be directed into at least two different orientations; those organized into regular patterns can be used to represent information.

## 19.17 MAGNETIC RECORDING

A section of magnetic tape consists of a backing medium covered by a thin film of ferromagnetic material. This film represents a large number of magnetons distributed in some uniform density. Since these magnetons are ferromagnetic and the directions of their axes are dictated by exchange forces, these directions are stable. Nevertheless, they can be manipulated by strong magnetic fields! Once a direction has been established by an imposed field, it is held constant by an exchange force. If the directions of these film magnetons are either dispersed at random or all lie in the same direction, their alignment pattern carries no useful information. However, if they are organized so that "bars" of one (plus) direction are imposed on a background made up of the opposite (minus) direction, a bar pattern can be made to support useful information.

In Fig. 19.19, the voltage applied to a recording head is reversed rapidly in a series of plus/minus pulses as the tape passes under the head. The pulse pattern of the driving voltage is translated into a magnetic field extending across the recording gap, and this field is then used to organize the desired bar pattern. If the head shown is envisioned as a read head instead of as a record head, a voltage pulse will be generated each time the bar pattern changes from plus to minus or from minus to plus. The pattern of these voltage pulses represents recovered information.

# 20
# SIZE EFFECTS

Very small structures exhibit electronic/magnetic characteristics that are significantly different from those of larger structures fabricated from the same material. This is true not only because the ratio of surface to mass increases as structural size decreases but because very small particles do not represent enough total energy to support wall energies.

## 20.1 SURFACE EFFECTS

Both the magnetons and carriers present in a material are exposed to two significantly different electric/magnetic environments. One of these environments is provided by the material's bulk, whereas the other is supplied by a kind of "skin" that is adjacent to its surfaces.

The bulk of a material represents a "continuum" extending in three-dimensional space. Each molecular particle observes an environment that is established by its neighbors. Although neighboring molecules may well be different, the mixture of molecules in each unit-volume is more or less a constant. If a material is crystalline, the continuum represents a regular structure in which each chemical bond is taken care of by some kind of relationship between neighbors and in which each molecule is assigned a fixed position whose location is established by the nature of their chemical bonds.

Near surfaces, molecular environments are defined either by the absence of neighboring molecules or by the presence of molecules in one direction that are significantly different than those observed in some other direction. A surface represents a massive departure from a continuum in which bonds are either left dangling (and are therefore available for adsorption processes) or assume relationships that are completely different from those seen in the continuum. Because the molecular nature of a material requires that electric distances be measured in terms of molecular spacings, surface effects are usually confined to a thin layer only a few molecules deep.

Both bulk and surface effects enter into those processes that establish the electric/magnetic characteristics of a body of material. If a body is very large as measured in terms of molecular spacings, bulk effects dominate and surface effects may not be detectable. On the other hand, if a body is small enough, surface effects are a real consideration and may even dominate.

## 20.2 FINE-WIRE LONGITUDINAL MAGNETORESISTANCE

Fine wires exposed to axial magnetic fields experience variations in electric resistance with changes in field strength. These variations are the result of *FINE-WIRE LONGITUDINAL MAGNETORESISTANCE*. This phenomenon is based on the helical paths followed by conduction electrons in the presence of axial magnetic fields.

An electron traveling at constant speed in a uniform magnetic field traverses a circular path with a radius of curvature inversely proportional to field intensity and proportional to speed. As illustrated by Fig. 2.3, this radius is called a "magnetron radius."

A magnetron trajectory can be superimposed on any other trajectory. As shown in Fig. 2.4, the effective paths of conduction electrons are of a spiral or helical nature, circling flux lines, if their components of motion are present both in the direction of an ambient field and at right angles to that ambient field. A net movement of carriers responding to current flow provides axial movement, whereas scatter mechanisms (with resulting random trajectories) can provide net tangential movements.

The electric resistance of fine wires is made up of two components. One of these is a result of conduction electrons interacting with crystal lattice ions in the bulk, whereas the other results from conduction electrons interacting with surfaces. If a magnetic field is directed along the axis of a wire of very small diameter, low-field helixes of large diameter force carrier electrons out toward the wire's rim. Here they are exposed to a maximum of surface effects as well as bulk effects. Under these circumstances, the electric resistance as seen by carriers is significantly affected by scattering caused by carrier/surface interactions.

As the strength of an axial field increases, contracting helical diameters pull the carriers away from wire surfaces to resistive exposures primarily determined by bulk effects. Under these circumstances, the magnetoresistance of fine wires decreases over that range of increasing field strength in which the majority of carriers are decoupled from surface effects. As magnetic fields further increase in magnitude, helical tightening increases the lengths of paths traversed by carriers, thereby increasing the carrier interactions with bulk effects. Magnetoresistance then increases after reaching some minimum value.

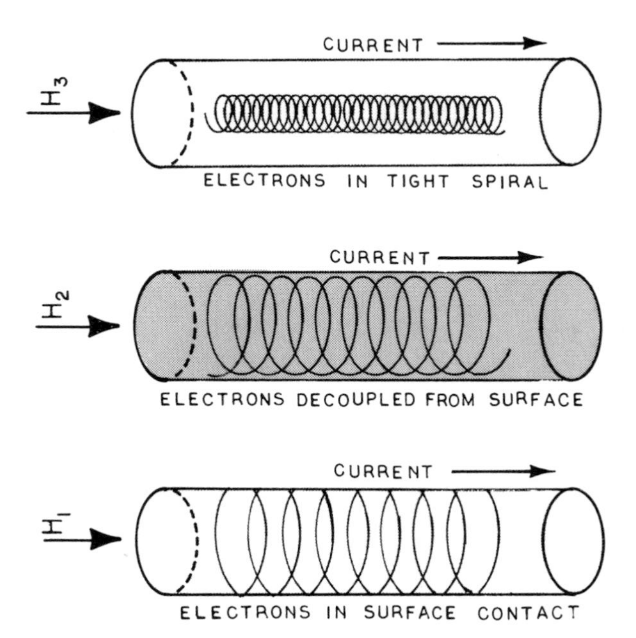


Fig. 20.1 Spiral paths in fine wire: Electron carriers traveling in an axially magnetized wire try to travel in spiral paths. If the diameter of the spiral would otherwise be larger than that of the wire, the spiral path is interrupted by collisions with the wire's surface discontinuity.

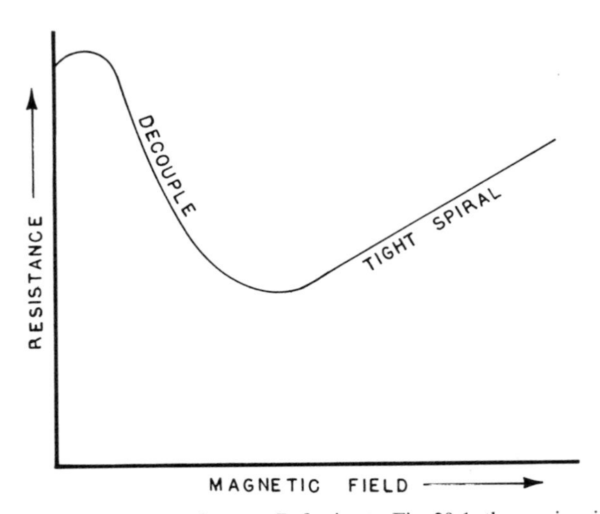


Fig. 20.2 Longitudinal magnetoresistance: Referring to Fig. 20.1, the carriers in a conducting wire are exposed to an axial magnetic field flow in a helical path. The resistance of such a wire is increased if the helix has a diameter greater than the wire diameter because the carriers interact with the discontinuity of the wire's surface. Furthermore, the resistance is increased if the helix is in a very tight spiral because of an increase in the carrier's mean-free path. Such a conductor has minimal resistance when the helix diameter is just a little less than the wire diameter.

In Figs. 20.1 and 20.2, the three conditions of "surface interaction," "surface decouple," and "spiral tightening," are each shown under the influence of different magnetic field strengths.

In the *WYDER EFFECT,* variations in thermal conductivity of fine wires follow the same general pattern as variations in electric conductivity.

## 20.3 FERROMAGNETIC PARTICLE SIZE EFFECTS

Two ferromagnetic phenomena are based on particle size. In the first, domain walls are lost below a critical size and particles become single domains. In the second, exchange forces are lost below a second, smaller critical size and materials become paramagnetic.

When a great many small particles are brought together in circumstances underwhich the condition of any one of them does not unduly influence the others, bulk effects closely track individual particle effects. Crystals smaller than critical size cannot support wall disaligning forces against aligning exchange forces. Such crystals exist as single domains in the absence of immersion fields. All of the magnetons present in a single-domain crystal are aligned in one direction, this direction being that of one of the crystal axes. In such circumstances, directional stability can be achieved around any one crystal axis but not in between axes.

Immersion fields of appropriate magnitude and direction can be used to move an alignment, against crystal anisotropic forces, from one crystal axis to another. Furthermore, it can be observed that immersion fields that exercise anisotropic forces near crystal axes do not experience hysteretic effects; that is, when a domain axis is displaced slightly by an imposed field, it returns automatically to its nearest crystal axis after the imposed field is removed.

As single-domain crystals are reduced in volume, they reach a second critical size below which they cannot support the aligning exchange forces against the disaligning thermal force. This size is a function of the ratio of crystal volume to temperature. Because temperature is a random phenomenon, the magnetism in particles approximating critical size is unstable. In this case, a particle changes back and forth between domain and no-domain structures. In the no-domain state below this critical size, ferromagnetic materials are said to experience *SUPERMAGNETISM*.

The change of magnetism experienced by both single-domain critical size and by supermagnetic critical size resembles a Curie transformation to some degree. Nevertheless, these are different phenomena and based on completely different principles.

## 20.4 SUPERMAGNETIC EFFECT

Utilizing bodies assembled from supermagnetic particles, a "susceptance thermometer" can be constructed whose transfer characteristics resembles those in Fig. 5.2. In the *SUPERMAGNETIC EFFECT*, however, the transfer characteristics are a function of particle-size distribution and not of a Curie transform.

## 20. 5 DOMAIN ROTATION

In minute samples of very thin ferromagnetic crystalline films, the particles are too small to support wall energy and magnetons do not have enough room to orient themselves in a direction normal to the plane of the film. Under these circumstances, but still in response to exchange forces, all magnetons are aligned in the same direction, parallel to the plane of the film. The particular direction assumed by the magnetons is one of some finite number determined by crystalline anisotropic forces.

If these magnetons are exposed to a magnetic field directed parallel to the plane of the film, they will rotate in response to that field and align themselves in the crystallographic direction that is most nearly coincident with the direction of the imposed field (see Fig. 20.3). Even when the imposed field is removed, the magnetons will remain in the new direction,

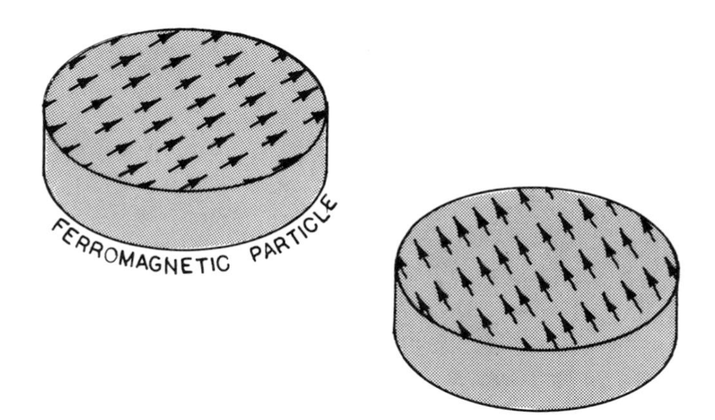


Fig. 20.3 Domain rotation: Single ferromagnetic domains can be formed in very small portions of thin films where the magneton orientations are stable along any one of the crystal axes. An imposed magnetic field pulse can move the magneton axes to the crystal axis most nearly oriented in the field direction. Here the new configuration is stable in the absence of any imposed field.

locked in place by crystalline anisotropic forces. These forces provide a mechanism equivalent to a mechanical "detent." Figure 20.3 shows two particles that have been left in "detent" directions after an aligning field has been removed.

Since magnetons can be rotated from one crystal axis to another faster than walls can be moved, this phenomenon of domain rotation has been used in computer memories where speed of response is a primary objective.

## 20.6 SONDHEIMER EFFECT

The *SONDHEIMER EFFECT* is observable in certain thin films exposed to magnetic fields directed at right angles to the plane of the film. The magnetoresistance of such films exhibits alternating plus and minus variations with increasing field strength, depending on the magnitude of the field. Those regions of field strength where variations are maximum occur in minimal fields and disappear completely in intense fields.

All galvanomagnetic effects are susceptible to Sondheimer variations.

## 20.7 ISOTHERMAL ELECTROMAGNETIC MAGNETORESISTIVE EFFECT

In the configuration of Fig. 9.11, small-diameter wires carrying electric currents are exposed simultaneously to axial magnetic fields and transverse electromagnetic radiation. When the carrier magnetron and wire diameters coincide, and the radiant and carrier-cyclotron frequencies are the same, the resistance of the wire is minimal. This frequency-sensitive minimum is here called the *ISOTHERMAL ELECTROMAGNETIC MAGNETORESISTIVE EFFECT* (IEMMR).

In detecting the IEMMR Effect, the wire temperature must be kept constant; otherwise, the absorption of electromagnetic energy by the wire causes an increase in resistance that will mask the effect.

This phenomena exists as a result of electromagnetic energy being added to carriers in magnetron orbit. The added increment of energy reduces the amount of energy required by the voltage driving these same carriers in their magnetron/axial paths.

# 21 STRONG MAGNETIC FIELD EFFECTS

Electronic carriers tend to circle flux lines when they move in magnetic fields. Under the circumstances discussed in Sec. 2.8 and shown in Fig. 2.4, these circular paths are extended into helices with radii, pitch distances, and frequencies of circulation. The radius referred to here is the magnetron radius described in Sec. 23.1, and the frequency is the cyclotron frequency described in Sec. 23.2.

In relatively weak magnetic fields, helical diameters are fairly large, circumferential traverses fairly long, and cyclotron frequencies fairly low. In strong magnetic fields, on the other hand, the pitch distances and cyclotron wave lengths are small enough for them to be able to interact with any periodic characteristic of a propagating medium that is similarly dimensioned.

Electrons traveling through solid materials experience characteristic mean free paths between collisions with lattice elements. After each collision, they embark in a new direction with a new velocity. If magnetic fields are present, the mechanisms of collisions and helical paths interact. If the mean free path represents a small percentage of a circumferential traverse, the periodicity of electron circulation is inconsequential, and magnetron path effects are manifested as galvanomagnetic effects. On the other hand, if magnetic fields are strong enough, magnetron circumferences are small, and an electron can complete a significant portion of a path revolution between collisions. (In metals, an electron has a magnetron radius on the order of $10^{-3}$ centimeters in a l-tesla field.) In these circumstances, the periodicity of an electron traverse can interact with other periodic functions that are characteristic of the environment provided by the solid material. These functions can be either temporal or spatial in nature.

Since strong fields provide the opportunity for periodic interactions, the resulting phenomena are here classified as "Strong Magnetic Field Effects."

## 21.1 AZ'BEL-KANER RESONANCE

In the Skin Effect of Sec. 12.8, electrons carrying a high-frequency alternating current flow very near the surface of a conductor where the depth of penetration is described by Eq. 12.3. If a magnetic field is applied in a direction parallel to the surface of such a conductor, the electrons will be forced into a helical path, as shown in Fig. 21.1, with a magnetron orbit as determined by Eq. 23.1. Since the diameter of this orbit can be greater than the skin depth, conduction electrons in orbit contribute to current flow for only a portion of an orbit's circumference. In such a configuration, there is both a time and a space periodicity in which an orbit circumference is included within the skin depth.

An *AZ'BEL-KANER RESONANCE* is then experienced as an interaction between the periodicity of electrons moving in response both to an alternating current flow and the periodicity of these same electrons in helical magnetron orbit. When conduction electrons are exposed to such interaction, the resistance they encounter is much greater if the orbit frequency and the frequency of the alternating current are either the same or share a harmonic relationship.

In a second interaction, as shown by Fig. 21.1, electrons near the surface in a cyclotron traverse can interact with a tangentially directed electromagnetic field. When this happens, the radiant energy is absorbed just as if it had encountered a resonant condition.

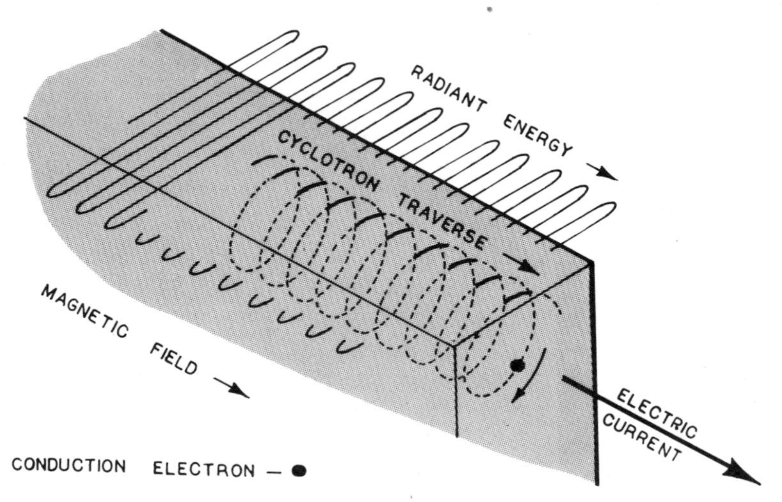


Fig. 21.1 Az'Bel-Kaner Resonance: Radiant energy interacts with electronic carriers in cyclotron traverses when both energy and traverses have the same wave length.

## 21.2 MAGNETOACOUSTIC RESONANCE

A *MAGNETOACOUSTIC RESONANCE* is an interaction between the periodicity of the alternating regions of compression and dilation established by sound wave propagation and the periodicity of electrons in magnetron orbit.

If a longitudinal sound wave is propagated through a metal, the alternating regions of compression and dilation cause conduction electrons to be moved in a sinusoidally varying trajectory that is equivalent to an alternating electric current. If a magnetic field is then applied at right angles to the direction of sound propagation, the moving conduction electrons are forced into helical orbits, as described by Fig. 2.4 and amplified in Sec. 21.1

When conduction electrons are exposed to such combined forces, the energy that causes them to move—in this case, the sound wave—encounters more propagation resistance if the orbit frequency and the frequency of the sound wave are either the same or share a harmonic relationship.

## 21.3 STRONG-FIELD MAGNETORESISTANCE

The term *magnetoresistance* describes a condition where the electric resistance of a conductor is changed when it is placed in a magnetic field. In Sec. 17.16, the general characteristics of magnetoresistance are divided into two parts. The first is a reduction in carrier forward velocity since carriers are here forced to move sideways as well as forward. The second is a reduction in the effective cross-sectional area of a conductor since carriers are crowded to one side. In this phenomenon, conduction electrons move along relatively small portions of their magnetron paths as these are indicated by Fig. 2.4.

In *STRONG-FIELD MAGNETORESISTANCE,* on the other hand, conduction electrons move through relatively large portions of magnetron paths. In this case, path lengths are extended by orbit, but the electrons are not necessarily crowded to one side. The fact that the electrons are following significant portions of helical paths and that these paths have periodicity of both pitch and cyclotron frequency makes it possible for these periodicities to interact with any structural periodicities to which they may be exposed. Since the strength of the magnetic field determines the periodicity of the helical path, the later can be changed by changing the magnetic field. The resistivity of a material will then experience periodic variations as changing magnetic fields sweep past conditions of periodic interaction.

In the *SHUBNIKOV-DE HASS EFFECT*, the resistivities of metals experience periodic variations in response to changing magnetic fields, while in the *MAGNETOPHONON RESONANCE EFFECT*, the same response is observed in weakly doped semiconductors.

## 21.4 ANISOTROPIC STRONG-FIELD MAGNETORESISTANCE

Many crystals are anisotropic, that is, as shown in Fig. 17.4, their structures exhibit different spatial periodic patterns when viewed in different directions. Each of these spatial patterns interacts differently with any of the helical carrier paths to which they may be exposed. A polar plot of strong-field magnetoresistance as a function of crystal orientation will then display an anisotropic pattern, with sharp peaks separated by regions of low magnetoresistance. This magnetoresistive anisotropy is a direct result of the crystalline anisotropy.

## 21.5 STRONG-FIELD MAGNETOSUSCEPTANCE

A conduction electron in spiral traverse exhibits periodicities of both time (cyclotron frequency) and space (spiral pitch). Section 21.1 discusses conduction electron interactions with both ac current and electromagnetic radiation; Sec. 21.2, with acoustic waves; Sec. 21.3, with structural patterns; and Sec. 21.4, with anisotropic structural patterns. These, however, are not the only periodic phenomena with which helical path interactions are possible.

For instance, the various relationships that exist between elements in lattice structures (electrons, atoms, molecules, etc.) are changed in quantized steps where these quantifications are described in terms of either phonon (vibrational energy) or magnon (oscillatory magnetic change) frequencies. Whenever the periodicity of helical paths interacts with either phonon or magnon frequencies, resonant effects may result.

In the *DE HASS-VAN ALPHEN EFFECT*, the ability of a metal to support changing magnetic fields (its susceptibility) experiences periodic variations in response to changing strong magnetic fields. This is a direct result of magnon frequency and helical path interactions. This phenomenon is best observed at low temperatures because the structural activities encountered at high temperatures tend to blur any periodic effects.

# PART IV MAGNETONS MOVING UNDER LOOSE CONSTRAINTS AS IN A VACUUM OR GAS

When an electrically charged particle is able to get far enough away from its neighbors, as in a vacuum or gas, its actions become fairly independent of its environment. If it starts on a curved path, it has a fair chance of completing that path. Even if it is deflected from its original trajectory by colliding with a gas ion, it may well complete the new trajectory without interference from some other ion.

Part IV discusses a number of the magnetic consequences that may be expected when magnetons are able to move without much interference from, or interaction with, other particles in the same environment, as in a vacuum or gas.

# 22 IONIC CURRENTS

Charged particles have opportunities to travel over significant portions of their magnetron trajectories when they move through gasses immersed in magnetic fields. They may not complete every trajectory necessarily, however. The mechanism that can interrupt a magnetron trajectory is a collision between one of these particles and a gas molecule.

Such collisions may do nothing more than deflect the moving charged particles into other trajectories, but it is also possible for them to ionize the gas molecules if they are violent enough. After an ionizing collision, three charged particles must be accounted for: the original particle, the gas ion, and the freed electron. The movements of ionizing particles before collisions, the movements of gas ions, the movements of freed electrons, and the movements of ionizing particles after collisions are all subject to Lorentz forces.

The interactions of ionizing/ionized particles with magnetic fields when these particles move through gas environments are the basis for a number of phenomena.

## 22.1 AURORA

The *aurora borealis* is a glow in the sky that is sometimes observed near the North Pole. It occurs when atmospheric molecules deionize after having been ionized by solar activity. (The same phenomenon in the southern hemisphere is called the *aurora australis*.)

The sun emits a great deal of energy in many different forms. Among these forms are high-speed charged particles of various kinds that are ejected in all directions. Those particles that approach the earth encounter the earth's magnetic field.

Figure 2.4 illustrates what happens when moving charged particles interact with flux lines. The particles are induced to circle the flux lines in helical paths, the net trajectory of which coincides with the flux lines. Any extraterrestial charged particles that encounter the earth's magnetic field from any direction are redirected into flux paths. As all of the earth's flux paths exit the earth from one pole and reenter at the other, interacting charged particles end up at one or the other pole.

Those high-speed charged particles that enter the atmosphere of the earth near the poles encounter atmospheric molecules. Collisions between such solar particles and gas molecules can ionize the latter, and electrons are knocked loose from some of them. Since ionized molecules represent more energy than deionized molecules, there is a tendency for electrons and gas ions to recombine. In those cases when ions recapture their missing electrons, electromagnetic energy is emitted. The term *aurora* is used to describe the visible portion of this electromagnetic emission.

As an auxiliary benefit to all life forms, the earth's magnetic field acts as a "roof" that protects the central portion of the globe from the effects of continuous bombardment by these solar-emitted, high-speed particles.

## 22.2 SINGING FLAMES

Flames exposed to oscillating magnetic fields sometimes emit audible audio notes at oscillating frequencies.

Flames consist of high-temperature, gaseous products that accompany certain reaction processes. When high temperatures separate orbit electrons from certain gas molecules, rich mixtures of free electrons and positive ions are created. Flames thus make both electrons and positive ions available to act as carriers in the support of electric currents. At the same time, flames generate thermal gradients that cause convection currents to flow in the gasses of which they are composed. Since electric carriers are a part of these gasses, convection currents also represent electric currents.

The electric/convection currents of flames exposed to magnetic fields are subject to Lorentz forces. Depending on the nature of the current/field interactions, the imposition of a magnetic field can either dim or brighten a flame. Gas pressures of flames interacting with oscillating magnetic fields are modulated by Lorentz forces acting on convection currents. Since gas-pressure modulation represents sonic vibrations, flames can serve as "transducers" to convert electric energy into sonic energy. Such flames are here called *singing flames*.

## 22.3 BLOWOUT MAGNET

*BLOWOUT MAGNETS* are used to extinguish arcs. They accomplish this task by lengthening an arc to the point at which the voltage gradient extending along the arc is less than the ionizing potential of the gas that is otherwise necessary to support the arc.

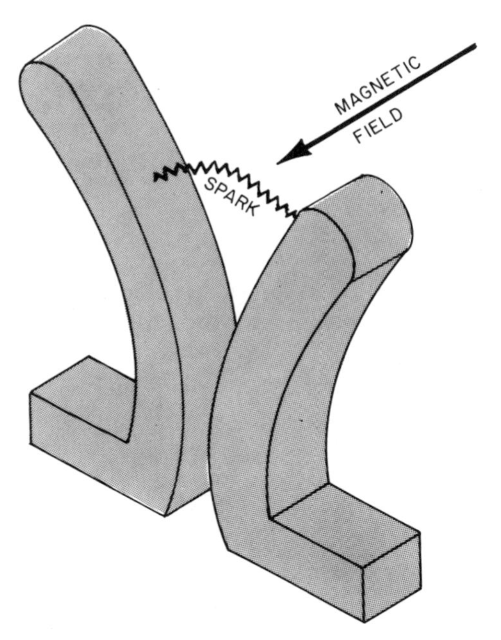


Fig. 22.1 Blowout magnet: Lorentz forces acting on the carriers in a spark can be used to extend its length to the point of extinguishing it.

An *ARC* is a discharge of electric current through a gas. The discharge may be supported by either one of two (or both) mechanism. In the first, carrier electrons are emitted from electrode surfaces if voltage gradients at these surfaces exceed the work functions of the surfaces. In the second, voltage gradients ionize gas molecules. After ionization, free electrons and remanent positive ions are available to act as current carriers. The voltage drop across a gas-supported arc is approximately equal to the ionization potential of the molecules in the arc.

Regardless of how they are supported, arcs represent electric currents and, as such, respond to Lorentz forces. At the very least, then, arcs are bent, and hence their lengths are increased, when they are exposed to transverse magnetic fields. This current/field interaction is commonly used to extinguish arcs that form when relay contacts open high-voltage circuits. As shown by Fig. 22.1, if relay contacts are paralleled by electrodes forming a "V"-shaped air gap, an arc representing the interrupted voltage originates at the bottom of the "V." A transverse magnetic field is then used both to bend the arc and to drive it along the electrode surfaces toward the open end of the "V." Both mechanisms significantly increase the length of the arc. As the same voltage is stretched along an increasingly lengthened path, the arc is extinguished at that point at which voltage gradient within the gas is reduced below the ionization potential of the gas molecules.

## 22.4 MAGNETOSPHERE

The *MAGNETOSPHERE* is that portion of the earth's environment dominated by magnetic fields. It is the vector sum of the earth's geomagnetic field and the magnetic fields accompanying the energy emanating from the sun.

As a superheated body subject to violent thermal and radioactive perturbations, the sun throws off vast quantities of plasma that consists of about equal parts of electrons and protons. Although this plasma is ejected from the sun's surface in all directions, much of it moves away from the sun in a trailing mode, more or less in one direction, because of the sun's movement through space. This plasma migration is called the *SOLAR WIND*. So long as the solar wind's constituent electrons and protons move together in equal quantity, they do not generate magnetic fields. Any differences in drift velocities represent electric currents, however, and differences in concentration represent voltages capable of driving currents. In any case, flows of plasma currents are accompanied by magnetic fields.

The earth lies in the path of the solar wind. When the particles of the latter and their associated magnetic fields approach the earth, they interact with the geomagnetic field. In this interaction, both fields are modified. The shape and behavior of the geomagnetic field is partially determined, then, by the solar wind's passage.

The emanative activities of the sun are highly variable both temporally and spatially over the sun's surface. Since the sun rotates on its axis, the character of the solar wind is in a constant state of change. Since the earth also rotates on its axis, the interactions between solar wind and geomagnetic field also change continuously. Significant effects of these interactive variations are called *MAGNETOSPHERIC STORMS* in the solar wind and *MAGNETIC STORMS* in the geomagnetic field.

Other interactions between solar wind particles and the magnetosphere include the aurora, mentioned previously, and an electric current that flows around the earth through the atmosphere in an east-to-west direction.

## 22.5 MAGNETIC FOCUSING

The path of a charged particle passing through a toroidal magnetic field is bent by that field in a manner analogous to paths of light rays passing

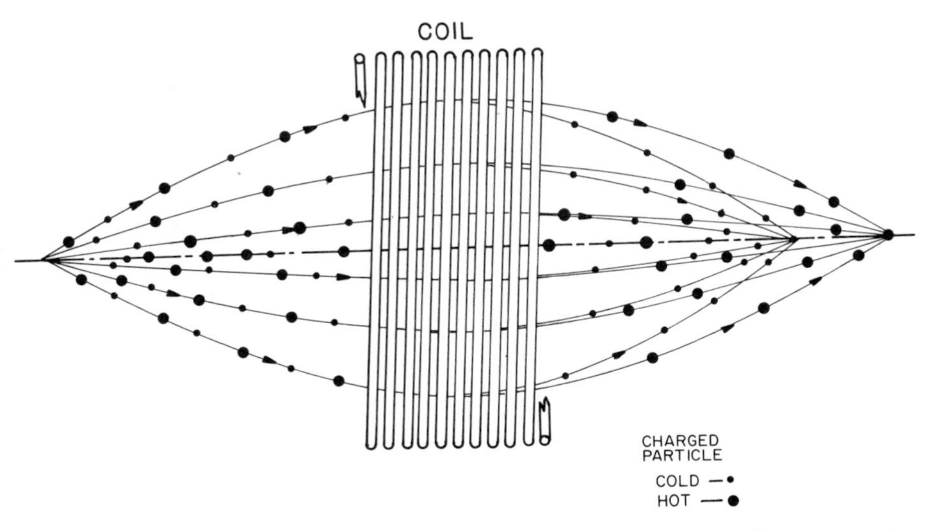


Fig. 22.2 Magnetic focusing: A coil functions as a lens to bring carriers of like speed and like mass to a single focal point.

through a glass lens. In fact, such a field functions as a magnetic lens. In this *MAGNETIC FOCUSING* (see Fig. 22.2), all of the charged particles, traveling at the same velocity, exit such a field at the same angle at which they entered. If they all originated at a point source, therefore, they will all be brought back to a common focal point. The faster a particle moves, the greater the distance to the focal point.

## 22.6 BETA RAY SPECTROMETER

Consider those circumstances in which a radioactive source emits electrons whose velocities (energies) extend over a wide range of possibilities. Using a magnetic lens, all electrons of a particular energy can be brought to a common focus by adjusting the strength of the focusing field appropriately. If a common magnetic strength is used, each velocity will bring its electrons to a different focal point along the lens's center line. Conversely, by choosing particular strengths, each velocity, in turn, can be brought to the same focal point.

In the *BETA RAY SPECTROMETER* of Fig. 22.3, all paths between the source and the focal point are blocked except those for electrons traveling at a particular velocity. By sweeping the focusing current and measuring the resulting detector current, it is possible to develop the spectrum of electron energies that emerge from a given source, that is, the relative quantities of electrons that are emitted at each energy level.

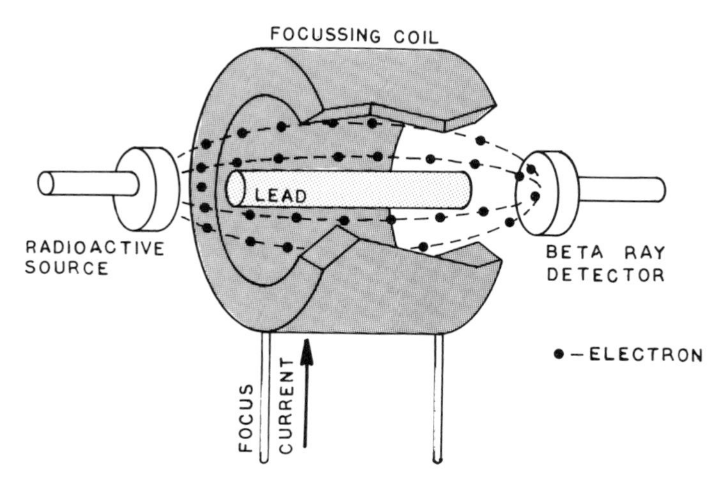


Fig. 22.3 Beta ray spectrometer: By sweeping the focusing current over a particular range, carriers of differing characteristics can be brought to the same focal point in time sequence.

## 22.7 STERN-GERLACH EXPERIMENT

The *STERN-GERLACH EXPERIMENT* proves the simultaneous existence of parallel and antiparallel magnetons in an aligning magnetic field by separating one from the other. In this experiment, a beam of metal ions is passed through a nonuniform magnetic field whose nonuniformity lies in the field direction. These spinning, monatomic nucleii are aligned by the field, some with it and some against, according to the strength of the field. In either event, one end of each is exposed to a stronger field than the other end. For a magneton aligned with the field, one end is attracted more by the stronger field than the other end is by the weaker field. For a magneton aligned against the field, on the other hand, one end is repelled more by the stronger field than the other end is by the weaker field. Those magnetons aligned with the field then experience a small accelerating force in the field direction, whereas those aligned against the field are accelerated in the opposite direction.

As shown by Fig. 22.4, if the beam continues through the nonuniform field far enough, the parallel magnetons are separated from the nonparallel magnetons. The former arrive at a target slightly displaced in the strong field direction, whereas the latter are slightly displaced in the weak field direction.

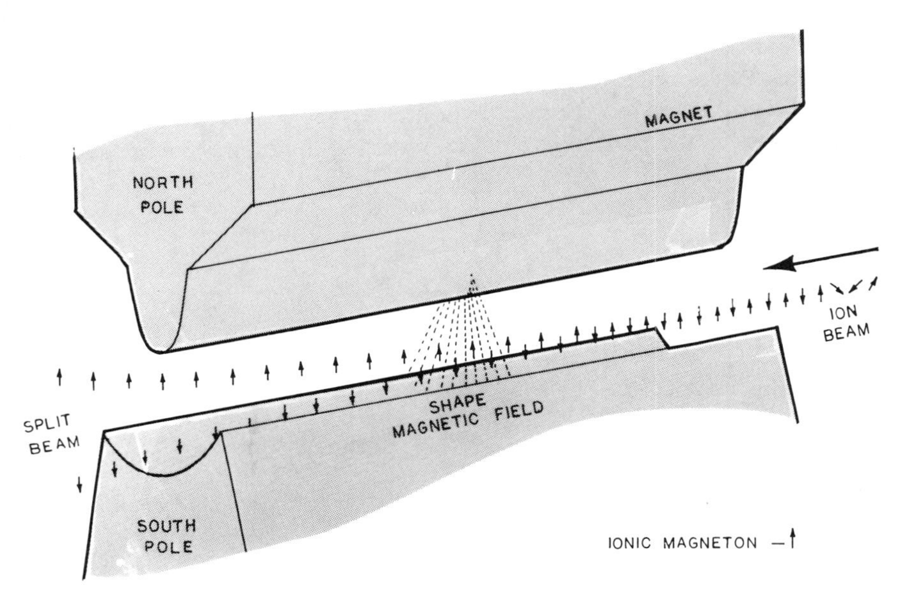


Fig. 22.4 Stern-Gerlach configuration: A shaped magnetic field can be used to separate beam magnetons that are oriented in opposite directions.

## 22.8 RABI CONFIGURATION

Experiments with the Rabi Configuration prove the existence of nuclear precessions by separating one type of nucleus from all other types by precessing that one and not the others. Rabi combined two Stern-Gerlach separators in series opposition so that parallel magnetons are deflected downward by the first and upward by the second to arrive at an ultimate target. The antiparallel magnetons are rejected in this process by their failure to pass various slits, or templates, inserted at strategic locations in their paths. Complete separation is not accomplished by this means, however, since molecules with differing characteristics might arrive at the same target having travelled different paths.

Rabi then introduced a region of uniform magnetic field ($H$) between the two deflection systems. He then superimposed an alternating field on those magnetons passing through this uniform field, as shown by Fig. 8.1. By choosing a strength of uniform field and an alternating frequency of appropriate values, certain magnetons passing through the uniform space are conditioned to precess. Since precession changes the vertical component of magnetic moment, precessing magnetons arrive at different target positions than nonprecessing magnetons. When $H$ and the precessing frequency are adjusted to minimize the ionic current arriving at the receiving target position, the particular characteristics of the precessing magnetons are established.

# 23 MAGNETRON EFFECTS IN GAS

A charged particle moving in a direction at right angles to a magnetic field attempts to follow a circular path. This path is called a *MAGNETRON ORBIT*. The radius of a magnetron orbit, the distance traveled in it, and the velocity of travel are factors that can be manipulated to accomplish various objectives.

## 23.1 MAGNETRON RADIUS

Charged particles traveling through magnetic fields follow magnetron paths in response to Lorentz forces. The radius of such paths is determined as follows:

MAGNETRON RADIUS $R = mv/eH$ (23.1)

where $R$ is the resulting radius of curvature when $m$ is the mass of a particle; $v$, its velocity; $e$, the electric charge on it; and $H$, the strength of the magnetic field.

A charged particle is accelerated by a voltage gradient. All charged particles that start with zero velocity and carry the same electric charge ultimately reach the same velocity when exposed to the same accelerating voltage. The momentum increment added to a charged particle, on the other hand, is proportional to the time over which the particle is exposed to an accelerating force. Since momentum is proportional to the product of mass times velocity ($mv$), all originally stationary particles (carrying the same electric charge) exposed to the same voltage gradient for the same length of time achieve the same momentum.

If both $e$ and $H$ in Eq. 23.1 are held constant, and if all charged particles are dropped through the same voltage distance, the magnetron radii will be proportional to particle mass. If only H is held constant and if all charged particles are dropped through the same voltage time, the magnetron radii will be proportional to particle charge.

## 23.2 CYCLOTRON FREQUENCY

When a particle travels in magnetron orbit, the radius of circular travel increases as particle velocity increases. Since this radius is proportional to velocity, a moving particle takes the same length of time to complete a magnetron circumference for every possible magnetron orbit. Each type of particle in magnetron orbit therefore circulates at a particular frequency, this being determined solely by field strength.

Each type of particle has its own unique frequency/field proportionality. For an electron, this cyclotron frequency is determined as follows:

*CYCLOTRON FREQUENCY* $$f = 2.8 \times 10^6 B \qquad (23.2)$$

where $f$ is an electron's cyclotron frequency when the magnetic field intensity is $B$, as measured in gauss.

## 23.3 CYCLOTRON

A *CYCLOTRON* is an instrument that utilizes oscillating voltages supplied at orbit frequencies to accelerate charged particles. In Fig. 23.1, two hollow "D"-shaped electrodes are driven by a voltage supplied at such a frequency that it always has the necessary polarity across the gap between the two "D"s to exert an accelerating force each time a chosen type of particle makes a gap crossing. Since each gap crossing provides exposure to an accelerating voltage, the velocity of a particle is increased by a fixed increment after completing each one-half cycle of cyclotron revolution. Each added increment of velocity results in an added increment of magnetron radius, and a particle moves in a more or less step-spiral path from the cyclotron's center to its outer edge. A particle moving from the center of such a device toward its outer rim travels with almost zero velocity at the beginning but reaches a very high speed by the time it reaches the outer edge.

Cyclotrons operate at constant frequency for particles of constant mass. The masses of particles change, however, when their velocities approach the speed of light. *SYNCROTRONS* are accelerating instruments whose frequencies are shifted to compensate for mass changes.

## 23.4 SPLIT-ANODE MAGNETOMETER

An electron beam is deflected by a magnetic field. The stronger the field, the more the deflection. If the beam moves with constant velocity, as

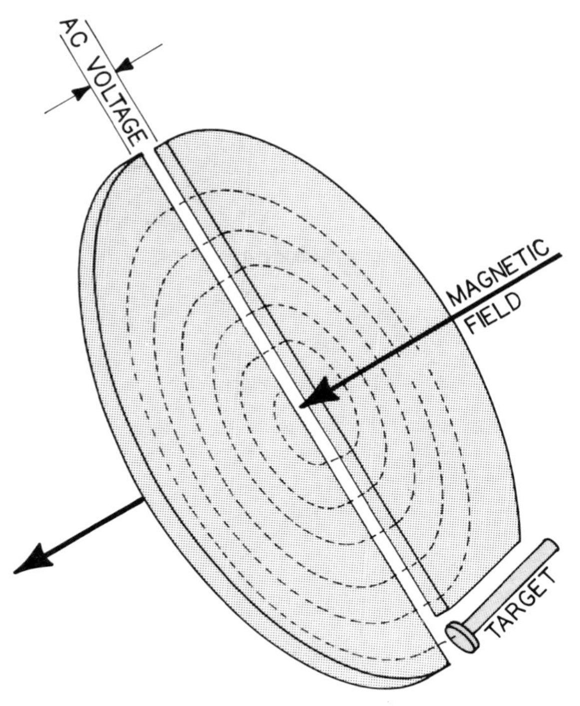


Fig. 23.1 Cyclotron: Oscillating voltages supplied at orbit frequencies are used to accelerate charged particles.

determined by a constant accelerating voltage, the amount of deflection can be a sensitive measurement of field strength, as determined by the split-anode magnetometer schematized in Fig. 23.2.

In orthogonal terms, consider an electron beam traveling from a cathode to an anode for some distance along an X axis. In this case, the anode is split into two parts with each part oriented in such a fashion that the electron beam current is equally divided between them. This equal division remains constant for any beam deflection in the Z direction. The presence of a magnetic field in the Z direction causes the beam to bend in the Y direction. As the beam does so, the balance of beam current between the two anode halves is changed. The ratio of beam current between the two halves of the anode can then be calibrated in terms of the strength of the deflecting field that extends in the Z direction.

Conversely, this configuration can be used to measure accelerating voltage when the device is immersed in a constant magnetic field.

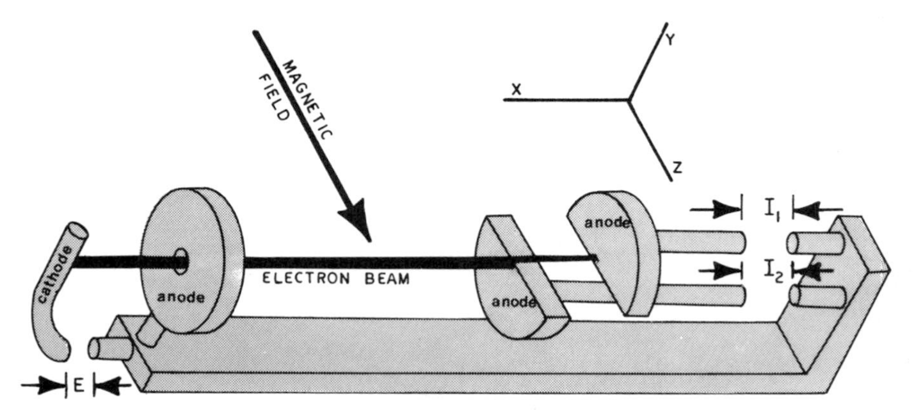


Fig. 23.2 Split-anode magnetometer: The movement of an electron beam caused by Lorentz or other forces can be detected as a change in the distribution of currents between the two halves of a split anode.

## 23.5 MAGNETRON

A *MAGNETRON* is a diode in which electrons traveling from a central cathode to a peripheral anode may or may not reach the anode, depending on how they are deflected by a magnetic field.

Figure 23.3 illustrates a configuration in which a straight-wire cathode is centered in a sleeve shaped anode, with both anode and cathode axes coincident to the direction of a magnetic field. When no magnetic field is present in such a device, an electron travels directly from the heated cathode to the cylindrical anode. With the application of a magnetic field, electrons travel in curved paths, as dictated by Eq. 23.1. At some critical field strength, the curvature is such as to cause the electrons to miss the anode and return toward the cathode. Constant current flows through this type of diode with increasing field strength until the latter reaches a critical value. At the critical point, the current falls abruptly to zero and remains there with energy further increase in field strength. If the field strength is held constant, the diode current is cut off below some critical accelerating voltage. In actuality, this critical point is unstable.

Electrons following curved paths create Lenz fields. When the current in a magnetron misses its anode and completes circular paths, the Lenz fields are stronger than they would be if the current were collected by the anode.

The magnetic field seen by an electric current is the sum of the imposed field and the Lenz field. There is, then, a critical magnetic field (and a

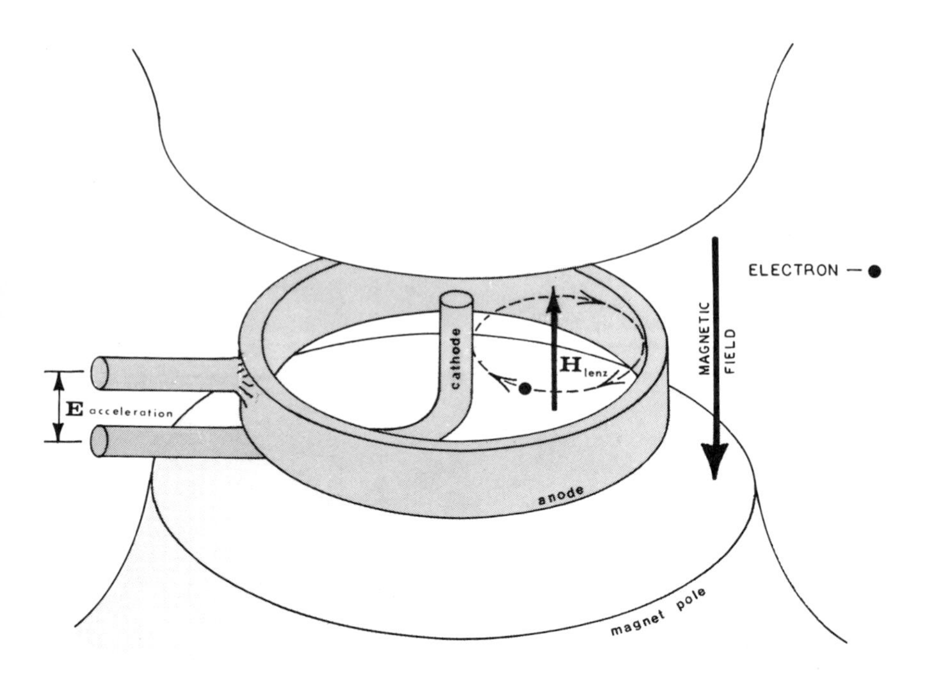


Fig. 23.3 Magnetron: Because of the Lenz field generated by electrons in circular orbit, the radius within which an electron in circular path just contacts an anode is unstable. As a result, high frequency oscillations occur at an interval between just making and just missing contact with the anode.

critical voltage) for which the anode current oscillates between cut-off and reconstitution. In this critical field, the anode current is first cut off by the combination of imposed field and driving voltage, but subtraction of the Lenz field from the imposed field reduces the field seen by the anode current and turns the diode current back on. As soon as the anode current is restored, the Lenz field is again reduced, whereupon the imposed field cuts off the current once more.

Since electrons have little mass and can be accelerated to high speeds, magnetron oscillations occur at high frequencies.

## 23.6 MAGNETRON PRESSURE GAGE

Electrons traveling through a gas experience periodic collisions with gas molecules. If the electrons travel fast enough, the molecules are ionized by these collisions. When these ions serve as carriers for an electric cur-

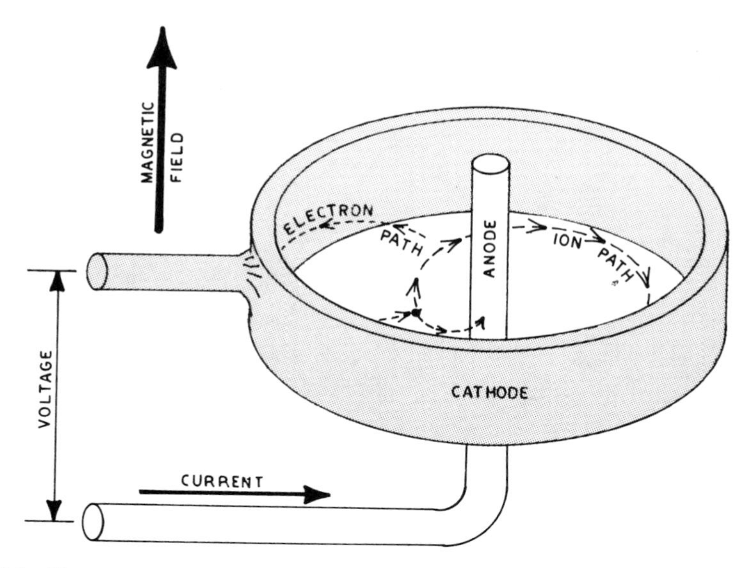


Fig. 23.4 Magnetron pressure gage: When an electronic carrier in a vacuum is configured to travel a circular path that misses an anode, no current flows through the device. When a gas is introduced into the evacuated space, electron collisions with gas molecules distort the circular paths and current flows between cathode and anode.

rent, the magnitude of this current can be used as a measure of the number of gas molecules present and hence a measure of gas pressure.

If the cathode and anode of Fig. 23.3 are interchanged so that the sleeve acts as the cathode and the central wire the anode, the device can be used to measure gas pressure, as shown in Fig. 23.4. In this inverted mode, the accelerating voltage and the magnetic field are adjusted to create circular paths for the electrons that travel at a high enough speed to ionize gas molecules. If there are no molecules present in such a device, the diode current will be zero. If gas molecules are present, some will be ionized. These ions are then accelerated by the electric field and move to the cathode. At the same time, some of the electrons—either those deflected by collisions or those knocked loose from the molecules—move to the anode. These two charged-particle movements represent an electric current whose magnitude serves a measure of gas pressure.

## 23.7 MASS SPECTROMETER

If all of the different types of molecules present in a complex gas are ionized, an analytical process based on Lorentz forces can be used to separate one type of ion from every other type.

Following the dictates of Eq. 23.1, an individual ion exposed to both electric and magnetic fields follows a circular path. If several different types of ions are present and each is exposed to common electric and magnetic fields, the radii of their various paths depend on their characteristic *m/e* ratio. Since *e* is an integer that can be set at 1, path radii represent mass differences. Since mass differences are more or less integer multiples of a proton's mass, paths traveled by different types of ions have discrete radii separated by forbidden zones.

If collector anodes are placed to intercept each path, the presence of an electric current on one collector identifies the presence of its characteristic ion. At the same time, the relative magnitudes of all collector currents serve as a measure of the proportion of each type of molecule present.

An *ASTON MASS SPECTROMETER* (see Fig. 23.5) achieves essentially the same result with one collector anode by creating all ions at a

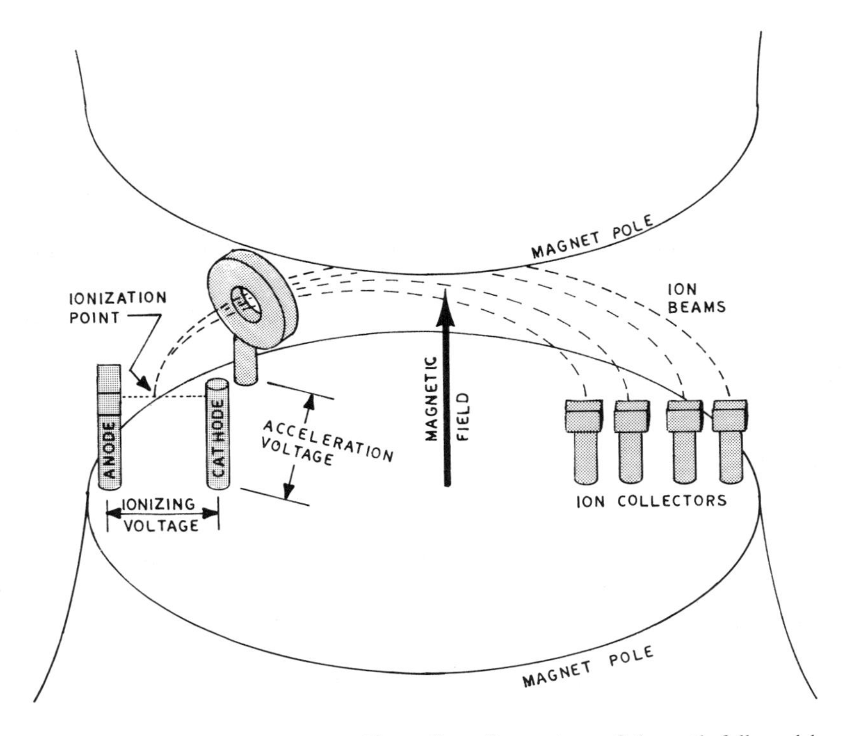


Fig. 23.5 Aston mass spectrometer: The radius of curvature of the path followed by a charged particle in a magnetic field is a function of field strength, particle charge, particle velocity, and particle mass. If all of these factors except one are held constant, particles can be separated according to their mass variations.

common point source and holding either the electric field or the magnetic field constant while varying the other field smoothly with time. The various ion currents are collected when their path's changing radii sweep them past the common anode. A time-varying plot of collector current then shows the proportion of each type of ion present in a gas sample.

A mass spectrometer has several functional parts: an ion source, a region of ion acceleration, an analyzer or separation region, and a detector–collection system. Ionization can be accomplished by bombarding molecules with electron beams. By holding both the ionizing current and the ionizing voltage constant and normalizing the gas pressure, it is possible to obtain reproducible ionization results with like samples of gas. Once ions have been created, they are collimated into a mono-energy beam and injected into the separation region.

In the *DEMPSTER MASS SPECTROMETER,* the ion path is 180 degrees, and the source and collector share the magnetic field of the separation region.

In the *SECTOR MASS SPECTROMETER,* the ion path is either 90 or 60 degrees, and the source and collector have their own individual magnetic fields.

In the *DOUBLE-FOCUSING,* or *CYCLOID, MASS SPECTROMETER,* the ion paths pass through two focus points. This makes it possible to collect more ions of any one type when the ion beam has a significant energy distribution caused by incomplete collimation at the source. In other words, all other factors being equal, this type of mass spectrometer has larger ion currents for a given sample.

In a *SOLIDS MASS SPECTROMETER,* very high voltage ionizing currents can be used to drive ions from the surface in the form of sparks. Once ionization is achieved, the performance is the same as described previously.

## 23.8 OMEGATRON

Following the dictates of Eq. 23.1, an alternating voltage of particular frequency can cause an ion to traverse a spiral path in a plane at right angles to a magnetic field's direction. Under these circumstances, the mass of an ion that reaches a collector is a function of the driving frequency. By varying this frequency, it is possible to "scan" the spectrum of masses in much the same fashion as described in Sec. 23.7. A device utilizing this principle is called an *OMEGATRON* (see Fig. 23.6).

## 23.9 TIME-OF-FLIGHT MASS SPECTROMETER

Ions injected into magnetic fields follow the helical paths illustrated in Fig. 2.4. In a *TIME-OF-FLIGHT MASS SPECTROMETER,* ions of dif-

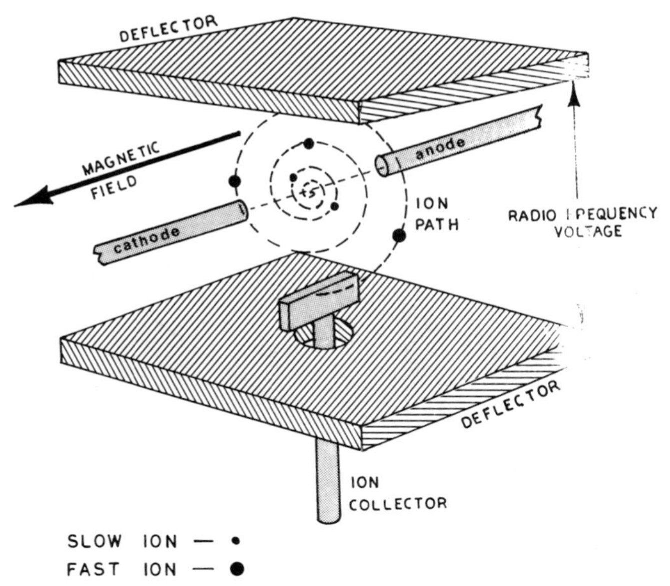


Fig. 23.6 Omegatron: Ions of one particular mass and one particular charge can be made to follow a unique spiral path that ends on a collecting electrode.

ferent mass travel with a different pitch in helical paths of the same diameter. Since the pitch is different, the time taken to complete a helical traverse is different, and ions of differing mass but originating at one point arrive at a target located at a second point at different times.

In the configuration of Fig. 23.7, means are provided to ionize gas molecules, an electric field is supplied to accelerate these ions, and a magnetic field is present to induce the ions into helical paths. Here the electric and magnetic fields are commonly directed. If a pulse of voltage is applied between the two accelerating electrodes for a period of time less than that required for the ions to move between the two electrodes, all types of ions achieve a common momentum, a common direction of movement, and a common radius of helical path. All ions then move from the point of inonization to the collector electrode in helical paths of the same radius. Since the pitch of helical path is different for each type of ion, however, the length of path and the time required to pass from the point of ionization to the collector electrode are also different for each type of ion. Following a pulse of accelerating voltage, ions of different types arrive at the collector at different times.

In this type of instrument, time-of-flight is inversely proportional to an ion's charge-to-mass ratio.

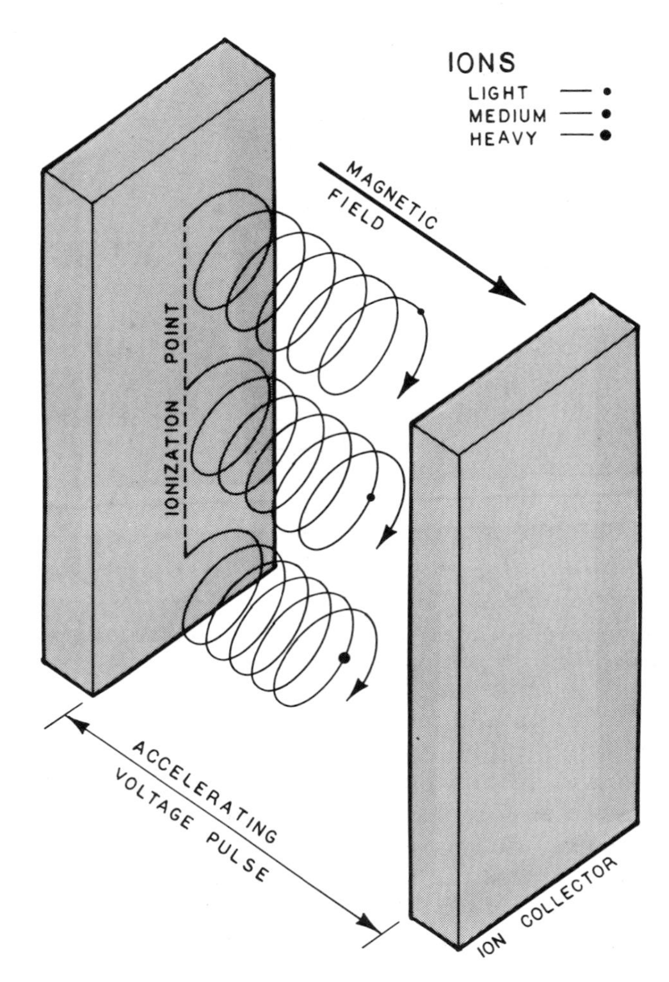


Fig. 23.7 Time-of-flight mass spectrometer: Ions of the same electric charge but of differing mass are caused to follow spiral paths whose pitch depends on mass. As the traverse length varies with spiral pitch, ions of differing mass arrive at a collecting anode at different times.

## 23.10 PHILIPS ION GAGE

In a *PHILIPS ION GAGE,* an electron beam is used to ionize gas molecules, and the ion current serves as a measure of gas pressure. If the ionizing electrons are forced into helical paths instead of following linear paths, the number of molecular collisions that occur between two electrodes is significantly increased. As a result of these helical paths, ionic currents are greater for the same gas pressure, and gage sensitivity is improved.

# PART V
# MAGNETONS MOVING IN ENVIRONMENTS WITH A VERY LOW ENERGY CONTENT

The environments to which moving electronic magnetons are exposed when moving through solid structures differ dramatically depending on whether or not there is an interchange of energy between the moving magnetons and those structures. Below some temperature, near absolute zero, there is a sudden change between interaction and no interaction. Part V discusses some of the supercooled magnetic phenomena that occur around and below this critical temperature.

When a unit of solid material has a low enough energy content, the thermal activities represented by that material are minimal. The ions that were shown in Fig. 17.1 do not vibrate around their nodal position very much, and free electrons can sneak through the interstices of a crystal without much ionic interaction. In fact, in the absence of an imposed magnetic field, there is a critical temperature below which the repulsion between orbiting and free electrons keeps the later moving in channels that avoid any interactions with the swinging nodal ions.

Under these circumstances, the electrons can move freely as long as they stay in those channels. The electrostatic channeling forces are fairly weak, however, and it doesn't take much to cause clashes between the free electrons and the still feebly swinging ions. An imposed magnetic field that is strong enough can force the free electrons out of their channel and back into contact with the ions; likewise, an increase in temperature can increase the ionic swings until they intrude upon the free electron's path.

If the free electrons and the nodal ions do have interactions, the magnetic results are pretty well covered by the previous chapters. If there is no interaction, however, and the free electrons stay in their own channels, the magnetic consequences are quite different. Part V discusses some of the magnetic consequences that may be expected when magnetons are able to move through crystalline structures without an energy interchange taking place with these structures.

# 24
# CHEMICAL ENVIRONMENT

Crystalline materials are said to be "superconducting" when they pass electric currents without an energy interchange taking place between current carriers and lattice structures. *SUPERCONDUCTION* is detected when the flow of electric current fails to generate an accompanying voltage gradient.

If all lattice elements are imagined to be static, Fig. 17.1 can be taken as a pictorialization of a superconducting relationship between carrier electrons and lattice components. If conduction electrons are exposed to disordered lattice components or to significant thermal activities, or if they are deflected by strong Lorentz forces, they are forced to interact with the lattice, and superconduction is lost. Superconduction depends on the ability of conduction electrons to move through regular lattice structures *without* interacting with those structures. Superconduction is possible because there are weak forces present that tend to keep the conduction electrons moving in linear paths away from the lattice elements.

Superconducting materials have *CRITICAL TEMPERATURES* below which they superconduct and above which they do not so conduct; that is, if the lattice components are quiet enough, the conduction electrons can "sneak" through the interstices without "noticing" the nodal elements. At a critical temperature, the lattice elements are active enough to move into the path of the carrier electrons and interactions do take place. In other words, there is a point at which the thermal energy of lattice activities overcomes the weak forces that attempt to keep the carrier electrons away from the lattice elements. Below that temperature, the evading forces dominate; above that temperature, thermal forces dominate.

Critical temperatures for all materials are very close to absolute zero. A critical temperature of 20°K is probably a near maximum for any material, but most materials have much lower maximums.

Superconductivity exists in the presence of weak magnetic fields but is destroyed by strong magnetic fields. There is, then, a *CRITICAL MAGNETIC FIELD* below which superconductivity exists and above which it does not exist. In the latter case, the weak forces that tend to keep the electrons on the straight and narrow are overcome by the Lorentz forces that bend electron beams into curved paths, thus causing collisions between conduction electrons and lattice elements. The critical magnetic field above which Lorentz collisions occur is a function of temperature. The relationship between the two is called *TUYN'S LAW*.

Superconductivity also exists in the presence of weak electric currents but is destroyed by strong electric currents. There then appears to be a *CRITICAL CURRENT* below which superconductivity exists and above which it does not exist. Actually, the critical nature of such a current is caused by its accompanying magnetic field, and the phenomenon is in reality an expression of the critical magnetic field effect.

Since the strength of the magnetic field that accompanies a given current is a function of the curvature of that current's path, the magnetic field seen by each segment of a conductor bent into a circle is greater than the field seen when the conductor is straight. In the *SILSBEE EFFECT*, the critical current for a superconducting loop of wire is proportional to the diameter of the loop.

Superconduction is a phenomenon that depends on lattice spacings as well as on a regularity in the periodicity of these spacings. As a result, both critical temperatures and critical magnetic fields are functions of mechanical strain.

Because heat is carried through superconductors by the same carrier electrons utilized by electric currents, heat transport phenomena are similarly affected by the environment seen by these electrons.

Although superconducting materials offer no resistance to the flow of direct currents, losses for time-varying currents are present and increase with frequency. At near infrared frequencies, these losses are more or less equivalent to those of nonsuperconductors.

All thermoelectric effects disappear in the superconducting state.

## 24.1 MEISSNER EFFECT

Superconducting materials exhibit perfect diamagnetism. Because the conduction electrons in a superconductor are not exposed to energy-absorbing constraints, they freely move to create Lenz fields that are exactly equal, but opposite, to imposed fields. Since the Lenz fields and imposed fields are always maintained in exact equality, no magnetic flux can pass through superconducting materials exposed to any field except when the materials are subjected to temperatures above their critical temperature. Even in this case, the flux is ejected as soon as these materials are cooled below their critical temperature. This ejection of magnetic flux from the

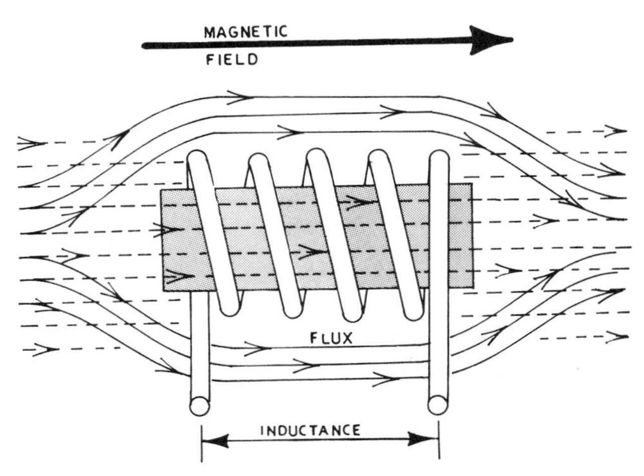


Fig. 24.1 Meissner device: Since the Lenz field caused by superconducting electrons exactly opposes the effects of an imposed field, a magnetic field cannot penetrate superconducting material.

interior of a superconductor is called the *MEISSNER EFFECT*. The magnetic field pattern in the presence of a Meissner device is shown in Fig. 24.1.

If a coil of wire is wrapped around a cylinder of superconducting material, no flux can pass through the coil since Meissner diamagnetism excludes all flux. If the critical temperature of the material is exceeded, however, the conventional flux passes through the material and hence through the coil. An impulse of voltage is then generated in the coil each time the critical temperature is passed in either direction. This voltage pulse is caused by the sudden change in flux through the coil that occurs at the critical temperature.

In the configuration of Fig. 24.1, adjusting the temperature of the material above or below its critical temperature is equivalent to turning the ambient flux through the coil on or off. If this switching mechanism (heating and cooling) is exercised rapidly enough, an alternating voltage is induced in the coil whose amplitude is proportional to the vector of the ambient magnetic field in the direction of the coil axis. Since superconduction is destroyed by a critical magnetic field intensity, this technique of magnetic field measurement is useful only for weak magnetic fields. In the very same configuration, the condition of any material, whether superconducting or nonsuperconducting, can be detected by a measure of coil inductance.

## 24.2 TYPES OF SUPERCONDUCTION

There are two basic types of superconducting materials. In Type I materials, diamagnetism duplicates an imposed field for weak fields and abruptly falls to some constant low value above a critical field ($H_c$ in Fig. 24.2); that is, the relation between diamagnetism and magnetizing force is linear below the critical field and equal to Pauli diamagnetism above. Here Meissner diamagnetism is much greater than Pauli diamagnetism (see Sec. 3.8).

In Type II materials, there are two critical magnetic fields. For fields less than the lower critical field, these materials behave the same as Type I materials. At magnetic field intensities in between the two critical fields, however, Type II materials function as if a part of them were superconductive and a part were not. In this case, diamagnetism is gradually lost with increasing field strength in a series of quantized steps. Above the higher critical field ($H_{c2}$ in Fig. 24.2), Type II materials become Pauli diamagnetic as do the Type I materials.

Type II materials functioning between the two critical fields are said to be in a *VORTEX STATE,* which consists of a number of diamagnetic vortices imbedded in a nondiamagnetic base. The number and size of these vortices depend on the intensity of the imposed magnetic field. All vortices are destroyed at the upper critical field.

All critical fields are functions of temperature. The upper critical field

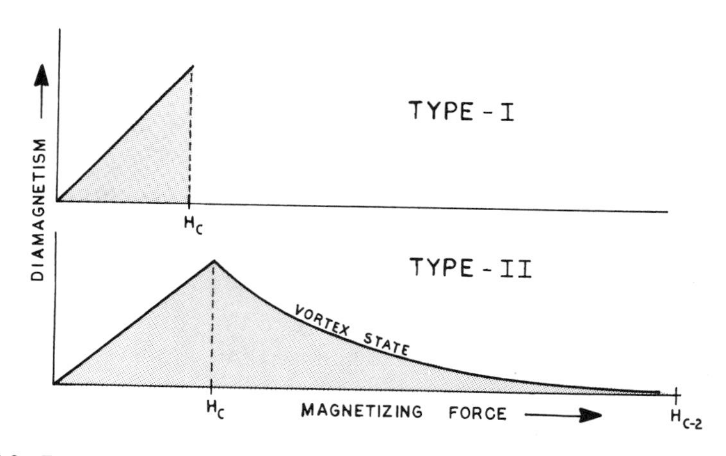


Fig. 24.2 Types of superconduction: Diamagnetism in superconducting materials can be accomplished by more than one mechanism. In the Meissner Effect, for example, electrons flow in one pattern, but diamagnetism is also accomplished as the sum of a large number of small vortices.

for some Type II materials occurs at a much higher temperature than can be reached by the superconductivity of any Type I material.

## 24.3 SPIN GLASS

The term *SPIN GLASS* describes that condition in which magneton axes freeze into a pattern of "directional disorder" below a critical temperature. This critical temperature is very close to absolute zero. The term "spin glass" is derived from an analogy with physical glass. In this case, the "glass" is a near solid in which the constituent molecules are frozen, more or less, into a pattern of "positional disorder."

The frozen directional disorder of spin glass contrasts with the condition typical of ferromagnetism, antiferromagnetism, etc., where exchange forces freeze magneton axes into patterns of orderly alignments, or that of paramagnetism, where disorder is dynamic. In short, spin glass might be thought of as frozen paramagnetism. The *KONDO TEMPERATURE* is the dividing line between spin glass freezing below and paramagnetism above.

Spin glass is found in some alloys in which a magnetic material, acting as an impurity, has been dissolved in a nonmagnetic host. Magneton interactions do occur in these alloys, but they are very weak because the magnetons are separated from each other by the diluent. Spin-glass behavior, then, occurs as a cooperative action among impurity magnetons.

*MICTOMAGNETISM* is a further phenomenon occurring in some spin-glasses in which clusters of magnetons freeze into an orderly pattern of either ferro- or ferrimagnetism. These clusters are of varying size and varying direction and are distributed more or less at random throughout volumes of what are otherwise the random orientations of the spin-glass magnetons.

## 24.4 KONDO EFFECT

In the *KONDO EFFECT,* a conducting material experiences a resistivity minimum at a very low temperature (only a few degrees above absolute zero).

In a dilute solution of magnetic ions in a nonmagnetic host metal, the widely dispersed magnetons interact through the intermediary effects of conduction electrons, which provide both the means of electric transport and the mechanism for magneton alignments. At the same time, the magnetons provide both a source of magnetism and a means of scattering the conduction electrons.

The *KONDO TEMPERATURE* (see Fig. 24.3) is a temperature of min-

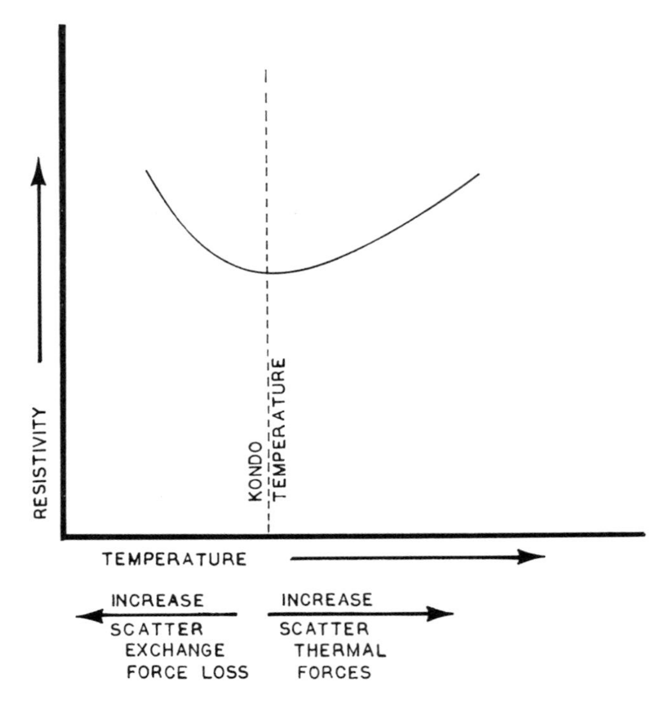


Fig. 24.3 Kondo Effect: The resistance of certain materials has a minimum value near, but not at, absolute zero.

imum resistivity. Above it, thermal activities induce disorder, whereas below it, the conduction electrons loose their ability to transmit exchange forces. This temperature is a function of magneton ion concentration and increases, more or less, as the one-fifth power of concentration.

## 24.5 ONNES EFFECT

When a wire in the form of a closed loop is placed between the pole pieces of a magnet, a portion of the magnetic field generated by that magnet passes through the loop. If the material from which the loop is fabricated is then cooled to a superconducting temperature, and if the loop in the superconducting state is removed from the vicinity of the magnet, the same magnetic flux will continue to pass through the loop! This magnetic flux has been trapped by the loop's superconducting state as a result of a mechanism for flux trapping here called the *ONNES EFFECT*.

To maintain this trapped flux, an initially induced electric current of an appropriate magnitude continues to flow in the superconducting ring. Since this current can be made to flow in either a clockwise or counterclockwise direction, depending on the direction of the originating magnetic field, the Onnes Effect has been used in computer memories, where one direction represents a "zero" and the other direction represents a "one."

## 24.6 CRYOTRON

A *CRYOTRON* is a configuration of two superconducting coils in which current in one winding controls the presence or absence of superconductivity in the other winding. Since such a device has two stable states, it has been used in computer memories, where one state represents a "zero" and the other state a "one."

## 24.7 HUEBENER-GOVEDNIK SWITCH

Tantalum superconducts when immersed in liquid helium. A noninductively wound tantalum coil can be used as a switching mechanism that

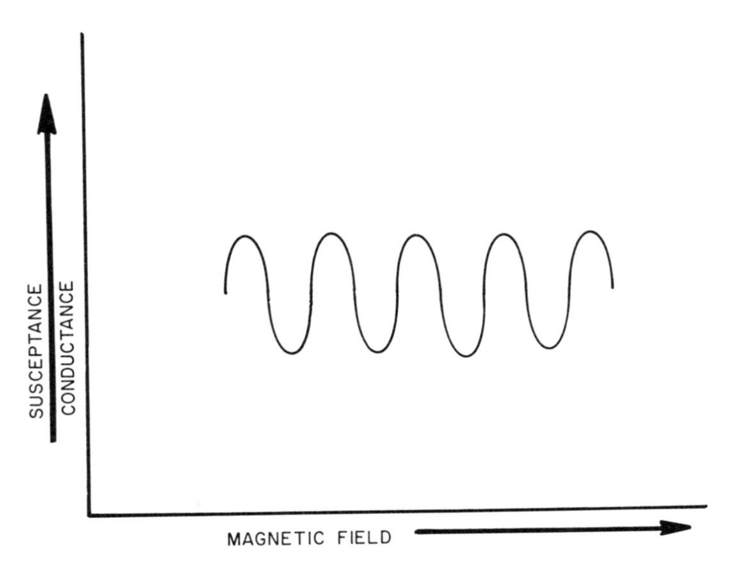


Fig. 24.4 Factor periodicity: The presence of positive charges at each crystal interstice represents spatial periodicity. Conduction electrons have periodicities in terms of magnetron radii, cyclotron frequencies, and helical path pitches. Since crystal periodicity is fixed in space, whereas electronic periodicity varies with magnetic field, interactions between the two are a function of field strength and result in periodicities of both conductance and susceptance.

presents zero impedance while it is superconducting and a fairly high resistance when it isn't. Switching between these two states can be accomplished by the presence or absence of the critical magnetic field. A tantalum coil of this kind can be assembled around a ferrite core along with a second coil to carry a current capable of supplying the critical field. Exposed to liquid helium temperatures, this device is operated by turning the energizing current on or off.

If the tantalum coil is shunted across the output of a thermocouple, the thermocouple voltage is shorted out while the tantalum superconducts. Thermocouple measurements in cryogenic applications are complicated by the relatively large spurious voltages generated in lead wires. A voltmeter will indicate the sum of the thermocouple voltage and the spurious voltages, but when the thermocouple is shorted out, it will indicate the spurious voltages only. The latter reading is then subtracted from the sum to yield an accurate measurement of the thermocouple voltage.

## 24.8 FACTOR PERIODICITY

In some materials, the diamagnetic susceptibility of conduction electrons shows a periodic variation at very low temperatures. When the applied magnetic field component perpendicular to the principle crystal axis is changed, the magnetic moment increases and decreases as shown in Fig. 24.4. The phenomenon is called the *DE HASS–VAN ALPHEN EFFECT*. Associated with this effect is a similar variation in conductivity called the *SHUBNIKOV–DE HASS EFFECT*.

# 25
# FLUX QUANTIZATION

In the Meissner Effect, the diamagnetism that results from superconductivity creates magnetic fields inside of a body that are exactly the same as, but of opposite sign to, those of imposed fields. As a result, the magnetic induction inside a superconducting body is theoretically zero. In actuality, an imposed field does extend into a very thin layer of material at the surface. The magnetic induction in this layer is called the *LONDON PENETRATION*. The intensity of magnetization varies throughout this layer, being maximum at the surface and falling off exponentially with depth below the surface.

If currents are induced in the London layer to create magnetic fields to oppose the imposed magnetic fields, the former will not exactly equal the latter as before because the magnetic fields inside the London layer can be established as only one possibility in a series of discrete values. The increment between these values is called a *FLUXOID*, or *flux quantum,* where one fluxoid is equal to $2 \times 10^{-15}$ webers. The fluxoid concept establishes that magnetic flux in a superconducting ring cannot be changed by any arbitrarily chosen value but only by integral multiples of one fluxoid.

The Meissner Effect is a bulk property, whereas the London penetration is a very shallow surface phenomenon. The quantized steps that occur in the London depth are very small in relation to what is possible in a Meissner Effect.

## 25.1 SQUID

Figure 25.1 shows a SQUID, a superconducting film deposited on an insulating cylinder in a configuration designed to limit magnetic phenomena to those that take place within the London penetration. As a result, all magnetic changes inside the coil shown take place in clearly defined quantized steps. Responding to magnetic field change, a voltage pulse is generated in the coil each time a quantized increment is completed.

Quantized flux modulators of this type can be miniaturized, with a wall

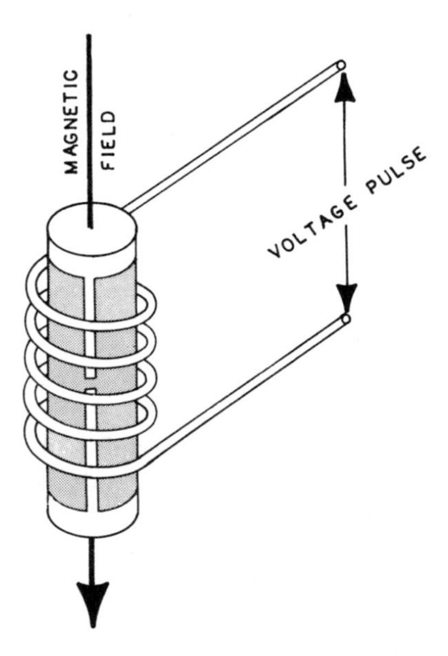


Fig. 25.1 Squid: Magnetization changes that occur in the very thin walls of cylinders constructed from superconducting materials occur in discrete, quantized steps.

thickness as little as 1 micron. Because of their small mass, switching frequencies as much as three orders of magnitude faster than those afforded by the Meissner Effect are possible.

The acronym "SQUID" stands for "Superconducting Quantum Interferometer Device." A number of different configurations are possible that are capable of detecting magnetic field strengths as small as one-hundred-thousandth of a gamma.

# 26
# TUNNELING

If two conductors are separated by an insulator, the insulating material acts as a barrier to the flow of conduction electrons from one conductor to the other. If the insulating layer is sufficiently thin, however, some electrons will pass through the barrier. The probability that some will do so is one consequence of the quantum theory of matter.

The passage of low-energy electrons through an insulating, higher-energy barrier without damage to the material of which the barrier is constructed is called *TUNNELING*. The components of a tunneling device are shown in Fig. 26.1.

When a tunnel junction is constructed from two metals separated by an insulating layer, its voltage/current relationship for low voltages is ohmic, with the current density proportional to the applied voltage. If at least one of the two conductors superconducts, on the other hand, two nonohmic phenomena are possible. These are the *GIAEVER EFFECT* and the *JOSEPHSON DC EFFECT*.

## 26.1 GIAEVER EFFECT

A *GIAEVER JUNCTION* is formed when one conductor of a two-conductor tunnel junction is superconducting. The *GIAEVER EFFECT* is characterized by a low-voltage region in which no current flows in response to imposed voltages and by high-voltage regions in which current/voltage relationships are ohmic. The Effect is based on tunneling by electrons functioning as single units.

As shown by Fig. 26.2, no current flows in a Giaever Junction until a critical voltage is exceeded. When an increasing voltage reaches the critical magnitude, the current jumps from zero to some value on the proportionality line. Actually, this is true only at a superconducting temperature of absolute zero. At temperatures above absolute zero, the zero current is slightly displaced by a small, constant current flowing across the junction. This zero-offset current is a result of electrons in the superconducting state being thermally excited through the barrier.

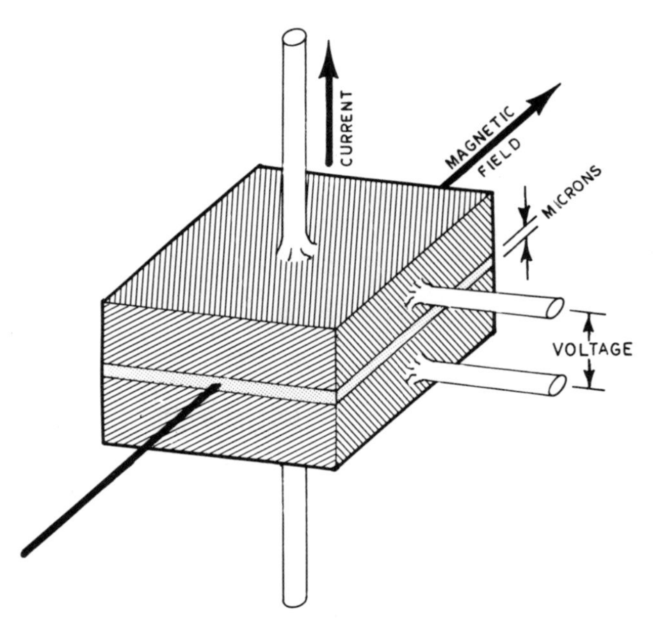


Fig. 26.1 Superconducting junction: Components of a tunneling device.

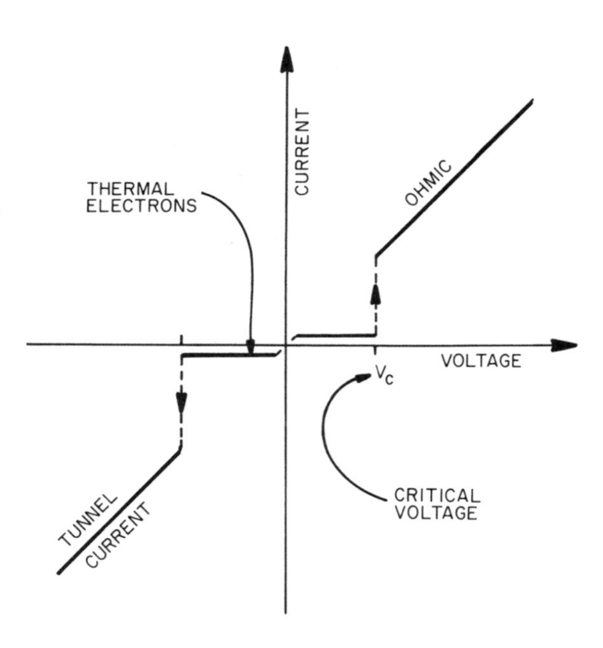


Fig. 26.2 Giaever junction transfer characteristics: Under certain conditions of magnetizing field and temperature, single electrons will pass through a barrier if the voltage is high enough.

## 26.2 JOSEPHSON DC EFFECT

A *JOSEPHSON JUNCTION* is formed in the presence of a magnetic field when both conductors of a tunnel junction are superconducting. As shown in Fig. 26.3, the *JOSEPHSON DC EFFECT* is characterized by a low-current region in which currents flow through these Josephson junctions without causing voltage drops across them.

The low-current regions of Josephson junctions have current maximums, that are a function of the strength of magnetic fields directed at right angles to the plane of the junction. This maximum, or critical, current that can flow through a Josephson junction without a voltage drop varies with changes in magnetic field strength. It is described by the following:

CRITICAL JOSEPHSON CURRENT $\qquad I_c = I_o\,(\sin kB)/kB \qquad$ (26.1)

where $I_c$ is the critical current that results from the applied magnetic field of strength $B$; $I_o$, the critical current in the absence of a magnetic field; $B$, the magnetic flux linking the junction area; and k, a constant.

The critical current of a single junction is thus a periodic function of

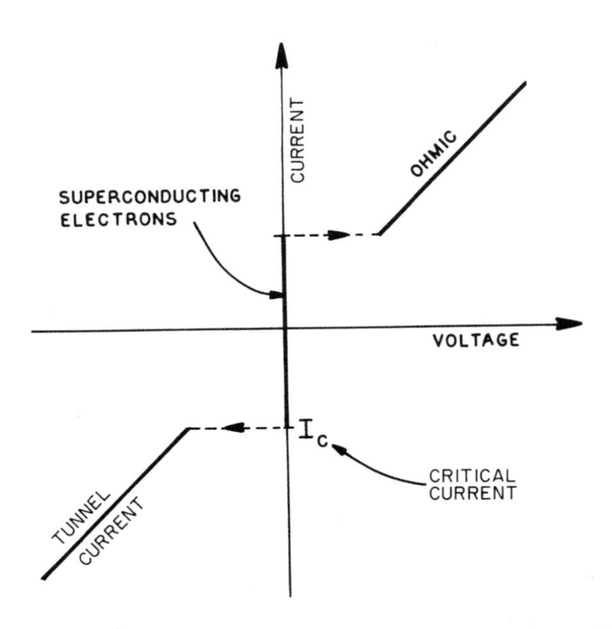


Fig. 26.3 Josephson junction transfer characteristics: Under certain conditions of magnetizing field and temperature, superconducting electron pairs pass through barriers without an associated voltage drop.

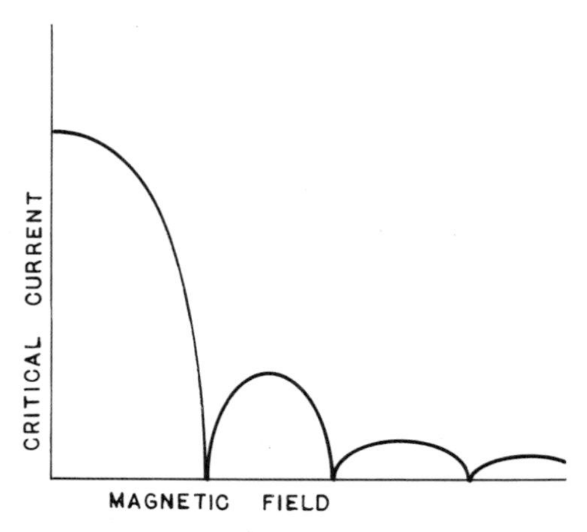


Fig. 26.4 Periodic current maximums: In a Josephson junction, the critical current relationship with field strength is a periodic function.

the magnetic flux linking that junction. In fact, Eq. 26.1 is analogous to the mathematical descriptions of an optical diffraction pattern associated with a single slit. (As a result, the plot of Fig. 26.4 is similar in form to the distribution of radiant energy caused by single-slit passage.)

Because the critical current in the presence of a magnetic field is a function of that magnetic field only, the field can be measured in terms of current.

If a Josephson junction is driven with a voltage that causes a current flow in excess of the critical current, this driving voltage must be returned to zero before the critical current can be re-established.

A number of Josephson junctions can be operated in parallel. Under these circumstances, the summed critical current is analogous to the interference achieved in multiple-slit optical interferometers.

The current density of Josephson junctions is a function of the phase difference of the superconducting electron pairs—called COOPER PAIRS—found on each side of a barrier. Transverse magnetic fields affect these phase differences.

## 26.3 JUNCTION MANIPULATION

Depending on whether superconduction takes place making use of single electrons or electron pairs, a particular junction can be operated in either the Giaever or Josephson mode. By changing performance back and forth between these two, it is possible to have either of two stable currents flowing in response to the same driving voltage. This change can be ac-

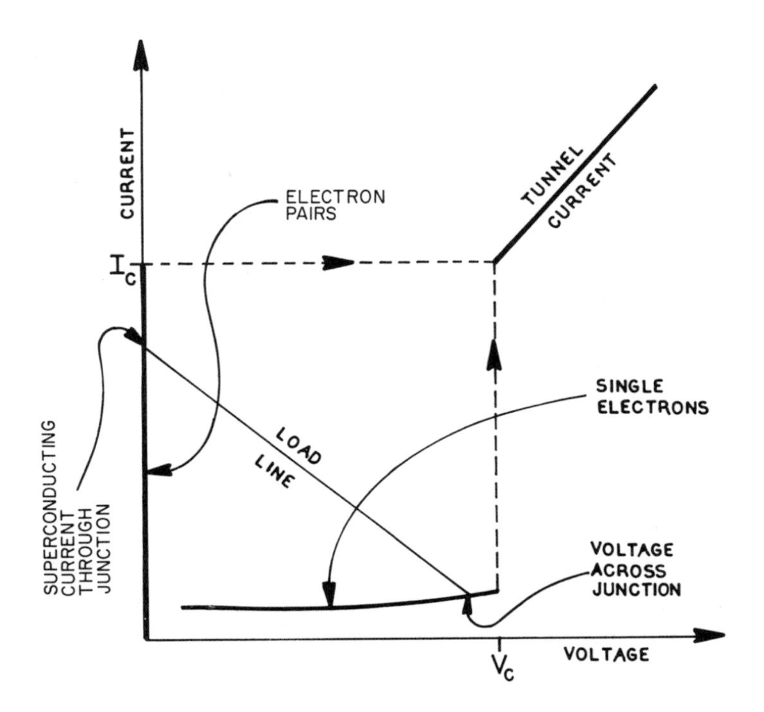


Fig. 26.5 Josephson-Giaever junction: By changing the magnetic field strength slightly, it is possible to switch the operating point of a superconducting junction back and forth between Josephson and Giaever functions, thereby providing two stable states that can be used in computer circuits to represent either ones or zeros.

complished by switching one of the two conductors from superconducting to nonsuperconducting. Switching can also be accomplished by the manipulation of magnetic fields. As shown in Fig. 26.5, the existence of two stable states achieved by combining Giaever and Josephson phenomena has potential use in digital logic circuits and computer memories.

The magnitude of the voltage jump experienced in a *GIAEVER–JOSEPHSON JUNCTION* is a function of the capacitance of the junction. Since capacitance is proportional to area, the smaller the area, the smaller the voltage jump associated with the junction and the faster the junction can be made to switch from one state to the other.

## 26.4 JOSEPHSON AC EFFECT

When a voltage is used to drive current through the combination of a resistive load and a Josephson junction, there will be no voltage drop across the junction as long as the current is less than the critical current.

On the other hand, an alternating current flows between the two superconductors of a Josephson junction when they are maintained at a finite, constant voltage difference. The frequency versus voltage relationship across a Josephson junction is expressed as follows:

JOSEPHSON FREQUENCY $\quad f = kV \quad$ (26.2)

where $V$ is the imposed voltage differential; $k$, a constant; and $f$, the Josephson frequency of the resulting alternating current. (This frequency is equal to 483.5976 ± 0.0012 megahertz per applied microvolt.)

This relationship is independent of temperature, magnetic field strength, and the character of the junction materials. It is an accurate method for determining the *STANDARD VOLT*.

## 26.5 JOSEPHSON RF EFFECT

A series of finite steps are induced into the voltage/current relationship when a Josephson junction is irradiated with *RF* energy. These steps appear as finite, discrete voltages for which the Josephson frequency is equal to an integer multiple of the *RF* frequency.

# GLOSSARY OF TERMS

## MAGNETIC FUNDAMENTALS

AMPERE'S LAW Magnetic fields surround moving charged particles.

BIOT-SAVART'S FIELD A cylindrically shaped field surrounding the linear path of an electric current.

BOSANQUET'S LAW A magnetizing force drives magnetic flux through magnetic circuits against circuit reluctance.

ENERGY INTERDEPENDENCE A quantity of material always strives to achieve the condition of minimum engery for the sum of all energy components: thermal, electrical, mechanical, chemical, and electromagnetic.

FIELD DIRECTION The Ampere field has a direction that depends both on the polarity and the direction of movement of the charged particles.

FIELD STRENGTH The Ampere field has a strength that depends both on the magnitude and the velocity of movement of the charged particles.

FLUX-QUANTIFICATION The magnitudes of magnetic fields can be changed only by discrete increments.

GROUP OPPOSITION Magnetons aligned in groups associate in patterns of group opposition, thereby minimizing local fields in one minimum energy configuration.

HYSTERESIS LOOP Displayed by those magnetic phenomina that have a range of stable configurations.

LARMOR'S PRECESSIONS Magnetons can be made to precess about the direction of an aligning magnetic field.

LENZ'S LAW A change in the path of a charged particle generates a magnetic field that opposes the force of change.

LORENTZ'S FORCE A force that acts on a moving charged particle orthogonally both to the magnetic field direction and to the movement direction.

LORENTZ'S VOLTAGE A voltage that is generated in conductors exposed to changing magnetic fields.

MAGNETIC MOMENT This is the force with which magnetized bodies tend to align themselves in magnetic fields.

MAGNETOMECHANICAL HYSTERESIS Changes in mechanical strain that result from changes in magnetization track the magnetic hysteresis loop.

MAGNETONS All electrons, protons, neutrons, and nuclei spin; those with electric charge function as magnetons, that is, as microminiature permanent magnets.

PAIRS OPPOSITION Magnetons associate in pairs with their fields in direct opposition, thereby completely cancelling local field effects in one minimum energy configuration.

RANDOM OPPOSITION Magnetons exposed to thermal jostling assume random orientations, thereby cancelling local fields in one minimum energy configuration.

REFLECTED IMPEDANCE Currents flowing in response to Lorentz voltages impose loads on the sources of such voltages.

TIME CONSTANTS Time is always expended when a change is imposed on one magnetic condition until stability is established at some other condition.

TOROIDAL FIELD IN A COIL Biot-Savart fields bent into circles assume a toroidal shape.

TOROIDAL FIELD IN A MAGNETON The magnetic fields surrounding spinning charged particles are toroidal in shape.

## ELECTRONIC MAGNETOCONDUCTIVE PHENOMENA

ELECTRONIC MAGNETOCONDUCTION The conduction of a material depends on the environment seen by the constituent carriers. If this environment is changed by a magnetic field, the conduction will be changed.

FERRIMAGNETIC CURIE POINT CHANGE Conduction in a magnetic semiconductor changes over range of Curie point change from ferrimagnetism to paramagnetism.

FERRIMAGNETIC DOMAIN CONFIGURATION Conduction in a magnetic semiconductor depends on domain configuration.

FERROMAGNETIC CURIE POINT CHANGE Conduction changes over range of Curie point change from ferromagnetism to paramagnetism.

FERROMAGNETIC DOMAIN CONFIGURATION Conduction depends on domain configuration. In the Gauss Effect, the domain configuration is changed by an imposed magnetic field. In the Elastoresistance Magnetic Effect, the domain configuration is changed by strain.

MAGNETRON TRAJECTORY The conduction of carriers in spiral paths depends on the pitch of the spiral.

MAGNETRON TRAJECTORY/CYCLOTRON FREQUENCY INTERACTIONS The conduction of carriers in spiral paths is influenced by exposure to electromagnetic radiation (or other energy form) of cyclotron frequency.

MAGNETRON TRAJECTORY/SURFACE EFFECT INTERACTIONS The conduction of carriers is decreased if magnetron paths interact with the surfaces of conductors.

PARAMAGNETIC MAGNETON ALIGNMENTS Conduction depends on magneton orientations; if magnetons are forced into an orderly pattern, the conduction is greater than under a disorderly pattern.
TUNED MAGNETIZATION The conduction of a body of material decreases when the constituent magnetons are caused to precess. (Such precession can result from electromagnetic radiation.)
SPIN WAVE INTERACTION The conduction of a body of material is decreased when the constituent magnetons are stimulated into a spin wave by some form of energy.

## THERMAL MAGNETOCONDUCTIVE PHENOMENA

FERROMAGNETIC CURIE POINT CHANGE Conduction changes over range of Curie point change from ferromagnetism to paramagnetism.
FERROMAGNETIC DOMAIN CONFIGURATION Conduction depends on domain configuration, which can be changed either by imposed magnetic fields or by mechanical deformations.
PARAMAGNETIC MAGNETON ALIGNMENTS Conduction depends on magneton orientations; if magnetons are forced into an orderly pattern; the conduction is greater than under a disorderly pattern.
THERMAL MAGNETOCONDUCTION The conduction of a material depends on the environment seen by the constituent carriers. If this environment is changed by a magnetic field, the conduction will be changed.
TUNED MAGNETIZATION The conduction of a body of material decreases when the constituent magnetons are caused to precess.

## MAGNETON PRECESSION PHENOMENA

ABSORPTION Energy stimulating precessions is absorbed; if it is exposed to material whose magnetons are precessing, it is absorbed more than if there were no precessions.
CHEMICAL SHIFTING The actual field seen by a magneton is the sum of an imposed field and local fields. The Larmor frequency that would otherwise be established by the imposed field is shifted to that of the local field.
DIMENSION magnetostrictive dimensions are changed by precessions.
DIRECTIONAL COHERENCE magnetons align with, or against, imposed magnetic fields.
ELECTROMAGNETIC PUMPING Precessions are stimulated by electromagnetic radiation of Larmor frequencies.

LARMOR FREQUENCY Proton resonance: 42.58 B megahertz (B in tesla); Electron resonance: 28.0 B gigahertz (B in tesla)

MAGNETIC PUMPING Precessions are stimulated when an imposed magnetic field direction is changed cyclically at a Larmor frequency.

MAGNETIZATION Magnetization is reduced by precessions. The inductance of a circuit containing a ferrite element is reduced when the magnetons in that element precess.

MAGNETONS Electrons, protons, neutrons and nuclei all spin; those with electric charge function as magnetons, that is, as microminiature permanent magnets.

MAGNETON PRECESSION PHENOMENA Magnetons can be caused to precess. The characteristics of a material that contains precessing magnetons are different from those of a material without precessions.

OPTICAL TRANSMISSION The transmission of polarized electromagnetic radiation is reduced by precessions.

PERMITTIVITY Permittivities are decreased by precessions.

PRECESSION Magnetons precess about new directions when an imposed field direction is changed.

RELAXATION TIMES Longitudinal: time to assume new field alignment; Transverse: time for precessions to die down.

RESISTIVITY Resistivities are increased by precessions.

## HALL-RELATED PHENOMENA

ADIABATIC NEGATIVE HALL EFFECT (electrons alone) With Ettinghausen thermal gradient and with or without Hall current

ADIABATIC POSITIVE HALL EFFECT (HOLES ALONE) With Ettinghausen thermal gradient and with or without Hall current

ETTINGHAUSEN EFFECT Thermal gradients are established across conductors when hot carriers are deflected more than cold ones.

ETTINGHAUSEN RESISTANCE The increase in the resistance of a conductor when heat is allowed to flow as a result of an Ettinghausen thermal gradient.

GALVANOMAGNETIC RESISTANCE The reduction in effective cross-sectional area and the increase in length of traverse when carriers are crowded to one side of a conductor.

HALL EFFECT Voltages are generated when carriers, either holes or electrons, are deflected to one side by Lorentz forces acting on electric currents.

HALL RESISTANCE The increase in the resistance of a conductor when a Hall current flows.

HALL REVERSAL (electrons and holes acting together) The condition of zero Hall voltage in an intrinsic semiconductor when the effect of the electron carriers balances that of the holes.

ISOTHERMAL NEGATIVE HALL EFFECT (electrons alone) With Ettinghausen heat flow and with or without Hall current

ISOTHERMAL POSITIVE HALL EFFECT (holes alone) With Ettinghausen heat flow and with or without Hall current

## NERNST-RELATED PHENOMENA

ADIABATIC GALVANOMAGNETIC RESISTANCE The reduction in effective cross-sectional area and the increase in length of traverse when carriers are crowded to one side of a conductor.

ADIABATIC NEGATIVE NERNST EFFECT (electrons alone) With Righi-Leduc thermal gradient and with or without Nernst current

ADIABATIC POSITIVE NERNST EFFECT (holes alone) With Righi-Leduc thermal gradient and with or without Nernst current

NERNST EFFECT Voltage gradients are established when carriers, either holes or electrons, are deflected to opposite sides of conductors by Lorentz forces during heat flow.

NERNST RESISTANCE The resistance of a conductor to heat flow increases when a Nernst current flows.

NERNST REVERSAL (electrons and holes together) The condition of zero Righi-Leduc thermal gradient in an intrinsic semiconductor when the effects of electron carriers balance those of holes.

RIGHI-LEDUC EFFECT Thermal gradients are established across conductors when hot carriers are deflected more than cold ones by Lorentz forces during heat flow.

## TWISTED MAGNETOSTRICTIVE ROD PHENOMENA

BOBECK EFFECT A rod is helically magnetized by twisting it after axial magnetization. A voltage is generated between its two ends when its axial magnetization is changed by an increment applied in one direction. The polarity of the voltage so created indicates the direction of helical magnetization.

INVERSE WERTHEIM EFFECT An axially magnetized rod twists when a current passes through it.

INVERSE WIEDEMANN EFFECT A rod is circularly magnetized when it carries an electric current as a result of an accompanying Biot-Savart field. It is axially magnetized when it is twisted.

WERTHEIM EFFECT A voltage is generated between the two ends of an axially magnetized when it is twisted.

WIEDEMANN EFFECT A current-carrying rod twists when it is axially magnetized. The vector combination of circular magnetization and axial magnetization forces the magnetons into a helical pattern, and the rod twists to follow the helix.

## CYCLOTRON/MAGNETRON PHENOMENA

CYCLOTRON A device in which particles are accelerated at cyclotron frequencies.

CYCLOTRON FREQUENCY Magnetron trajectories are accomplished at cyclotron frequencies.

MAGNETRON A diode in which electrons pass from a cathode to an anode in circular paths. The condition of transmission is unstable at the point at which the path just barely brushes the anode.

MAGNETRON PRESSURE GAGE A diode in which the electrons' circular paths normally miss the anode. If an electron collides with a gas molecule, however, the circular paths reach the anode after deflection, and a current flows that is proportional to the number of collisions (i.e., the number of gas molecules present).

MAGNETRON TRAJECTORY A charged particle travels in a circular path when it is exposed to a magnetic field, the radius being proportional both to mass and to velocity and inversely proportional both to electric charge and field strength.

MASS SPECTROMETERS Particles of different mass or different electric charge are separated by circular path manipulations.

PERIODICITY OF ALTERNATING CURRENT Interactions with pitch periodicity and interactions with cyclotron periodicity.

PERIODICITY OF ELECTROMAGNETIC RADIATION Interactions with pitch periodicity and interactions with cyclotron periodicity.

PERIODICITY OF LATTICE NODES Interactions with pitch periodicity and interactions with cyclotron periodicity.

PERIODICITY OF PHONON WAVES Interactions with pitch periodicity and interactions with cyclotron periodicity.

SPIRAL PATH (with a pitch of a given periodicity) Charged particles travel in spiral paths when components of motion both in the field direction, and at right angles to the field direction, are present.

# INDEX

W9-BLZ-930

# Nasreen's Secret School

A TRUE STORY FROM AFGHANISTAN

BY JEANETTE WINTER

Beach Lane Books

NEW YORK * LONDON * TORONTO * SYDNEY

*To the courageous*
*women and girls of Afghanistan*

# Author's note

The Global Fund for Children, a nonprofit organization committed to helping children around the world, contacted me about basing a book on a true story from one of the groups they support.

I was immediately drawn to an organization in Afghanistan that founded and supported secret schools for girls during the 1996–2001 reign of the Taliban.

The founder of these schools, who requested anonymity, shared the story of Nasreen and her grandmother with me. Nasreen's name has been changed.

Before the Taliban seized control of Afghanistan,

- 70% of schoolteachers were women
- 40% of doctors were women
- 50% of students at Kabul University were women.

After the Taliban seized control of Afghanistan,

- girls weren't allowed to attend school or university
- women weren't allowed to work outside the home

- women weren't allowed to leave home without a male relative as chaperone
- women were forced to wear a *burqa* that covered their entire head and body, with only a small opening for their eyes.

There was no singing or dancing or kite flying. Art and culture, in the birthplace of the immortal poet Rumi, was banished. The colossal Bamiyan Buddhas, carved into the side of a mountain, were destroyed. Years of isolation and fear had begun.

But there was also bravery from citizens who defied the Taliban in many ways, including supporting the secret schools for girls.

Even now, after the fall of the Taliban in Afghanistan in 2001, danger remains. Still, schools are bombed, set on fire, and closed down. Still, there are death threats to teachers. Still, girls are attacked or threatened if they go to school.

And STILL, the girls, their families, and their teachers defy the tyranny by keeping the schools open.

Their courage has never wavered.

My granddaughter, Nasreen, lives with me in Herat,
an ancient city in Afghanistan.
Art and music and learning once flourished here.

Then the soldiers came and changed everything.
The art and music and learning are gone.
Dark clouds hang over the city.

Poor Nasreen sat at home all day,
because girls are forbidden to attend school.
The Taliban soldiers don't want girls to learn about the world,
the way Nasreen's mama and I learned
when we were girls.

One night, soldiers came to our house

and took my son away,
with no explanation.

We waited many days and nights for his return.

Finally, Nasreen's frantic mama went searching for him,
even though going out alone in the streets
was forbidden for women and girls.

The full moon passed our window many times
as Nasreen and I waited.

Nasreen never spoke a word.
She never smiled.
She just sat, waiting for her mama and papa to return.

I knew I had to do something.

I heard whispers about a school—
a secret school for girls—
behind a green gate in a nearby lane.
I wanted Nasreen to attend this secret school.
I wanted her to learn about the world, as I had.
I wanted her to speak again.

So one day, Nasreen and I hurried down the lanes
until we came to the green gate.
Luckily, no soldier saw us.

I tapped lightly.
The teacher opened the gate,
and we quickly slipped inside.

We crossed the courtyard to the school—
one room in a private house,
filled with girls.

Nasreen took a place at the back of the room.
*Please Allah, open her eyes to the world,*
I prayed as I left her there.

Nasreen didn't speak to the other girls.
She didn't speak to the teacher.
At home, she remained silent.

I was fearful that the soldiers would discover the school.
But the girls were clever.
They slipped in and out of school at different times,
so as not to arouse suspicion.
And when boys saw soldiers near the green gate,
they distracted them.

I heard of a soldier who pounded on the gate,
demanding to enter.

But all he found was a room filled with girls
reading the Koran, which was allowed.
The girls had hidden their schoolwork,
outwitting the soldier.

One of the girls, Mina,
sat next to Nasreen every day.
But they never spoke to each other.
While the girls were learning,
Nasreen stayed inside herself.

My worry was deep.

When school closed for the long winter recess,
Nasreen and I sat by the fire.
Relatives gave us what food and firewood
they could spare.
We missed her mama and my son more than ever.
Would we ever know what had happened?

The day Nasreen returned to school,
Mina whispered in her ear.

And Nasreen answered back!

With those words,
her first since her mama went searching,
Nasreen opened her heart to Mina.

And she smiled for the first time
since her papa was taken away.

At last, little by little, day by day, Nasreen learned to read, to write, to add and subtract.

Each night she showed me
what she had discovered that day.

Windows opened for Nasreen
in that little schoolroom.

She learned about the artists and writers
and scholars and mystics who, long ago,

made Herat beautiful.

Nasreen no longer feels alone.
The knowledge she holds inside
will always be with her,
like a good friend.

Now she can see blue sky
beyond those dark clouds.

As for me, my mind is at ease.
I still wait for my son and his wife.
But the soldiers can never close the windows
that have opened for my granddaughter.

Insha'Allah.

The phrase *Insha'Allah* on page 38 means "God willing."

The Global Fund for Children (www.globalfundforchildren.org) is a nonprofit organization committed to advancing the dignity of children and youth around the world. The Global Fund for Children pursues its mission by making small grants to innovative community-based organizations working with some of the world's most vulnerable children and youth.

BEACH LANE BOOKS
An imprint of Simon & Schuster Children's Publishing Division
1230 Avenue of the Americas, New York, New York 10020

Book design by Ann Bobco
The text for this book is set in Bernhard Modern BT.
The illustrations for this book are rendered in acrylic paint.
Manufactured in China
32 34 36 38 40 39 37 35 33
Library of Congress Cataloging-in-Publication Data
Winter, Jeanette.
Nasreen's secret school : a true story from Afghanistan / Jeanette Winter.—1st ed.
p. cm.
ISBN: 978-1-4169-9437-4 (hardcover : alk. paper)
1. Girls' schools—Afghanistan—Juvenile literature. 2. Girls—Education—Afghanistan—Juvenile literature. 3. Taliban—Juvenile literature. 4. Afghanistan—Social conditions—21st century—Juvenile literature. I. Title.
LC2410.A3W56 2009
371.823'4209581—dc22
2009008285
0623 SCP